Communications in Computer and Information Science 1408

More information about this series at http://www.springer.com/series/7899

Daniel Alejandro Rossit ·
Fernando Tohmé · Gonzalo Mejía Delgadillo (Eds.)

Production Research

10th International Conference
of Production Research - Americas, ICPR-Americas 2020
Bahía Blanca, Argentina, December 9–11, 2020
Revised Selected Papers, Part II

 Springer

Editors
Daniel Alejandro Rossit (iD)
Universidad Nacional del Sur
and CONICET
Bahía Blanca, Argentina

Gonzalo Mejía Delgadillo (iD)
Universidad de la Sabana
Chía, Colombia

Fernando Tohmé (iD)
Universidad Nacional del Sur
and CONICET
Bahía Blanca, Argentina

ISSN 1865-0929   ISSN 1865-0937 (electronic)
Communications in Computer and Information Science
ISBN 978-3-030-76309-1   ISBN 978-3-030-76310-7   (eBook)
https://doi.org/10.1007/978-3-030-76310-7

This Springer imprint is published by the registered company Springer Nature Switzerland AG
The registered company address is: Gewerbestrasse 11, 6330 Cham, Switzerland

Preface

This CCIS volume includes selected articles presented at the 10th International Conference of Production Research - Americas (ICPR-Americas 2020), held in virtual form in Bahía Blanca, Argentina, during December 9–11, 2020. This conference was organized by the Americas chapter of the International Foundation of Production Research (IFPR). The aim of this conference was to exchange experiences and encourage collaborative work among researchers and practitioners from the Americas and the Caribbean region.

The first ICPR-Americas conference took place in November 2002 in St. Louis, Missouri, USA, under the general theme of "Production Research and Computational Intelligence for Designing and Operating Complex Global Production Systems". This conference generated very positive responses from the attendees. From then on, this ICPR regional meeting was held every other year. The second version of ICPR-Americas was held in August 2004 in Santiago, Chile, with the conference theme "Information and Communication Technologies for Collaborative Operations Management". The third edition of ICPR-Americas, held in Curitiba, Brazil, had the general theme of "Rethinking Operation Systems: New Roles of Technology, Strategy and Organization in the Americas' Integration Era". This event sought to promote a deep discussion about the role of production engineering in the Americas' integration process. In 2008, the venue of ICPR-Americas was Sao Paulo, Brazil, at the Universidade de São Paulo. The conference had the theme "The Role of Emerging Economies in the Future of Global Production: Creating New Multinationals". Bogotá, Colombia, hosted the fifth conference in 2010 on the subject of "Technologies in Logistics and Manufacturing for Small and Medium Enterprises". The sixth edition (2012) was organized around the topic "Production Research in the Americas Region: Agenda for the Next Decade" by the Universidad de Santiago de Chile. In Lima, Peru, the 2014 edition of ICPR-Americas addressed the general theme of "Towards Sustainable Eco-Industrialization through Applied Knowledge". The eighth edition was held in Valparaíso, Chile, in 2016, and more recently, in 2018, the ninth conference was held in Bogotá, Colombia, under the topic "Improving Supply Chain Management through Sustainability".

ICPR-Americas 2020 was the first edition held in virtual mode, due to the COVID-19 pandemic. Thanks to the participation and commitment of the attendees, the conference went on successfully, allowing many young researchers to participate in an international conference, in a year in which such opportunities were scarce. The ICPR-Americas meeting space provided them with the possibility of sharing their work, as well as exchanging ideas and points of view, all in the cordial atmosphere that characterizes the ICPR-Americas conferences.

ICPR-Americas 2020 was organized by a local committee from the Universidad Nacional del Sur, Argentina, supported by the university administration and the IFPR.

The conference received 280 submissions, all subjected to peer review. The refereeing process followed a single-blind procedure. The reviews were in the charge of a panel of experts and invited external reviewers (outside the Program Committee). Each submission had an average of three independent reviews and each reviewer was assigned, on average, two submissions. The best 53 articles were selected to be part of this CCIS volume.

March 2021

Daniel Alejandro Rossit
Fernando Tohmé
Gonzalo Mejía Delgadillo

Organization

General Chair

Daniel Rossit   Universidad Nacional del Sur, Argentina

Program Committee Chair

Fernando Tohmé   Universidad Nacional del Sur, Argentina

Organizing Committee

Nancy López   Universidad Nacional del Sur, Argentina
Antonella Cavallin   Universidad Nacional del Sur, Argentina
Adrián Toncovich   Universidad Nacional del Sur, Argentina
Ernesto Castagnet   Universidad Nacional del Sur, Argentina
Diego Rossit   Universidad Nacional del Sur, Argentina
Mariano Frutos   Universidad Nacional del Sur, Argentina
Daniel Carbone   Universidad Nacional del Sur, Argentina
Luciano Sívori   Universidad Nacional del Sur, Argentina
Adrián Castaño   Universidad Nacional del Sur, Argentina
Martín Safe   Universidad Nacional del Sur, Argentina
Marisa Sánchez   Universidad Nacional del Sur, Argentina
Agustín Claverie   Universidad Nacional del Sur, Argentina
Fernando Nesci   Universidad Nacional del Sur, Argentina
Carla Macerates   CONICET, Argentina

Program Committee

Adrián Toncovich   Universidad Nacional del Sur, Argentina
Mariano Frutos   Universidad Nacional del Sur and CONICET, Argentina
Antonella Cavallin   Universidad Nacional del Sur, Argentina
Marisa Analía Sánchez Universidad Nacional del Sur, Argentina
Martín Safe   Universidad Nacional del Sur and CONICET, Argentina
Diego Gabriel Rossit Universidad Nacional del Sur and CONICET, Argentina
Máximo Méndez Babey Universidad de las Palmas de Gran Canaria, Spain
Héctor Cancela   Universidad de la República, Uruguay
Pedro Piñeyro   Universidad de la República, Uruguay
Sergio Nesmachnow Universidad de la República, Uruguay
Adrián Ferrari   Universidad de la República, Uruguay

Marcus Ritt	Universidade Federal do Rio Grande do Sul, Brazil
Marcelo Seido Nagano	Universidade de São Paulo, Brazil
José Framiñan	Universidad de Sevilla, Spain
Rubén Ruiz	Universidad Politécnica de Valencia, Spain
Begoña González Landín	Universidad de las Palmas de Gran Canaria, Spain
Ricardo Aguasca Colomo	Universidad de las Palmas de Gran Canaria, Spain
Roger Ríoz-Mercado	Universidad de Nueva León, Mexico
Enzo Morosini Frazzon	Federal University of Santa Catarina, Brazil
Ciro Alberto Amaya Guio	Universidad de los Andes, Colombia
Cihan Dagli	Missouri University of Science and Technology, USA
Shimon Nof	Purdue University, USA
Bopaya Bidanda	University of Pittsburgh, USA
Sergio Gouvea	Pontifícia Universidade Católica do Paraná, Brazil
Edson Pinheiro	Pontifícia Universidade Católica do Paraná, Brazil
Fernando Deschamps	Pontifícia Universidade Católica do Paraná, Brazil
Cecilia Montt Veas	Pontifícia Universidade Católica de Valparaíso, Chile
Óscar C. Vásquez	Universidad de Santiago de Chile, Chile
Pedro Palominos	Universidad de Santiago de Chile, Chile
Luis Ernesto Quezada Llanca	Universidad de Santiago de Chile, Chile
Dusan Sormaz	Ohio University, USA
Gursel Suer	Ohio University, USA
José Ceroni	Pontifícia Universidade Católica de Valparaíso, Chile
Karen Y. Niño	Universidad Militar Nueva Granada, Colombia
Nubia Velasco	Universidad de los Andes, Colombia
Gonzalo Mejía Delgadillo	Universidad de la Sabana, Colombia
Jairo Rafael Montoya Torres	Universidad de la Sabana, Colombia
William Javier Guerrero Rueda	Universidad de la Sabana, Colombia
Leonardo Jose Gonzalez Rodriguez	Universidad de la Sabana, Colombia
Luis Alfredo Paipa Galeano	Universidad de la Sabana, Colombia
Alfonso Tullio Sarmiento Vasquez	Universidad de la Sabana, Colombia
Vícto Viana Céspedes	Universidad de la República, Uruguay
Diego Ricardo Broz	Universidad Nacional de Misiones, Argentina
Aníbal Blanco	CONICET, Argentina
Alberto Bandoni	Universidad Nacional del Sur, Argentina
José Luis Figueroa	Universidad Nacional del Sur, Argentina
Mauricio Miguel Coletto	Universidad Nacional de Río Negro, Argentina
Alejandro Olivera	Universidad de la República, Uruguay
Sandra Robles	Universidad Nacional del Sur, Argentina
Daniela Alessio	Universidad Nacional del Sur, Argentina
Frank Werner	Otto von Guericke University Magdeburg, Germany
María Clara Tarifa	Universidad Nacional de Río Negro and CONICET, Argentina

Fernanda Villarreal	Universidad Nacional del Sur, Argentina
Valentina Viego	Universidad Nacional de Río Negro and CONICET, Argentina
Lorena Brugnoni	Universidad Nacional del Sur and CONICET, Argentina
Jorge Lozano	Universidad Nacional del Sur and CONICET, Argentina
Guillermo Crapiste	Universidad Nacional del Sur and CONICET, Argentina
Facundo Iturmendi	Universidad Nacional de Río Negro, Argentina
Ana Maguitman	Universidad Nacional del Sur and CONICET, Argentina
Santiago Maiz	Universidad Nacional del Sur, Argentina
Héctor Chiacchiarini	Universidad Nacional del Sur and CONICET, Argentina
Gabriela Pesce	Universidad Nacional del Sur, Argentina
Susana Moreno	CONICET, Argentina
María Teresa González	Universidad Nacional del Sur and CONICET, Argentina
Ignacio Costilla	Universidad Nacional del Sur and CONICET, Argentina
Adrián M. Urrestarazu	Universidad Nacional del Sur, Argentina
Elda Monetti	Universidad Nacional del Sur, Argentina
Fabio Miguel	Universidad Nacional del Río Negro, Argentina
Diego Hernán Peluffo-Ordóñez	Yachay Tech, Ecuador
Pedro Ballesteros Silva	Universidad Tecnológica de Pereira, Colombia
Leandro Leonardo Lorente Leyva	Universidad Técnica del Norte, Ecuador
Katty Alicia Lagos Ortiz	Universidad Agraria del Ecuador, Ecuador
José Medina-Moreira	Universidad de Guayaquil, Ecuador
Miguel Heredia	Instituto Tecnológico de Gustavo A. Madero, Tecnológico de Nacional de México, Mexico
Israel David Herrera Granda	Universidad Técnica del Norte, Ecuador
Andrés Fioriti	CONICET, Argentina
Fernando Delbianco	Universidad Nacional del Sur and CONICET, Argentina
Claudio Delrieux	Universidad Nacional del Sur and CONICET, Argentina
Katyanne Farias	École des Mines de Saint-Etienne, France
Ana Carolina Olivera	Universidad Nacional de Cuyo and CONICET, Argentina
Francesco Pilati	University of Trento, Italy
Yanina Fumero	INGAR, Argentina
Gabriela Corsano	INGAR, Argentina
Marco Cedeño Viteri	Universidad Tecnológica de Chile INACAP, Chile

Marcela C. González Araya	Universidad de Talca, Chile
Leandro Rodriguez	Universidad Nacional de San Juan and CONICET, Argentina
Juana Zuntini	Universidad Nacional del Sur, Argentina
Natalia Urriza	Universidad Nacional del Sur, Argentina
María Angélica Viceconte	Universidad Nacional del Sur, Argentina
Marianela De Batista	Universidad Nacional del Sur, Argentina
Pau Fonseca i Casas	Universtitat Politècnica de Catalunya, Spain
Ariel Behr	Universidade Federal do Rio Grande do Sul, Brazil
José Fidel Torres Delgado	Universidad de los Andes, Colombia
Sepideh Abolghasem Ghazvini	Universidad de los Andes, Colombia
Alex Ricardo Murcia Cucaita	Universidad de los Andes, Colombia
Adriana Lourdes Abrego Perez	Universidad de los Andes, Colombia
Camil Martinez	Universidad de los Andes, Colombia
Jorge Luis Chicaiza Vaca	Dortmund Technical University, Germany
Fernando Daniel Mele	Universidad Nacional de Tucumán, Argentina
Humberto Heluane	Universidad Nacional de Tucumán, Argentina
Jorge Marcelo Montagna	INGAR, Argentina
Melisa Manzanal	Universidad Nacional del Sur and CONICET, Argentina
Gustavo Ramoscelli	Universidad Nacional del Sur, Argentina
José María Cabrera Peña	Universidad de Las Palmas de Gran Canaria, Spain
Francisco Javier Rocha Henríquez	Universidad de Las Palmas de Gran Canaria, Spain
Maarouf Mustapha	Universidad de Las Palmas de Gran Canaria, Spain
Carlos Hernández Hernández	Universidad de Las Palmas de Gran Canaria, Spain
Dagoberto Castellanos Nieves	Universidad de La Laguna, Spain
Martha Ramírez Valdivia	Universidad de la Frontera, Chile
Sri Talluri	Michigan State University, USA
Dmitry Ivanov	Berlin School of Economics and Law, Germany
Pietro Cunha Dolci	Universidade Santa Catarina do Sul, Brazil
Carlos Ernani Fries	Universidade Federal de Santa Catarina, Brazil
Claudemir Tramarico	Universidade Estadual Paulista, Brazil
Eduardo Ortigoza	Universidad Nacional de Asunción, Paraguay
Liang Gao	Huazhong University of Science and Technology, China
Carlos Contreras Bolton	Universidad de Concepción, Chile
Marcela Filippi	Universidad Nacional de Río Negro, Argentina
María Beatriz Bernabé Loranca	Benemérita Universidad Autónoma de Puebla, Mexico
Fernando Espinosa	Universidad de Talca, Chile

Patrick Hirsch	University of Natural Resources and Life Sciences, Austria
Erfan Babaee Tirkolaee	Mazandaran University of Science and Technology, Iran
Albert Ibarz	Universidad de Lerida, Spain
Darla Goeres	Montana State University, USA
Diane Walker	Montana State University, USA
Gabriela Vinderola	Universidad Nacional del Litoral and CONICET, Argentina
Pedro Rizzo	INTA, Argentina
Lorena Franceschinis	Universidad Nacional del Comahue and CONICET, Argentina
Vítor Alcácer	Instituto Politécnico de Setúbal, Portugal
Marcela Ibañez	Universidad de la República, Uruguay
Diego Passarella	Universidad de la República, Uruguay
Krzysztof Polowy	Poznan University of Life Sciences, Poland
Marta Glura	Poznan University of Life Sciences, Poland
Axel Soto	Universidad Nacional del Sur and CONICET, Argentina
Carlos Lorenzetti	Universidad Nacional del Sur and CONICET, Argentina
Rocío Cecchini	Universidad Nacional del Sur and CONICET, Argentina
Eduardo Xamena	Universidad Nacional de Salta and CONICET, Argentina
Diego Rodriguez	Universidad Nacional de Salta, Argentina
Silvio Gonnet	Universidad Tecnológica Nacional and CONICET, Argentina
Diego Pinto-Roa	Universidad Nacional de Asunción and CONACYT, Paraguay
Ángel Augusto Roggiero	Univesidad Nacional de Cuyo, Argentina
Claudia Noemí Zárate	Universidad Nacional de Mar del Plata, Argentina
Alejandra María Esteban	Universidad Nacional de Mar del Plata, Argentina
Carlos Vecchi	Universidad Nacional del Nordeste, Argentina
Ângela de Moura Ferreira Danilevicz	Universidade Federal do Rio Grande do Sul, Brasil
Germán Rossetti	Universidad Nacional del Litoral, Argentina
Marcelo Tittonel	Universidad Nacional de La Plata, Argentina
César Pairetti	Universidad Nacional de Rosario, Argentina
Victor Albornoz	Universidad Técnica Federico Santa María, Chile
Simone Martins	Universidade Federal Fluminense, Brazil
Antonio Mauttone	Universidad de la República, Uruguay
Débora Ronconi	Universidade de São Paulo, Brazil
Libertad Tansini	Universidad de la República, Uruguay
Carlos Testuri	Universidad de la República, Uruguay

Javier Dario Fernandez   Universidad Pontificia Bolivariana, Colombia
  Ledesma
Carlos Romero   Technical University of Madrid, Spain
Julio Arce   Universidade Federal do Paraná, Brazil
Pete Bettinger   University of Georgia, USA
Pascal Forget   Université du Québec à Trois-Rivières, Canada
Sergio Maturana   Pontificia Universidad Católica de Chile, Chile
  Valderrama
Gilson Adamczuk Oliveira   Universidade Tecnológica Federal do Paraná, Brazil
Juan José Troncoso   Universidad de Talca, Chile
  Tirapegui
Luis Diaz-Balteiro   Technical University of Madrid, Spain
Marcio Pereira da Rocha   Universidade Federal do Paraná, Brazil
Jamal Toutouh   Massachusetts Institute of Technology, USA
Maico Roris Severino   Universidade Federal de Goiás, Brazil
Joaquín Orejas   Universidad Nacional de Río Cuarto, Argentina

Contents – Part II

Simulation

Machine Learning and Big Data

Contents – Part I

Metaheuristics and Algorithms

Industry 4.0 and Cyber-Physical Systems

Smart City

A Benders Decomposition Approach for an Integrated Bin Allocation and Vehicle Routing Problem in Municipal Waste Management

Arthur Mahéo[1], Diego Gabriel Rossit[2(✉)], and Philip Kilby[3]

[1] Monash University, Melbourne, Australia
arthur.maheo@monash.edu
[2] INMABB, DI, Universidad Nacional del Sur (UNS)-CONICET,
Bahía Blanca, Argentina
diego.rossit@uns.edu.ar
[3] Australian National University, Canberra, Australia
Philip.Kilby@data61.csiro.au

Abstract. The municipal solid waste system is a complex reverse logistic chain which comprises several optimisation problems. Although these problems are interdependent – i.e., the solution to one of the problems restricts the solution to the other – they are usually solved sequentially in the related literature because each is usually a computationally complex problem. We address two of the tactical planning problems in this chain by means of a Benders decomposition approach: determining the location and/or capacity of garbage accumulation points, and the design of collection routes for vehicles. We also propose a set of valid inequalities to speed up the resolution process. Our approach manages to solve medium-sized real-world instances in the city of Bahía Blanca, Argentina, showing smaller computing times in comparison to solving a full MIP model.

Keywords: Municipal solid waste · Reverse supply chain · Integrated allocation-routing problem · Benders decomposition algorithm · Valid inequalities

1 Introduction

Regardless of their size, city councils have the duty to provide efficient service to their constituents. Municipal Solid Waste (MSW) management is a crucial example of such a service. It has direct economic and a social impacts; poor collection service can be both expensive and unsanitary. In this paper, we will focus on a less traditional MSW design called Garbage Accumulation Points (GAP). Instead of providing a "door-to-door" pickup of garbage, constituents have to drop their garbage at specific facility – the GAPs. These facilities can range from collective bins to recycling centres. When using GAPs, MSW management comprises the following design decisions:

© Springer Nature Switzerland AG 2021
D. A. Rossit et al. (Eds.): ICPR-Americas 2020, CCIS 1408, pp. 3–18, 2021.
https://doi.org/10.1007/978-3-030-76310-7_1

- The design of a pre-collection network, which consists in defining the location and capacity of GAPs.
- The design and schedule of routes for collection vehicles.

The geographical distribution of GAPs affects the actual route that the collection vehicles must perform. Additionally, the storage capacity of these sites will define the visit frequency in order to avoid overflow. Finally, the availability and type of vehicles[1] affect the distribution and capacity of the GAPs in a (global) optimal solution. Thus, there is a trade-off between the cost of the installation of GAPs and the routing cost; solving both simultaneously is often beneficial [12]. However, solutions in the literature [7,10,20] often address each separately. This is due to the complexity of tackling MSW as a whole. Indeed, only solving the design of routes is tantamount to solving a Vehicle Routing Problem (VRP) [23], a well-known NP-hard problem.

In this paper, we propose the following contributions to the field of MSW management with GAPs:

- A novel mathematical model which combines the allocation of bin arrangements to GAPs and defining collection routes (Sect. 3.1).
- A Benders decomposition-based approach to tackle the resulting problem (Sect. 4).

The problem we are tackling is an "inventory routing problem" [4]. The resulting formulation is a mixed-integer program (MIP) which is still too large to be tractable. However, we can see it as a combination of two problems:

1. a routing problem, similar to a vehicle routing problem [23]; and,
2. an allocation problem, similar to a nonlinear resource allocation problem [3] – in which the used amount of resource (bin) should be minimized, though not limited.

This *natural* decomposition lead us to use Benders decomposition [1], a well-suited method for problems with this structure. Benders decomposition works by solving such problems in an iterative fashion. First, it solves the *difficult* part to generate a candidate solutions. It then checks this solution against the dual of the *easy* part. From the dual solution, it either terminates, when the dual solution's objective value is equal to an incumbent; or, it generates constraints, called "Benders cuts," which are added to the difficult part and the problem solved anew.

However, we cannot use standard Benders decomposition because the subproblem contains integer variables. Therefore, we use a framework called Unified branch-and-Benders-cut (UB&BC) [17]. This framework is based on a modified Branch-and-Cut (B&C) with callbacks from a commercial solver. In the callbacks, it derives dual information and an upper bound for the subproblem. Using these, it terminates the branch-and-bound tree with a set of *open solutions* – whose objective function value falls below the best upper bound. To find

---

[1] Mainly capacity, but could also be cost.

the global optimum, the B&C is followed by a post-processing phase where the framework solves those open solutions to integer optimality.

We test our model on a real-world use case: the city of Bahía Blanca, Argentina (Sect. 6). Although the city currently uses a door-to-door collection service, they are interested in switching to GAPs. We simulated instances using data from a survey [5] and provide optimal allocation and routing for a variety of scenarios. Finally, we present our conclusions and future directions in Sect. 7.

2   A Working Example

In this section we will present a short example to illustrate the workings of our solution approach. We will use a toy instance shown in Fig. 1a. It comprises: a two-day-horizon, two GAPs, and two vehicles. We also consider two types of waste bins with a storage capacities of $1.1\,\mathrm{m}^3$ and $1.73\,\mathrm{m}^3$ respectively.

We will now show the iterations the solution algorithm takes.[2] At each iteration, we will report:

- the master solution (routing cost, which includes GAP allocation per vehicle per day);
- the objective function value of the LP relaxation of the subproblem (lower bound);
- the heuristic value (upper bound); and,
- the total cost of the solution.

A graphical representation of the iterations is provided in Figs. 1b to 1d.

Iteration 1. The first solution uses one vehicles on two days and one vehicle the second day only.

$$v_{0,0} : (0,2) \rightarrow (2,0)$$
$$v_{0,1} : (0,2) \rightarrow (2,0)$$
$$v_{1,0} : (0,1) \rightarrow (1,0)$$

The routing cost is: 494.2. The LP relaxation has an objective function value of 7.96 while the heuristic has a value of 10.48. We add a Benders cut to the master problem and continue.

Iteration 2. The second solution uses two vehicles with different routes during one day:

$$v_0 : (0,2) \rightarrow (2,0)$$
$$v_1 : (0,1) \rightarrow (1,0)$$

The routing cost is: 381.8. The LP relaxation has an objective function value of 7.06 while the heuristic has a value of 10.48. We add a Benders cut to the master problem and continue.

---

[2] We use the complete problem (**M1**) augmented with valid inequalities Eqs. (2) to (4). The Benders cuts we generate are "optimality cuts" given by Eq. (8b).

Iteration 3. The third solution found uses a single vehicle with the same route on both days, given by:

$$v_0 : (0,1) \rightarrow (1,2) \rightarrow (2,0)$$

The routing cost is: 316.8. The LP relaxation has an objective function value of 8.58 while the heuristic has a value of 10.48. We add a Benders cut to the master problem and continue.

At this point, the B&C will finish as no improving solution can be found, we can progress to the post-processing.

(a) Location of the depot and the two GAPs (green circles) on the toy instance.

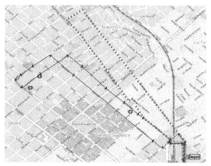

(b) It. 1: The first (blue) vehicle uses its route both days, while the second (red) vehicle only operates on the first day.

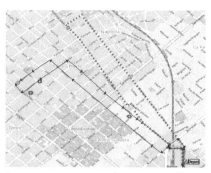

(c) It. 2: Both vehicles operate during the first day.

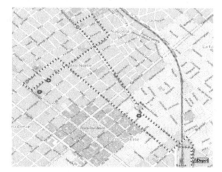

(d) It. 3: Only one vehicle operates during one day

Fig. 1. (a) Location of the depot and the two GAPs (green circles) on the toy instance. (b) It. 1: The first (blue) vehicle uses its route both days, while the second (red) vehicle only operates on the first day. (c) It. 2: Both vehicles operate during the first day. (d) It. 3: Only one vehicle operates during one day. (Color figure online)

Post-processing. At the start of the post-processing phase, the UB&BC orders solutions according to their lower bound values. In this case, it will process the solutions in reverse order: 3, 2, 1.

Starting with the solution found in Iteration 3, we solve the subproblem to integer optimality. This gives an optimal value of 10.48 – the same as the heuristic. Being an integer value, it can be used to update the upper bound to: 347.28 (routing + delivery).

Then, the framework verifies that the remaining open solutions' values are lower that the new-found upper bound. Both solutions found at Iterations 1 and 2 exceed the best upper bound and are thus skipped.

Our approach has managed to find the optimal solution to the problem. It did so as an integrated algorithm which solved the routing and allocation problems at once.

3   A Mathematical Model of MSW

3.1   Model Formulation

The model has the following sets:

- I: the set of potential GAPs.
- $L = \{l_0, l_1, \ldots, l_{|L|}\}$: is an ordered set of vehicles. We consider a homogeneous and finite fleet of vehicles.
- T: the set of days in the time horizon, which coincides with a week (seven days).
- R: the set of possible visit combinations.
- U: the set of all bin arrangements that can be installed in a GAP.

A potential GAP $i \in I$ is a predefined location in an urban area in which bins can be installed. We define the superset: $I^0 = I \cup 0$, where 0 is the depot from which vehicles start and finish their daily tours, and where the collected waste is deposited. We also define a special notation for the set of edges given a set of nodes: $E(\cdot)$, such that: $E(I) = \{ (i, j) \mid i \in I, j \in I, i \neq j \}$. Bin arrangements are set of bins that are feasible to install in a GAP, respecting the space limitation.

We now define the parameters of the model:

- Q: vehicle capacity.
- c_{ig}: travel time between i to g.
- s_i: service time of GAP i.
- b_i: waste generation per day at GAP i.
- cap_u: capacity of bin arrangement u.
- cin_u: adjusted cost of installing bin arrangement u for the time horizon T.
- α: cost per kilometre of transportation.
- β_r: maximum number of days between two consecutive visits of the visit combination r.
- a_{rt}: 1 if day t is included in visit combination r.
- TL: time limit of the working day.

Notice that cin_u is an *adjusted cost*. This is because we are considering two different level of decision and cost:

1. a strategic decision that involves purchasing and installing the bin arrangements that will last probably for several years; and,
2. a tactical decision which involves the transport costs of the routing schedule [19].

Therefore, the cost assigned to a bin arrangement (cin_u) includes a proportional part of the purchase/installation cost and the maintenance cost. With regards to parameters a_{rt} and β_r, they can be introduced with an example:

Example 1. Let time horizon T be a week – i.e., $T = \{t_1, t_2, t_3, t_4, t_5, t_6, t_7\}$. Then, one possible visit combination $r* \in R$ is $\{t_1, t_3, t_5, t_7\}$. In this case, we have: $a_{r*t_1} = a_{r*t_3} = a_{r*t_5} = a_{r*t_7} = 1$, and, conversely: $a_{r*t_2} = a_{r*t_4} = a_{r*t_6} = 0$. Thus, the maximum number of days between two consecutive visits that this combination has is two days: $\beta_{r*} = 2$, and the chosen bin arrangement for this GAP must be able to store the waste generated in two days. (A similar consideration is performed in [12]).

Finally, we define the following decision variables:

- x_{iglt}: binary variable set to 1 if vehicle l performs the collection route between GAPs i and g on day t, 0 otherwise.
- v_{iglt}: continuous variable representing the load of vehicle l along the path between GAP i and g on day t.
- m_{ir}: binary variable set to 1 if visit combination r is assigned to GAP i, 0 otherwise.
- n_{ui}: binary variable set to 1 if bin arrangement $u \in U$ is used for GAP i, 0 otherwise.

We can now present the mathematical model for the MSW management problem:

$$\min \quad \sum_{\substack{i \in I \\ u \in U}} n_{ui}\, cin_u + \alpha \sum_{i,g \in E(I^0)} c_{ig} \left(\sum_{\substack{l \in L \\ t \in T}} x_{iglt} \right) \qquad \textbf{(M1)}$$

$$s.t. \quad \sum_{u \in U} n_{ui}\, cap_u \geq \sum_{r \in R} b_i m_{ir} \beta_r \qquad \forall\, i \in I \quad (1a)$$

$$\sum_{u \in U} n_{ui} = 1 \qquad \forall\, i \in I \quad (1b)$$

$$\sum_{r \in R} m_{ir} = 1 \qquad \forall\, i \in I \quad (1c)$$

$$\sum_{\substack{g \in I^0, g \neq i \\ l \in L}} x_{iglt} - \sum_{r \in R} a_{rt} m_{ir} = 0 \qquad \forall\, t \in T,\, i \in I \quad (1d)$$

$$\sum_{i \in I^0, i \neq q} x_{iqlt} - \sum_{g \in I^0, g \neq q} x_{qglt} = 0 \qquad \forall\, q \in I^0,\, l \in L,\, t \in T \quad (1e)$$

$$\sum_{i \in I} x_{0ilt} \leq 1 \qquad \forall\, l \in L,\, t \in T \quad (1f)$$

$$\sum_{i,g \in E(I^0)} (c_{ig} + s_i)\, x_{iglt} \leq TL \qquad\qquad \forall\, l \in L,\, t \in T \quad (1\text{g})$$

$$v_{iglt} \leq Q\, x_{iglt} \quad \forall\, (i,j) \in E(I^0), l \in L,\, t \in T \quad (1\text{h})$$

$$\sum_{i \in I^0, i \neq g} v_{iglt} + b_g \sum_{r \in R} (m_{gr}\, \beta_r) \leq \sum_{i \in I^0, i \neq g} v_{gilt} + Q \left(1 - \sum_{i \in I^0, i \neq g} x_{iglt} \right)$$
$$\forall\, g \in I,\, l \in L,\, t \in T \quad (1\text{i})$$

$$v \geq 0; n, x, b \in \mathbb{B}$$

The objective function is the sum of the routing cost and the adjusted cost of installing bins. Equation (1a) limits the maximum amount of garbage that can be accumulated in a GAP to the installed capacity of the bin arrangement. Equation (1b) enforces that one bin arrangement has to be chosen for each GAP. Equation (1c) establishes that one visit combination is assigned to each GAP. Equation (1d) ensures that each GAP is visited by the collection vehicle the days that corresponds to the assigned visit combination. Equation (1e) ensures that if a vehicle visits a GAP, it leaves the GAP on the same day. Equation (1f) states that every vehicle can be used at most once a day. Equation (1g) guarantees that a tour does not last longer than the allowable time limit associated with the working day of the drivers. Equation (1h) limits the total amount of waste collected in a tour to the vehicle capacity. Equation (1i) establishes that the outbound flow after visiting a GAP equals the inbound flow plus the waste collected from that GAP and, thus, also forbids subtours.

3.2 Valid Inequalities

The model presented above for the MSW (**M1**) is still a difficult problem. In particular, it contains a lot of *symmetric solutions*. Two solutions are said to be symmetric if they have the same objective function value but different variable assignments. Consider the following: during a given day, two trucks undertaking the same collection route would have the same cost. There is no way for the solver to omit one of them.

One way to address this issue is to add Valid Inequalities (VIs) to the model. A VI is a constraint that reduces the feasible polytope of the problem without removing every optimal solution. We decided to focus on VIs for the routing part of the problem because the allocation part is *easy* in comparison. For examples of VIs in the context of vehicle routing problems, we refer the interested reader to [6].

One thing to remember is that our graph is asymmetric. Therefore, we do not need to address symmetries in routes with the same GAPs. We have developed the following valid inequalities to remove as much symmetry from the optimal solutions as possible.

Empty Start. A vehicle must start its tour unloaded. This prevents solutions with different *delivery plans* – when a vehicle finishes its collection tour below full capacity, we can consider another solution where the vehicle starts with any amount less than the difference.

$$v_{0glt} = 0, \forall\, g \in I,\ l \in L,\ t \in T \tag{2}$$

Vehicle Ordering. We impose that a vehicle with index l can only leave the depot if the vehicle with index $l - 1$ has. In the case where a solution does not use all available vehicles, we can consider swapping an unused vehicle with a used one. For brevity, we define: $L' = L \setminus \{0\}$, as the set of vehicles minus the first one.

$$\sum_{i \in I} x_{0ilt} \leq \sum_{i \in I} x_{0ipt}, \forall\, l \in L',\ p = l - 1,\ t \in T \tag{3}$$

Furthest Visit. We assign the furthest GAP from the depot to the first vehicle. Because each GAP must be visited at most once a day, so does the furthest. Because only one vehicle can visit each GAP on a given day, we can forbid others vehicle than the first vehicle – using L' defined above – to visit the furthest GAP.

$$\sum_{i \in I, t \in T} x_{iglt} = 0, \forall\, l \in L', g = \underset{i \in I}{\operatorname{argmax}}\, c_{0i} \tag{4}$$

4 A Resolution Approach Based on Benders Decomposition

The classic Benders decomposition was devised in [1] for addressing large MIPs that have a characteristic *block diagonal* structure. In summary, it starts by decomposing the *original problem* into a *master problem* and a *subproblem*. The master problem is a relaxation of the original problem used to determine the values of a subset of its variables. It is formed by retaining the *complicating variables*, and projecting out the other variables and replacing them with an *incumbent*. The subproblem is formed around the projected variables and a parameterised version of the complicating variables. By enumerating the extreme points and rays of the subproblem, the algorithm defines the projected costs and the feasibility requirements, respectively, of the complicating variables. Because this enumeration is seldom tractable, the algorithm proceeds in the following manner:

1. It solves the (relaxed) master problem to optimality, which yields a *candidate solution*.
2. This candidate solution is used as a parameter in the subproblem.
3. The resulting problem is solved to optimality and, using LP duality, a set of coefficients are retrieved.

4. These coefficients are used to generate a constraint, called a "Benders cut," which is added to the master problem.
5. If the objective function value of the subproblem is equal to the incumbent value in the master problem, the algorithm stops. Otherwise, it repeats from point 1. using the master problem with the additional constraint.

One key limitation of the classic Benders decomposition is that the subproblem cannot contain integer variables. This is because of point 3. above: the method needs to use LP duality, which is not well-defined for MIPs. We use a recent framework called Unified branch-and-Benders-cut (UB&BC) [18] to bypass this issue. This new framework operates by using a modified B&C where, at each integer node, it:

1. solves the LP relaxation of the subproblem to get a lower bound and generate Benders cuts; and,
2. uses a heuristic to determine if the master solution is feasible and, if yes, a valid, global upper bound.

The second point is key: by maintaining a valid upper bound, the framework ensures that no optimal solution is removed during the search. However, this leads to having a set of *open solutions* after finishing the B&C tree – solutions whose objective function value falls between the lower and upper bound. Thus, the UB&BC finishes by a *post-processing phase* during which subproblems associated with open solutions are solved to integer optimality. The combination of maintaining a global upper bound and using a post-processing phase enables the framework to find an optimal solution.

As stated in Sect. 3.1, the problem addressed in this work comprises two characteristic decision-making problems in MSW. On the one hand, the allocation of bins in the GAPs and, on the other, the design and schedule of routes for the collection vehicles. This division can be exploited by applying Benders decomposition. The bins allocation equations are moved to the subproblem while the master problem aims at designing the schedule and routes of the collection vehicles.

4.1   Creating the Subproblem

The subproblem, which aims at allocating the bins of each GAP, is an integer programming problem:

$$q(\overline{m}) = \min \qquad \sum_{\substack{i \in I \\ u \in U}} n_{ui} \, cin_u \qquad\qquad\qquad \textbf{(SB)}$$

$$s.t. \qquad \sum_{u \in U} n_{ui} \, cap_u \geq b_i \sum_{r \in R} \overline{m_{ir}} \, \beta_r \qquad \forall \, i \in I \qquad (5a)$$

$$\sum_{u \in U} n_{ui} = 1 \qquad\qquad\qquad \forall \, i \in I \qquad (5b)$$

$$n \in \mathbb{B}$$

We define the positive continuous variables δ_i and unrestricted continuous variables γ_i, the dual variables of Eqs. (5a) and (5b) respectively. The dual formulation of the LP relaxation of (**SB**), which will be used to generate cuts, is then:

$$q^{LP}(\overline{m}) = \max \quad \sum_{i \in I} \left(\gamma_i - \delta_i b_i \sum_{r \in R} (\overline{m_{ir}} \beta_r) \right) \tag{LP}$$

$$\text{s.t.} \quad \gamma_i - \delta_i \sum_{u \in U} cap_u \leq \sum_{u \in U} n_{ui} \qquad \forall i \in I \tag{6a}$$

$$\delta, \gamma \geq 0$$

Heuristic for the Subproblem. In order to apply Benders decomposition when the subproblem has integer variables, an efficient method for solving the subproblem is required. Therefore, we devised a rounding heuristic procedure based on the LP relaxation of the subproblem:

1. We solve the LP relaxation of (**SB**). The (relaxed) solution will contain n_{ui} with fractional values.
2. We estimate the *joint fractional capacity* K_i^f of each GAP using:

$$K_i^f = \sum_{u \in U} n_{ui} \, cap_u \tag{7}$$

3. We define a feasible (non-fractional) bin arrangement $u \in U$ for each GAP by finding the bin arrangement with minimal cost among those with storage capacity larger than K_i^f. It is guaranteed that there will always be a bin arrangement which respects this rule since considering Eq. (7) and Eq. (5b) implies that: $K_i^f \leq cap_{u*}, \forall\, i \in I$, where $u* = \underset{u \in U}{\operatorname{argmax}}\{cap_u\}$.

4.2 Stating the Master Problem

The master problem retains the same constraint structure as (**M1**) but the bin allocation part is replaced by an incumbent variable q. Let us consider the set of extreme points ($\mathcal{O}$) and extreme rays ($\mathcal{F}$) of the LP relaxation of (**SB**). These generate the optimality (8b) and feasibility (8a) cuts, respectively. Therefore, the master problem is:

$$\min \quad \alpha \sum_{i,g \in E(I^0)} c_{ig} \left(\sum_{\substack{l \in L \\ t \in T}} x_{iglt} \right) + q \tag{MPB}$$

$$\text{s.t.} \quad \text{Eqs. (1c) to (1i) and Eqs. (2) to (4)}$$

$$\sum_{i \in I} \left(\gamma_i^f - \delta_i^f b_i \sum_{r \in R} (\beta_r m_{ir}) \right) \leq 0 \qquad \forall\, f \in \mathcal{F} \tag{8a}$$

$$\sum_{i \in I} \left(\gamma_i^f - \delta_i^o b_i \sum_{r \in R} (\beta_r m_{ir}) \right) \leq q \qquad \forall\, o \in \mathcal{O} \qquad (8b)$$

$$v \geq 0; q \in \mathbb{R}; x \in \mathbb{B}$$

5 Literature Review

Allocation of bins and routing problems have been thoroughly studied as separate problems in the MSW related literature [16]. Comprehensive reviews of the study of these problems separately can be found in [20] and [10], respectively. However, the number of works considering integrated approaches is more scarce.

Among the works that consider an integrated approach, [14] presented a study case of the Tunisian city of Sousse, considering uncertainty in waste generation at GAPs. They proposed a transformed formulation to handle stochastic waste generation and solved the problem in a heuristic fashion: first they applied the k-means clustering algorithm to group the GAPs into sectors and later they applied an exact model solved with CPLEX to determine both the number of bins and the collection route of each sector. They consider that all GAPs are to be collected daily.

Another example is [12], which proposed an integrated approach where the bins allocation problem is solved jointly with the routing schedule. The authors compare different methods to solve this problem based on a Variable Neighbourhood Search (VNS) algorithm, for solving the problem hierarchically – i. e., first solving the bin allocation and then the routing and *vice versa* – and integrated approaches. They found that integrated approaches overcome hierarchical ones.

A further complex approach is presented in [13] for solving an integrated model that aims to simultaneously locate GAPs, size the storage capacity of each GAP (allocate bins) and set the weekly collection schedule and routes in the context of collaborative recycling problem. They solved this problem with an Adaptive Large neighbourhood Search algorithm based on Hemmelmayr's implementation [11]. They performed a sensitivity analysis for several of the parameters, such as available vehicle capacities, visiting schedules or GAPs' storage capacities.

Although integrated models have proven to efficiently handle the trade-off [7, 9], others works have tackled these problems in a sequential fashion. For example, [8] solved the GAP location with while considering that bins of incompatible types – i. e., that cannot be collected by the same vehicle – are not located in the same GAP. Then they applied an heuristic zoning algorithm to define the routes while minimising the number of required vehicles and the total distance. Finally, [21] solved a multi-objective bins allocation problem in which one of the objectives was to minimise the required collection frequency.

Differently from the tactical problem that we address in this paper, other authors have used Bender's decomposition approaches to deal with optimization problems of the strategic level of the MSW logistic chain, mainly considering stochastic parameters (which is a traditional application area of Bender's decomposition). For example, [22] used Bender's decomposition to model a logistic chain of MSW in which organic waste in send from sources into treatment plants to generate power. Uncertainty is considered in waste generation, and power price and demand. Another case are [15], who applied Bender's decomposition to optimize the location and capacity selection of waste transfer stations when considering uncertainty in the operational cost of the stations.

6   Computational Experiments

The instances used for these tests are based on simulated scenarios of the city of Bahía Blanca, Argentina. Although this city still has a door-to-door collection system, the local government and citizens are interested in studying more efficient collection systems that allow them to reduce the high logistic costs. In this sense, a community bins-based will simplified the required collection logistic [2,5]. Particularly the location of 76 GAPs in a central neighbourhood of the city and the generation rate were obtained from a recent field work in a central neighbourhood of the city [5]. We consider a *homogeneous VRP* by using the standard collection vehicle of Bahía Blanca, a 20 m^3 bin-tipper truck. The GAPs can hold three types of commercial bin with purchasing and maintenance cost (cin_u) 2.76, 3.53 and 5.24 monetary units (m.u.) and capacities (cap_u) 1.1, 1.73 and 3.1 m^3, respectively. Information about the travel time between GAPs was estimated with Open Source Routing Machine^3 using the approach proposed by Vázquez Brust [24]. The algorithms are implemented in Python 3.5, and we use a UB&BC framework called BRANDEC^4 v0.7. For the benefit of the scientific community, we open-sourced the instances used for the experimentation^5 and implementation. The solver used is CPLEX v12.7 in its default configuration, we disable heuristics when running the UB&BC. We ran the experiments on a computer with Intel Gold 6148 Skylake CPU@2.4 GHz and a 4 GB RAM limit.

We divide the computational experimentation in two parts. In Sect. 6.1 we deal with small instances in order to asses the value of the proposed valid inequalities in the resolution approach. Then, in Sect. 6.2 we explore the performance of the proposed Benders approach in comparison to full MIP when solving more complex instances.

6.1   The Value of Valid Inequalities

In order to explore the impact of valid inequalities in the resolution process we construct instances composed by five GAPs and the Depot. These sample of five GAPs were selected with QGIS Random Selection tool. We consider scenarios with two vehicles and two bin arrangements per GAP, the instances are formatted as "$I/T/n$," where n is the instance number. Figure 2 report the results of solving the resulting problem with:

---

^{3} http://project-osrm.org/.
^{4} https://gitlab.com/Soha/brandec.
^{5} http://doi.org/10.13140/RG.2.2.19210.49604.

MIP CPLEX using (**M1**);
MIP + VIs CPLEX using (**M1**) augmented with VIs (2) to (4);
BD our Benders approach; and,
BD + VIs our Benders approach augmented with VIs (2) to (4).

We ran five iterations of each configuration and report the minimum solve time. We can see in Fig. 2 that the VIs are necessary to have reasonable solve times. Both the MIP and our Benders approach benefit from them. When instances grow in size, that is when the time horizon is greater than two days, version with VIs do not manage to solve most instances. This experiment is not enough to tell for certain whether the Benders approach is better than MIP.

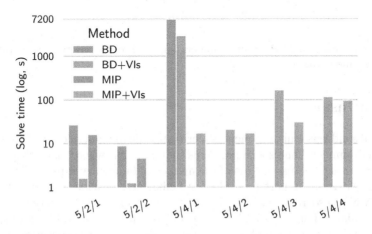

Fig. 2. Results of using different methods, with or without VIs, to solve a set of reduced instances.

6.2 Using More Bin Arrangements

The complexity of the problem grows with the size of the instance becoming increasingly time consuming for the algorithm. Indeed, we now use a larger set of bin arrangements (set U) for the GAPs in order to have a more complex subproblem with a larger number of binary variables. In Fig. 3 we can see the time taken to find the optimal solution on a set of instances by our approach (BD) vs. solving the full MIP model with CPLEX (MIP). All the solutions have the valid inequalities (2) to (4) included. Our approach is, on average, more than one order of magnitude faster than using a full MIP. We are able to solve instances up to 7 GAPs, which the MIP cannot.

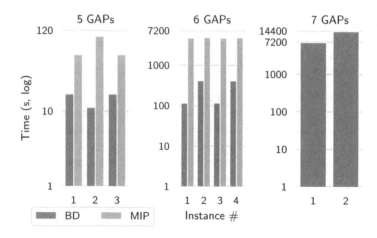

Fig. 3. Comparison of MIP vs. our Benders approach on a variety of scenarios. We report the time in log scale.

7   Conclusion

Municipal solid waste management is a critical issue in modern cities. Besides the direct environmental and social problems that can arise when it is mishandled, it usually represents a large portion of the municipal budgetary expense. Therefore, intelligent decision support tools that can efficiently provide this service to the citizens while also reducing the cost of the system can be a major asset for decision makers. This work addresses two common tactical problems that arise in the reverse logistic chain of solid waste: the design of a pre-collection network and the routing schedule of collection vehicles. These problems, which are usually solved individually in the related literature, are interdependent in the sense that the solution to one of the problem affects the other.

This work proposed an integrated approach that solves both problems simultaneously, making the trade-off an intrinsic element of the model. In particular, a new MIP formulation, valid inequalities and resolution approach based on Benders decomposition, using unified branch-and-Benders-cut, were proposed. Since the subproblem contains integer variables, we devised a heuristic for solving the bin allocation problem. Our approach was able to solve real-world instances in the city of Bahía Blanca. Tests performed on small instances showed the competitiveness of the valid inequalities to speed up the resolution process. Then, the resolution of larger instances showed that the proposed Benders approach was more competitive than full MIP. While not yet able to solve full-size instances, our approach holds promises to scale beyond a traditional MIP approach.

Future work includes expanding computational experiments with larger real-world instances to test the scalability of the approach. Another research line is to consider an allocation-first routing-second method. In that case, the master problem would be comparatively simpler than the subproblem. Such an approach would require efficient vehicle routing heuristics to work. We could also explore heterogeneous fleet of vehicles. Indeed, the city of Bahía Blanca already owns a fleet of vans of small capacity for spot operations.

Acknowledgements. The second author was supported by the Australia–Americas PhD Research Internship Program co-financed by the Australian Academy of Science, the Ministry of Foreign Affairs and Worship of Argentina and the Universidad Nacional del Sur.

References

1. Benders, J.F.: Partitioning procedures for solving mixed-variables programming problems. Numerische mathematik **4**(1), 238–252 (1962)
2. Bonomo, F., Durán, G., Larumbe, F., Marenco, J.: A method for optimizing waste collection using mathematical programming: a Buenos Aires case study. Waste Manag. Res. **30**(3), 311–324 (2012)
3. Bretthauer, K.M., Shetty, B.: The nonlinear resource allocation problem. Oper. Res. **43**(4), 670–683 (1995)
4. Campbell, A., Clarke, L., Kleywegt, A., Savelsbergh, M.: The inventory routing problem. In: Crainic, T.G., Laporte, G. (eds.) Fleet Management and Logistics. CRT, pp. 95–113. Springer, Boston, MA (1998). https://doi.org/10.1007/978-1-4615-5755-5_4
5. Cavallin, A., Rossit, D.G., Herrán Symonds, V., Rossit, D.A., Frutos, M.: Application of a methodology to design a municipal waste pre-collection network in real scenarios. Waste Manag. Res. **38**(1), 117–129 (2020)
6. Dror, M., Laporte, G., Trudeau, P.: Vehicle routing with split deliveries. Discrete Appl. Math. **50**(3), 239–254 (1994)
7. Ghiani, G., Laganà, D., Manni, E., Musmanno, R., Vigo, D.: Operations research in solid waste management: a survey of strategic and tactical issues. Comput. Oper. Res. **44**, 22–32 (2014)
8. Ghiani, G., Manni, A., Manni, E., Toraldo, M.: The impact of an efficient collection sites location on the zoning phase in municipal solid waste management. Waste Manag. **34**(11), 1949–1956 (2014)
9. Ghiani, G., Mourão, C., Pinto, L., Vigo, D.: Chapter 15: routing in waste collection applications. In: Arc Routing: Problems, Methods, and Applications, pp. 351–370. SIAM (2015)
10. Han, H., Ponce Cueto, E.: Waste collection vehicle routing problem: literature review. PROMET-Traffic Transp. **27**(4), 345–358 (2015)
11. Hemmelmayr, V.C.: Sequential and parallel large neighborhood search algorithms for the periodic location routing problem. Eur. J. Oper. Res. **243**(1), 52–60 (2015)
12. Hemmelmayr, V.C., Doerner, K.F., Hartl, R.F., Vigo, D.: Models and algorithms for the integrated planning of bin allocation and vehicle routing in solid waste management. Transp. Sci. **48**(1), 103–120 (2013)
13. Hemmelmayr, V.C., Smilowitz, K., de la Torre, L.: A periodic location routing problem for collaborative recycling. IISE Trans. **49**(4), 414–428 (2017)
14. Jammeli, H., Argoubi, M., Masri, H.: A bi-objective stochastic programming model for the household waste collection and transportation problem: case of the city of Sousse. Oper. Res., 1–27 (2019)
15. Kŭdela, J., et al.: Multi-objective strategic waste transfer station planning. J. Cleaner Prod. **230**, 1294–1304 (2019)

16. Lu, J.W., Chang, N.B., Liao, L., Liao, M.Y.: Smart and green urban solid waste collection systems: advances, challenges, and perspectives. IEEE Syst. J. **11**(4), 2804–2817 (2015)
17. Mahéo, A.: Benders and its sub-problems. Ph.D. thesis, Australian National University, Canberra, Australia (2020)
18. Mahéo, A., Belieres, S., Adulyasak, Y., Cordeau, J.F.: Unified branch-and-Benders-cut for two-stage stochastic mixed-integer programs. Les Cahiers du GERAD G-2020-54 (2020)
19. Nagy, G., Salhi, S.: Location-routing: issues, models and methods. Eur. J. Oper. Res. **177**(2), 649–672 (2007)
20. Purkayastha, D., Majumder, M., Chakrabarti, S.: Collection and recycle bin location-allocation problem in solid waste management: a review. Pollution **1**(2), 175–191 (2015)
21. Rossit, D.G., Toutouh, J., Nesmachnow, S.: Exact and heuristic approaches for multi-objective garbage accumulation points location in real scenarios. Waste Manag. **105**, 467–481 (2020)
22. Saif, Y., Rizwan, M., Almansoori, A., Elkamel, A.: Municipality solid waste supply chain optimization to power production under uncertainty. Comput. Chem. Eng. **121**, 338–353 (2019)
23. Toth, P., Vigo, D.: The vehicle routing problem. SIAM (2002)
24. Vázquez Brust, A.: Ruteo de alta perfomance con OSRM. Rpubs by RStudio (2018). https://rpubs.com/HAVB/osrm

Design of a *Nanostores'* Delivery Service Network for Food Supplying in COVID-19 Times: A Linear Optimization Approach

Daniela Granados-Rivera[✉] [iD], Gonzalo Mejía, Laura Tinjaca, and Natalia Cárdenas

Sistemas Logísticos Research Group, Faculty of Engineering, Universidad de La Sabana, Campus Universitario del Puente Común Km 7, Autopista Norte de Bogotá D.C., Chía, Cundinamarca, Colombia
{danielagrri,gonzalo.mejia,lauratiro,
karencarva}@unisabana.edu.co

Abstract. This work aims the implementation of a facility location mathematical model to improve the fruit and vegetable (F&V) demand coverage using the delivery service in the traditional channel (*nanostores*) and considering the changes caused by the COVID-19. It is a relevant topic due to the constraints related to social distance employed by many governments to avoid spread and ensure food security. The mathematical model has the objective to enhance the spatial distribution of *nanostores* for increasing the accessibility and availability of F&V through the delivery service. We used Chía and Cajicá as case studies, two towns of the Sabana Centro Region in Colombia's central part. The results indicate that implementing the delivery service in less than 40% of *nanostores* is achieved the total geographical coverage of both towns, improving the critical zones such as peripherical areas. However, to enhance the demand coverage (food tons), the delivery service is necessary for more than 50% of all *nanostores*.

Keywords: Delivery · Facility location problem · Food logistics · Nanostores · Corner stores · COVID-19

1 Introduction

From the beginning of the COVID-19 emergency, there have been more than 14 million cases and 591 thousand deaths worldwide. The virus has spread in more than 200 countries, being Colombia one of them, where there are more than 182 thousand cases and 6 thousand deaths [1]. As a solution, many countries have ruled lookdowns to reduce the number of ill people. However, this strategy has generated significant impacts on economic and trade development [2, 3].

Those pandemic restrictions have meant a logistic challenge without antecedents and a disruption on supply chains (SC) [4]. That is why there is a significant risk on operations, including changes in demand, distribution, and supply [5]. The *"retail"* sector has been significantly affected since to guarantee food access; it had to implement quick

D. A. Rossit et al. (Eds.): ICPR-Americas 2020, CCIS 1408, pp. 19–32, 2021.
https://doi.org/10.1007/978-3-030-76310-7_2

solutions to respond to consumers' change behaviors due to the politics adopted by many governments during the quarantine [6].

Factors that have affected the demand have been social distance and mobility constraints [7]. About the food sector, those limitations have implied panic buying and increases in online purchases [8]. The last factor has caused many retailers to implement delivery service to enhance their sales, mitigating the COVID-19 effects [6]. The implementation of the delivery services is considering several studies in which distance is one of the most critical factors in food buying decisions [9]. That is why ensuring food access and food availability is essential to handle economic consequences by the constraints mentioned above. Thus, the delivery service is a great opportunity to deal with this problem [3].

Delivery services are a significant part of e-commerce, being particularly complex for retailers to have to increment technological environment as well as logistic operations for last-mile distribution [10]. Additionally, the management of diverse sales channels represents a challenge for people who only handle in-person sales. Therefore, the immediate transformation that retailers suffered by the sanitary emergency has involved different inefficiencies [11].

Considering the above, the design of a delivery services' network is a relevant strategic decision to impact food access with the current pandemic scenario. This analysis allows us to evaluate food supply for households [12] and guarantee food security. For these reasons, this study proposes a mathematical model of mixed-integer linear optimization (MILP) to design a delivery service network of *nanostores* to improve fresh food access in Chía and Cajicá, Colombia. *Nanostores* in Colombia have been the main supply way during the quarantine. They have increased up to 96% of their sales [13, 14], taking into account the growth in e-commerce [15]. Delivery service might ensure the *nanostores* survival despite economic implications by the COVID-19 [10]. Also, we selected fresh food, mainly fruits and vegetables (F&V), because they have a relevant role in guaranteeing food security, considering low consumption in the study area [16].

This work is organized as follows. Section 2 shows background related to the topic of this research. Section 3 presents the formal problem definition and the associated mathematical model. Section 4 describes a case study in two towns, Chía and Cajicá. Section 5 explains the main results of the implementation in the case study. Finally, Sect. 6 argues the conclusions of this study and suggests some future research.

2   Background

In this section, we give a brief review of concepts related to the development of this work.

2.1   Disruptions in SC by Pandemics

The pandemic effects on SC have been studied from a humanitarian logistic perspective [17, 18]. Nonetheless, the great part of these investigations has focused on improving the response capacity to supply provisions and medical resources. For example, He et al. [19] presented a new linear programming model for a fast response of healthcare services to deal with sanitary emergencies. On the other hand, Anparasan et al. [20] proposed an analytics model to assign medical resources in developing countries using cholera epidemic information in Haití in 2010.

However, there are few studies associated with economic and trade impacts on SC by pandemics. Chou et al. [21] have researched the effects of the SARS virus in 2004, finding that the bigger consequence for Taiwán was in the airline sector. Other authors were Calnan et al. [22], who analyzed the effect of Ebola virus in Africa on social and economic areas concluding that building a support system deals with the pandemic impacts.

With the COVID-19, there are new studies due to many countries closing their borders, generating significant impacts on trade [5]. Ivanov [5, 23] developed one of these studies analyzing the factors required to create resilient SC and forecast the long-term implications using simulation. That study was general, whereas Nicola et al. [8] described each economic area's impacts. For the food supply area, they have mentioned that the growth in delivery services is an important part of keeping the economy. Similarly, Li et al. [6] did a survey in Wuhan in which they confirmed that e-commerce prevails to the food supply. However, different countries have adopted these strategies without previous planning, considering it is an important part of strategic decision-making.

2.2 Location Models to Design Delivery Service Network

Implementing delivery services is a relevant strategic decision since it implies redesign SC [12]. This approach can be analyzed from mathematical models MILP, particularly the facility location problem (FLP). This model consists of selecting locations for one or more facilities to serve the consumers' demand [24]. For delivery service, the studies have focused on tactical decisions associated with distribution. For example, Cagliano et al. [25] used dynamic systems to analyze SC flow and optimize route distribution according to real-time traffic. The results showed that the use of technological tools improves the efficiency of last-mile distribution. Authors as Pan et al. [26], Chen et al. [27], and Aktas et al. [28] employed mathematical models of vehicle routing problems to reduce operational costs, distance and enhance distribution efficiency.

In the scope of this research, we only found two studies that deal with the implantation of delivery service in supply points. The first study was made by Zambetti et al. [12], who used a facility location model to allocate depots for intermediate companies that provide delivery services. The second study was done by Hong et al. [29], who assigned delivery depots and optimized the distribution routes. Nonetheless, no one of these models considered the delivery service assignment directly in sales points according to geographical coverage. For this reason, this work aims to deal with this assignment problem considering mobility constraints by the COVID-19.

3 Proposal

In this section, the formal problem definition of delivery service network design is presented and the mathematical model MILP associated with the problem.

3.1 MILP Model

The problem definition is as follows:

A set of consumers (households) $H = \{1...j\}$ is considered geographically distributed in the study area according to residential density. They must be served fresh foods. Each household $j \in H$ has a demand d_j that has to be fulfilled by a *nanostore* $R = \{1...i\}$ within a maximum coverage distance according to the sales channel (in-person or delivery) $V = \{1...k\}$ defined by the parameter a_{ijk}. Each *nanostore* $i \in R$ has a capacity Q_{ik} for each sales channel $k \in V$. We consider a revenue P_1 for each household $j \in H$ within the coverage range of each *nanostore* $i \in R$ and other revenue P_2 for each kilogram of household demand $j \in H$ supplied by *nanostore* $i \in R$. Additionally, there is a cost C_k to implement sales channel $k \in V$. Finally, we establish a maximum number N of *nanostores* that can have delivery service and a minimum demand percentage β to cover total F&V demand.

3.2 Model Formulation

The FLP model is formulated as follows:

Sets

H set of households; $j \in H$
R set of potential *nanostores* to implement delivery service; $i \in R$
V set of sales channels offers by *nanostores*; $k \in V$

Parameters

d_j average estimated demand for household j
Q_{ik} capacity for each *nanostore* i in the sales channel k
a_{ijk} a binary parameter in which is 1 if *nanostore* i covers household j by channel k and 0 otherwise
P_1 revenue for geographical coverage of one household
P_2 revenue per covered kilogram of F&V demand
C_k cost for assignment sales channel k
N maximum percentage of *nanostores* that can have delivery service
t total number of *nanostores*
β minimum percentage of total F&V demand to cover
M sufficiently large number for logical constraints

Decision variables

$$X_{ik} = \begin{cases} 1 & \textit{If nanostore } i \textit{ uses sales channel } k \\ 0 & \textit{Otherwise} \end{cases}$$

$$Y_{jk} = \begin{cases} 1 & \textit{If household } j \textit{ is covered by sales channel } k \\ 0 & \textit{Otherwise} \end{cases}$$

$Z_{ji} =$ Proportion of household j demand covered by *nanostore* i

The objective function is:

$$Max \sum_{j \in H} \sum_{k \in V} P_1 Y_{jk} + \sum_{i \in R} \sum_{j \in H} P_2 d_j Z_{ji} - \sum_{i \in R} \sum_{k \in V} C_k X_{ik} \tag{1}$$

Subject to:

$$\sum_{i \in R} a_{ijk} X_{ik} \geq Y_{jk} \quad \forall j, k; j \in H, k \in V \tag{2}$$

$$\sum_{i \in R} a_{ijk} X_{ik} \leq M Y_{jk} \quad \forall j, k; j \in H, k \in V \tag{3}$$

$$Z_{ji} \leq \sum_{k \in V} a_{ijk} X_{ik} \quad \forall i, j; i \in R, j \in H \tag{4}$$

$$\sum_{i \in R} Z_{ji} \leq 1 \quad \forall j; j \in H \tag{5}$$

$$\sum_{i \in R} \sum_{j \in H} d_j Z_{ji} \geq \beta \sum_{j \in H} d_j \quad \forall t; t \in T \tag{6}$$

$$\sum_{j \in H} d_j Z_{ji} \leq \sum_{k \in V} Q_{ik} X_{ik} \quad \forall i; i \in R \tag{7}$$

$$X_{ik} = 1 \quad \forall i, k = 1; i \in H, k \in V \tag{8}$$

$$\sum_{i \in R} X_{ik} \leq N * t \quad \forall k = 2; k \in V \tag{9}$$

$$Z_{ji} \geq 0 \quad \forall j, i; i \in R, j \in H \tag{10}$$

$$X_{ik}, Y_{jk} \in \{0, 1\} \quad \forall i, j, k; i \in R, j \in H, k \in V \tag{11}$$

The objective function (1) maximizes the profit of *nanostores* considering both revenues for geographic coverage and sale kilograms and the cost to implement delivery service. Constraint sets (2) and (3) establish that a sales channel can cover a household within a coverage range of a *nanostore* with that channel. Constraint set (4) ensures that a nanostore sales channel can only cover household demand if it has that channel within the coverage range. Constraint set (5) limits the coverage of household demand to no more than 100%. Constraint set (6) guarantees that the F&V demand proportion covered is more than or equal to a pre-establish value. Constraint set (7) avoids that *nanostores'* sales be more than their capacities. Constraints set (8) defines the in-person channel for all *nanostores*, whereas constraint set (9) restricts the number of *nanostores* that can have delivery service. Finally, constraint sets (10) and (11) correspond to the nonnegative and binary restrictions for variables decision.

4   Description of Case Study

For this work, we used data collected from the towns Chía and Cajicá. Both are Colombian towns that belong to the Cundinamarca department located in the Sabana Centro Region, to 10 and 17 km from northern Bogotá city (capital of Colombia), respectively. Chía has a population of 142,302 [35], while Cajicá has 62,862 [36]. Both towns are approximately 30% of the geographical territory of the region.

In this zone, malnutrition remains, illustrating it with the rate of born children with low weight, being approximately 11.2%, a worrying figure [30]. One of the significant factors is the insufficient consumption of F&V [16] since it is 190 g per person in Colombia, whereas OMS suggests 400 g [31]. Food supply in this area has relevant problems. For instance, transport costs are higher, and times to supply are longer because people have to go until Bogotá [32]. Also, the lack of inventory management planning made many *nanostores* unable to respond to the demand increases by the lockdown. They could not introduce new alternative sales channels to keep themselves during the pandemic [6].

4.1 Data Preparation

Demand points were established, locating the population density of the study area using Google Earth™. For Chía, we located 700 demand points, whereas for Cajicá was 495 points. With population information of both towns, we assumed that each point of Chía cluster 203.3 inhabitants, whereas Cajicá cluster 127 inhabitants. The demand per person was determined using the Fruit and Vegetable National Consumption Profile Report [37], in which for Chía and Cajicá, the average F&V intake was 107 g and 107.8 g, respectively. Therefore, we considered that total consumption per person was 0.2148 kg. We determined each demand point a weekly demand of 305.665 kg for Chía and 190.948 kg for Cajicá.

On the other side, the number of potential *nanostores* in both towns was determined by Google Earth™ and Google Maps™. For Chía, we found 144 *nanostores,* and for Cajicá 104. Distances were calculated using the Haversine equation, which measures the distance between two points in a sphere according to latitude and longitude [33]. To consider road structure and other geographical conditions, we employed a multiplying factor [34]. The coverage ranges were for in-person channel 0.5 km, while for delivery service channel 2 km.

We used a survey made to 74 *nanostores* owners to define the capacity. Table 1 shows the results for buying frequency and quantity. We observed that 51% of owners supplied themselves every 2 or 3 days, and 22% bought between 151 and 180 kg every time they go to buy. That is why the capacity was taken by the result buying average quantity given by both data mentioned before. With these results, we obtained that the weekly capacity for *nanostores* in Chía was 672 kg, and in Cajicá 415 kg. We supposed an additional capacity of 15% when is enabled the delivery service in a *nanostore*.

The revenue for one household's geographical coverage was COP 82,500 (USD 22.1 approximately), determined according to the weekly value of the basic food basket in Colombia [35]. The revenue associated with the F&V sales of each *nanostore* was COP 3,500 (USD 1 approximately) per kilogram, which was taken from the Daily Report

Table 1. Results of purchase frequency and quantity according to the survey applied *nanostores'* owners

Purchase frequency	%	Quantities	%
2–3 times per week	51%	More than 180 kg	34%
Daily	27%	151–180 kg	22%
Once per week	16%	31–60 kg	19%
4–6 times per week	5%	91–120 kg	11%
Less than once per week	1%	61–90 kg	7%
		Less than 30 kg	4%
		121–150 kg	4%

of Prices of Corabastos (the main central market in Colombia) [36]. For the cost of implementing a sales channel, we assumed that in-person, the cost was COP 0 because this service is available for all *nanostores*. The cost was COP 714,000 (USD 192) for delivery service, considering the average time travel from a *nanostores* to a household in the maximum distance of coverage and the minimum salary for Colombia in 2020 [37].

5   Results

The mathematical model was implemented in GAMS software. We tested numerous instances with the variation of the maximum number of *nanostores* N that can have delivery service and the minimum percentage of households' F&V demand to cover β. The parameter N changes were made between 20% and 70% of the total number of *nanostores* t in Chía and Cajicá.

La Fig. 1 illustrates the real F&V demand coverage in the different instances with β of 0.1 and 0.5. We observed that the percentage of demand coverage varied, especially by *nanostores* N enabled to have delivery service. The minimum percentage of demand to cover β did not affect significatively the results. For instance, for β between 5% and 40% the obtained curve is the same as the shown one in Fig. 1 with β equal to 0.1.

The behavior mentioned above is due to the demand coverage depends on the *nanostores'* capacity, and the delivery service increases that capacity. It means, the more *nanostores* with delivery service, the more capacity to cover households' demand. We noticed that the covered demand stabilized from enabling the delivery service in 40% of total *nanostores* with a minimum demand to cover β up to 40%. This result is relevant that more than 40% of nanostores with delivery service are not necessary to improve the demand coverage. However, to cover more than 50% of the households' F&V demand, the delivery service has to enable at less 60% of the total current number of *nanostores*. That implementation would imply higher costs for *nanostores* owners to provide that service. It is important to highlight that we did not consider other sellers of F&V such as supermarkets, wholesalers, central markets, among others. Therefore, the solution that covers 50% of the demand by *nanostores* can assume as a good solution because the other sellers might supply the remaining demand.

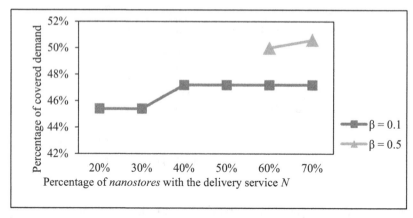

Fig. 1. Representation of the total percentage of households' demand covered according to variations in the number of *nanostores* enabled with the delivery service and the minimum percentage of F&V demand to cover.

On the other hand, the found pattern of the covered demand stability is confirmed with the results shown in Table 2. The number of enabled *nanostores* with delivery service is illustrated in each instance. It is observed that this number does not exceed 75 *nanostores* for β smaller than 40%. The last percentage does not affect the results until it is evaluated with 50%. The solution evidence that it is necessary to enable the delivery service in more than 75 *nanostores*.

Table 2. Number of enabled *nanostores* to provide the delivery service according to the changes in the parameters N and β

N	$\beta = 5\%$	$\beta = 10\%$	$\beta = 20\%$	$\beta = 30\%$	$\beta = 40\%$	$\beta = 50\%$
25%	9	9	9	9	9	
30%	9	9	9	9	9	
35%	75	75	75	75	75	
50%	75	75	75	75	75	
60%	75	75	75	75	75	148
75%	75	75	75	75	75	173

About the geographical coverage, Fig. 2 indicates uncovered areas which are located in peripherical zones of both towns, Chía and Cajicá (red points). These areas might have a higher COVID-19 spread risk because they have to go away from their homes to buy food. This coverage considers only the in-person channel for *nanostores* and the assumption that the household buying is made walking to supply points.

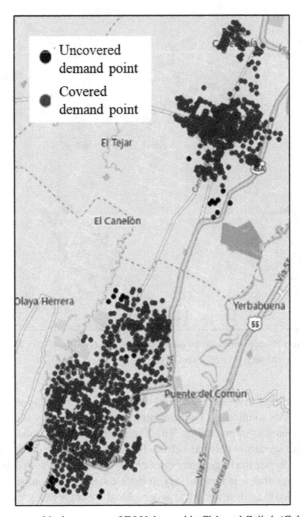

Fig. 2. Current geographical coverage of F&V demand in Chía and Cajicá. (Color figure online)

With the different instances, the geographical coverage improves. For example, for a solution with β equal to 40% and N equal to 30%, only it is enabled delivery service in 9 *nanostores* (See Table 2) distributed as shown in Fig. 3(a). Those locations obtain the geographical coverage represented in Fig. 4. We observed that this solution's delivery service provided the current scenario's uncovered demand points (see Fig. 2). That coverage guarantees that all demand points can access at least one *nanostore* for supplying by some sales channel. The F&V demand coverage is 45.4% in this solution.

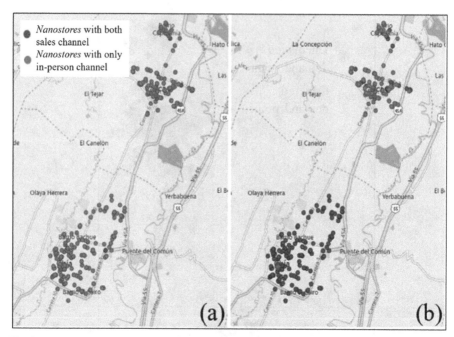

Fig. 3. Type of sales channel provided by each *nanostore* in scenario (a) with β of 40% and N of 30%, and in scenario (b) with β of 50% and N of 70%.

To analyze coverage in another scenario, we represented the solution with β of 50% and N of 70%. In this solution, 173 *nanostores* can provide delivery service (See Table 2) spatially distributed as shown in Fig. 3(b). The spatial coverage is the same as shown in Fig. 4 with the scenario analyzed before. It means that the geographical coverage does not enhance despite the larger number of enabled *nanostores* with delivery service. Those results justify that it is unnecessary to have a significant number of *nanostores* with delivery service to achieve complete geographical coverage of F&V demand points, as we mentioned above.

Nonetheless, this solution's demand coverage increases to 50.6%, 5% more than the solution shown in Fig. 3(a), improving F&V availability for both towns. Although the geographical accessibility does not change in this solution, the F&V availability raises to supply households' demand. Furthermore, this solution allows both towns to have more food supply options because there are more *nanostores* with both sales channels. However, having more *nanostores* with delivery service increases individual costs for each owner, as we analyzed before.

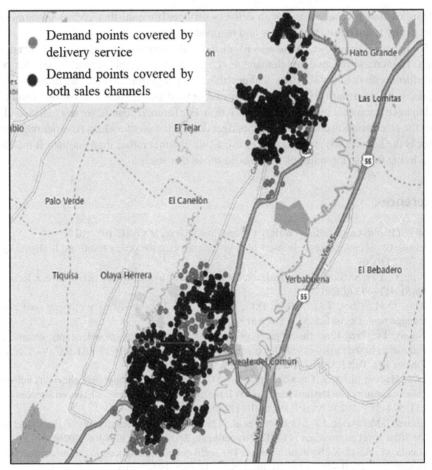

Fig. 4. Representation of geographical coverage for solutions in scenarios with β of 40% and N of 30%, and with β of 50% and N of 70% according to sales channel for both towns Chía y Cajicá

6 Conclusions and Future Works

This study proposed a mathematical model for designing a *nanostores* delivery service network considering the restrictions related to the COVID-19 virus, such as lockdown and social distance. The model allowed us to determine how many and which *nanostores* should provide delivery service to cover all demand points by some sales channel (in-person or delivery service).

The results showed that despite being able to have until to 70% of nanostores with delivery service, the optimal solution only assigned this channel for 40% of the total number of *nanostores*. With this percentage, the complete geographical coverage of demand points was achieved. This pattern was observed in Fig. 1, which indicates that it is unnecessary to have more than 10 *nanostores* with the delivery channel to improve F&V accessibility. However, to cover more than 50% of households' demand, it must have more than 140 *nanostores* with delivery service. It means that greater investment is necessary to increase the capacity and respond by delivery channel.

Considering the above, the results to select for decision-making depend on the approach. To increase revenue and demand coverage, we could analyze that the number of *nanostores* with delivery services changes although geographical coverage does not. To decide, it is important taking into account the limited budget that *nanostores'* owners tend to have because it may be a hard restriction. For future research, we suggest considering the size variation and the specific budget that each *nanostore* has. Another relevant factor is the households' willingness to use a sale channel rather than another. It means considering the buying patterns that we did not in this study.

References

1. World Health Organization (WHO). Coronavirus Disease (COVID-19) (2020)
2. Evans, O.: Socio-economic impacts of novel coronavirus: the policy solutions. BizEcons Q. **7**, 3–12 (2020)
3. Feng, C., Fay, S.: Store closings and retailer profitability: a contingency perspective. J. Retail. **96**(3), 411–433 (2020)
4. Araz, O.M., Choi, T.M., Olson, D.L., Salman, F.S.: Data analytics for operational risk management. Decis. Sci. **51**, 1316–1319 (2020)
5. Ivanov, D.: Predicting the impacts of epidemic outbreaks on global supply chains: a simulation-based analysis on the coronavirus outbreak (COVID-19/SARS-CoV-2) case. Transp. Res. Part E Logist. Transp. Rev. **136**, 101922 (2020)
6. Li, J., Hallsworth, A.G., Coca-Stefaniak, J.A.: Changing grocery shopping behaviours among chinese consumers at the outset of the COVID-19 outbreak. Tijdschr voor Econ. en Soc. Geogr. **111**, 574–583 (2020). https://doi.org/10.1111/tesg.12420
7. Chinazzi, M., Davis, J.T., Ajelli, M., et al.: The effect of travel restrictions on the spread of the 2019 novel coronavirus (COVID-19) outbreak. Science **368**, 395–400 (2020)
8. Nicola, M., Alsafi, Z., Sohrabi, C., et al.: The socio-economic implications of the coronavirus pandemic (COVID-19): a review. Int. J. Surg. **78**, 185–193 (2020)
9. Mejía-Argueta, C., Benitez-Perez, V., Salinas-Benitez, S., et al.: The importance of nanostore logistics in combating undernourishment and obesity. Eindhoven, The Netherlands (2019)
10. de Kervenoael, R., Schwob, A., Chandra, C.: E-retailers and the engagement of delivery workers in urban last-mile delivery for sustainable logistics value creation: leveraging legitimate concerns under time-based marketing promise. J. Retail. Consum. Serv. **54**, 102016 (2020). https://doi.org/10.1016/j.jretconser.2019.102016
11. Ailawadi, K.L., Farris, P.W.: Managing multi- and omni-channel distribution: metrics and research directions. J. Retail. **93**, 120–135 (2017). https://doi.org/10.1016/j.jretai.2016.12.003
12. Zambetti, M., Lagorio, A., Pinto, R.: A network design model for food ordering and delivery services. In: Proceedings Summer School Franceso Turco 2017-Septe, pp. 1–7 (2017)
13. El Tiempo. Cuarentena por coronavirus disparó ventas de tiendas de barrio y minimercados - Sectores - Economía (2020). https://www.eltiempo.com/economia/sectores/cuarentena-por-coronavirus-disparo-ventas-de-tiendas-de-barrio-y-minimercados-475932, Accessed 23 July 2020
14. La República. Aumenta ticket promedio en tiendas de barrio, pero baja frecuencia de compra (2020). https://www.larepublica.co/empresas/aumenta-ticket-promedio-en-tiendas-de-barrio-pero-baja-frecuencia-de-compra-2994292, Accessed 23 July 2020

15. El Tiempo. Ventas online pasaron del 6 al 30 por ciento en 2020 - Novedades Tecnología - Tecnología (2020). https://www.eltiempo.com/tecnosfera/novedades-tecnologia/ventas-onl ine-pasaron-del-6-al-30-por-ciento-en-2020-486816, Accessed 23 July 2020
16. World Health Organization. Diet, nutrition and the prevention of chronical diseases (2003)
17. Behl, A., Dutta, P.: Humanitarian supply chain management: a thematic literature review and future directions of research. Ann. Oper. Res. **283**(1–2), 1001–1044 (2018). https://doi.org/ 10.1007/s10479-018-2806-2
18. Balcik, B., Bozkir, C.D.C., Kundakcioglu, O.E.: A literature review on inventory management in humanitarian supply chains. Surv. Oper. Res. Manag. Sci. **21**, 101–116 (2016). https://doi. org/10.1016/j.sorms.2016.10.002
19. He, Y., Liu, N.: Methodology of emergency medical logistics for public health emergencies. Transp. Res. Part E Logist. Transp. Rev. **79**, 178–200 (2015). https://doi.org/10.1016/j.tre. 2015.04.007
20. Anparasan, A., Lejeune, M.: Data laboratory for supply chain response models during epidemic outbreaks. Ann. Oper. Res. **270**(1–2), 53–64 (2017). https://doi.org/10.1007/s10479-017-2462-y
21. Chou, J., Kuo, N.-F., Peng, S.-L.: Potential Impacts of the SARS outbreak on taiwan's economy. Asian Econ. Pap. **3**, 84–99 (2004). https://doi.org/10.1162/1535351041747969
22. Calnan, M., Gadsby, E.W., Kondé, M.K., et al.: The response to and impact of the ebola epidemic: towards an agenda for interdisciplinary research. Int. J. Heal. Policy Manag. **7**, 402–411 (2018). https://doi.org/10.15171/ijhpm.2017.104
23. Ivanov, D.: Viable supply chain model: integrating agility, resilience and sustainability perspectives—lessons from and thinking beyond the COVID-19 pandemic. Ann. Oper. Res. (2020). https://doi.org/10.1007/s10479-020-03640-6
24. Hutchison, D., Mitchell, J.C.: Invariant checking combining forward and backward traversal. Lect. Notes Comput. Sci. **9**(3), 414 (1973)
25. Cagliano, A.C., Gobbato, L., Tadei, R., Perboli, G.: ITS for e-grocery business: the simulation and optimization of Urban logistics project. Transp. Res. Procedia **3**, 489–498 (2014). https:// doi.org/10.1016/j.trpro.2014.10.030
26. Pan, S., Giannikas, V., Han, Y., et al.: Using customer-related data to enhance e-grocery home delivery. Ind. Manag. Data Syst. **117**, 1917–1933 (2017). https://doi.org/10.1108/IMDS-10-2016-0432
27. Chen, R.-C., Lin, C.-Y.: An efficient two-stage method for solving the order-picking problem. J. Supercomput. **76**(8), 6258–6279 (2019). https://doi.org/10.1007/s11227-019-02775-z
28. Aktas, E., Bourlakis, M., Zissis, D.: Collaboration in the last mile: evidence from grocery deliveries. Int. J. Logist. Res. Appl., 1–15 (2020). https://doi.org/10.1080/13675567.2020. 1740660
29. Hong, J., Lee, M., Cheong, T., Lee, H.C.: Routing for an on-demand logistics service. Transp. Res. Part C Emerg. Technol. **103**, 328–351 (2019). https://doi.org/10.1016/j.trc.2018.12.010
30. Sabana Centro Cómo vamos. Informe de Calidad de vida 2018, 98 (2018)
31. FAO, IFAD, UNICEF, et al.: The State of Food Security and Nutrition in the World 2019. Safeguarding against economic slowdowns and downturns. Rome (2019)
32. Secretaría Distrital de Desarrollo Económico. Línea de Base para la reformulación del Plan Maestro de Abastecimiento de Alimentos de Bogotá (2018)
33. Koroliuk, M., Connaughton, C.: Analysis of big data set of urban traffic data, 1–12 (2015)
34. Wang, H.: Consumer valuation of retail networks: evidence from the banking industry. SSRN Electron. J (2012). https://doi.org/10.2139/ssrn.1738084
35. Datos del DANE vs. realidad en las calles: ¿cuánto cuesta la canasta familiar para un colombiano? - YouTube. https://www.youtube.com/watch?v=kNWGcSbFVqk. Accessed 29 Apr 2021

36. CORABASTOS (2020) Boletin Diario de Precios. https://corabastos.com.co/sitio/historico App2/reportes/index.php. Accessed 01 Aug 2020

37. Salario mínimo para 2020 será de $877.802 - Ministerio del trabajo. https://www.mintrabajo. gov.co/prensa/mintrabajo-es-noticia/2019/-/asset_publisher/5xJ9xhWdt7lp/content/salario-m-c3-adnimo-para-2020-ser-c3-a1-de-877.802. Accessed 29 Apr 2021

A Decision Support Tool for the Location Routing Problem During the COVID-19 Outbreak in Colombia

Andrés Martínez-Reyes, Carlos L. Quintero-Araújo$^{(\boxtimes)}$,
and Elyn L. Solano-Charris

Operations and Supply Chain Management Research Group, International School
of Economic and Administrative Sciences, Universidad de La Sabana, Chía, Colombia
{andres.martinez9,carlosqa,elynsc}@unisabana.edu.co

Abstract. During the outbreak of coronavirus disease 2019 (COVID-19) in Bogotá, Colombia, some strategies for dealing with the increasing number of infected people and the level of occupation of intensive care units include the use of Personal Protective Equipment (PPE). PPE is a crucial component for patient care and a priority for protecting healthcare workers. For attending this necessity, the location of distribution centers within the city and the corresponding routes to supply the intensive care units (ICU) with PPE have an important role. Formally, this problem is defined as the Location Routing Problem (LRP). The LRP is an NP-Hard problem that combines the Facility Location Problem (FLP) and the Vehicle Routing Problem with Multiple Depots (MDVRP). This work presents a decision support tool based on a simheuristic method that hybridize an Iterated Local Search (ILS) algorithm with Monte Carlo simulation to deal with the LRP with uncertain demands. Realistic data from Bogotá (Colombia) was retrieved using Google Maps to characterize the geographical distribution of both potential facilities and ICUs, while demands were generated using the uniform probability distribution. Our preliminary results suggest the competitiveness of the algorithm in both the deterministic and the stochastic versions of the LRP.

Keywords: COVID-19 · Healthcare logistics · Location-routing · Decision support tool · ILS · Monte Carlo simulation

1 Introduction

In December 2019, a new coronavirus disease emerged characterized as a viral infection with a high level of transmission in Wuhan, China. Coronavirus 19 (COVID-19) is caused by the virus known as Severe Acute Respiratory Syndrome coronavirus 2 (SARS-CoV-2) established by the Coronaviridae Study Group of the International Committee on Taxonomy of Viruses (ICTV) [1]. As the cumulative numbers of confirmed cases have considerably increased worldwide,

D. A. Rossit et al. (Eds.): ICPR-Americas 2020, CCIS 1408, pp. 33–46, 2021.
https://doi.org/10.1007/978-3-030-76310-7_3

academics and practitioner are mainly concerned about modeling and prediction of COVID-19 [2,3].

In Colombia, according to the report of August 4th, 2020, 33.9% of the cases are located in Bogotá D.C. (capital city) with 113,548 confirmed cases, and the 89.2% of the total intensive care units are occupied [4]. As COVID-19 is predominantly caused by contact or droplet transmission attributed to relatively large respiratory particles which are subject to gravitational forces and travel only approximately one meter from the patient [5], personal protective equipment (PPE) has become crucial for protecting healthcare workers and alleviating the burden in the hospitals and controlling the epidemic [6].

Considering the city needs and that most of the studies about COVID-19 deal with prediction models and do not integrate their results to support decision making, e.g., estimating cities' implications, supplies and demand of material resources; in this paper, we study the location of distribution centers within the city and the corresponding routes to supply PPE to the Intensive Care Units (ICU) and, therefore, support decision making. The problem is formally defined as the Location Routing Problem (LRP). The LRP involves all decision levels in supply chain management and logistics, i.e., strategic, tactical and operational. From an operations research perspective, it can be seen as the combination of two well-known NP-Hard problems such as the Facility Location Problem and the Vehicle Routing Problem with Multiple Depots. Thus, the LRP is also NP-Hard.

From a practical point of view, there is a recent trend to create both easy-to-implement and powerful algorithms when solving complex problems. Accordingly, in this work, we propose a simheuristic method based on an ILS algorithm in which initial solutions are created using a random choice of depots combined with a biased randomized version of the nearest neighbor heuristic. Readers are referred to [7] as a key literature review of simheuristics for dealing with stochastic combinatorial optimization problems. Moreover, with the aim of providing an easy-to-use tool, we have implemented our algorithm as an Excel Macro, considering its advantages in data manipulation, consolidation and analysis. Our preliminary results have been compared to the ones obtained by GAMS/Cplex for the deterministic LRP, showing promising results. Besides, we have carried out a set of experiments for the LRP with Stochastic Demands by combining our Iterated Local Search (ILS) algorithm with Monte Carlo simulation (MCS). The results on the stochastic version demonstrate that using safety stocks as protective policies against uncertainty could improve not only expected costs but also the reliability of the obtained solutions.

The remainder of this paper is organized as follows: Sect. 2 presents the literature review; Sect. 3 introduces the description of the problem; Sect. 4 outlines the proposed approach; in Sect. 5 we analyze the obtained results; finally, Sect. 6 presents some conclusions and further research perspectives.

2   Literature Review

This section presents a brief literature review focusing on location routing problems and also on logistics applications for pandemics such as the current one generated by COVID-19.

2.1   Location Routing Problems

The LRP was initially proposed by [8]. Due to its complexity, the first studies on the LRP proposed to tackle it by separating the two related subproblems, i.e., the facility location problem and the vehicle routing problem. However, it has been demonstrated that such approach leads to sub-optimal solutions [9]. This complex problem can be used to support decision-making processes in different fields of application such as city logistics, humanitarian logistics, horizontal cooperation, among others [10, 11].

Considering its NP-Hard nature, heuristics and metaheuristics yield better results than classical optimization approaches, specially for large sized instances [12]. We refer to [13] for an overview on LRP problems. The deterministic version of the LRP has been widely studied while its stochastic counterpart has been scarcely analyzed. In the stochastic demands version, the main assumption is that demands are not known in advance, i.e., the real value of demands is revealed once the vehicle arrives at the customer. Thus, routes failures will occurs whenever the aggregated demand in a route exceeds the vehicle capacity [14].

Among recent works on the LRP with stochastic demands (LRPSD), Quintero et al. [14] proposed a simheuristic algorithm to deal with the LRPSD. The authors proposed three simulation processes to deal with: (i) the estimation of the right safety stock policy to protect against uncertainty, (ii) estimate stochastic costs and reliabilities of the proposed solution, (iii) refinement of the estimation of both stochastic costs and reliabilities. However, their work was tested using benchmark instances adapted from literature, while in this work we use realistic data from Bogotá, Colombia related to the current pandemic.

2.2   Logistic Approaches for Dealing with COVID-19

Mainly, most of the literature on COVID-19 is focused on prediction models (e.g., [2, 3, 15]). Torrealba-Rodriguez et al. [2] studied the prediction of cases of COVID-19 infection in Mexico. Shen [15] considered the logistic growth modelling of COVID-19 proliferation in China and its international implications. Wang et al. [3] predicted the global trend and the specific trends of Brazil, Russia, India, Peru and Indonesia.

However, just few works integrate the results of the prediction models to support decision making. For example, Loske [16] analyzed the impact of COVID-19 on transport volume in German food retail logistics, as well as its resulting implications. The author proposed a regression analysis to validate the interdependencies of COVID-19 and transport logistics in retail logistics, and considered the reallocation of health care capacity, repurposing of hospitals, and close collaboration between the government and the health care committee. Zhang et al. [17] deal with the scheduling of vehicles to transport infected people to isolated medical areas and solved it using a metaheuristic approach. Yu et al. [18] propose the design of a multi-objective reverse logistic network in epidemic outbreaks. More specifically, this work aims to determine the location of temporary facilities and transportation of the increased medical waste generated by a pandemic. Kaplan [19] combines predictive and prescriptive models for COVID-19 related decision making such as crowd-size restrictions, hospital surge planning, timing decisions (when to stop and possibly restart university activities), and scenario analyses to assess the impacts of alternative interventions, among other problems.

To the best of our knowledge there are no published works on the location of depots and the consequent routing of PPEs (maks, gloves and disposable suits) for caregivers (physicians, nurses, therapyst, etc.) located all long the city. Thus, the importance and novelty of this work.

3   Problem Description

Due to the COVID-19 pandemic and its associated effects, many governments have adopted lock-down mechanisms as a strategy to diminish the speed of contagion and have more prepared health systems, especially by having a higher number of available ICUs. Particularly, in Bogotá - Colombia, lock-down started in March 2020 but currently, we are facing a high number of new cases and deaths every day. Besides, the occupancy of ICUs is becoming critical with a 90% value. Thus, logistic approaches are a must to optimize the response of the system to the current pandemic.

In particular, we aim to analyze the efficient delivery of PPE (composed by masks, gloves, and disposable suits) required by medical teams (physicians, nurses, and therapists) to take care of COVID-19 patients. In this work, we propose to study the location of distribution centers within the city and the corresponding routes to supply the different ICUs that are habilitated to receive COVID-19 patients in Bogotá. This situation could be represented by the Location Routing Problem with Stochastic Demands (LRPSD) due to the nature of the field of application.

The LRP considered in this work is adapted from [20]. As stated by [21], the LRP belongs to the class of NP-hard problems, which means that it is not possible to find optimal solutions for large-sized instances in reasonable computing times. The LRP is defined in a directed graph $G = (V, A, C)$. V is a set of nodes comprising a subset I of m possible depot locations and a subset $J = V \backslash I$ of n customers. The cost of any arc $a = (i, j)$ in the arc set A is given by C_a. A capacity W_i and an opening cost O_i are associated with each depot site $i \in I$. Each customer $j \in J$ has a demand d_j. A set K of identical vehicles of capacity Q is available. When used, each vehicle incurs a fixed cost F and performs one single route. The following constraints must be taken into account:

- Each demand d_j must be served by one single vehicle.
- All nodes are allocated to an open depot.
- The number of depots within the set must guarantee that total demand can be serviced.
- Each route must begin and end at the same depot and its total load must not exceed vehicle capacity.
- The total load of the routes assigned to a depot must respect the capacity of the selected depot.

The goal of our problem is to determine the subset of distribution centers (DCs) to open, allocate ICUs to DCs, and planning the routes from DCs to serve ICUs, in order to minimize the total expected costs. The total expected costs of a solution include the fixed cost of opening facilities F, the costs of traversed arcs, the fixed cost of using vehicles and the cost of recursive actions (in case of route failures due to the stochastic demands). The mathematical model of the stochastic version can be found in [14].

Figure 1 depicts a complete solution for our LRP. Potential distribution centers locations are represented by squares while ICUs are represented by circles. This figure illustrates how, from an initial problem setting (top-left), a complete solution could be obtained by *(i)* selecting the distribution centers to be opened (top-right), *(ii)* assigning ICUs to open DCs (bottom-left) and, *(iii)* creating routes from each DC to its allocated ICUs (bottom-right), while satisfying all constraints.

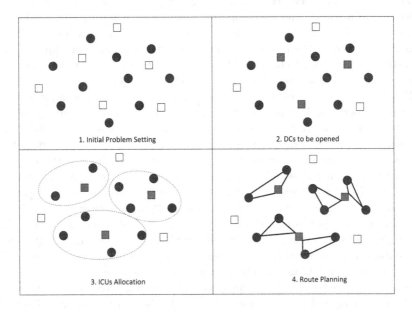

Fig. 1. A complete solution for our LRP

4 Solving Approach

To deal with the LRPSD, we have developed a simheuristic algorithm hybridizing an Iterated Local Search (ILS) algorithm [22] with Monte Carlo simulation. The procedure is driven by the ILS while the simulation is used to test the quality of the solution in the stochastic setting of the problem. Considering that ILS is a powerful local search-based metaheuristic to deal with deterministic problems, we need to use a kind of protection policy (safety stocks) to face uncertainty and get better results in the stochastic setting of the problem.

Safety stock is used when planing routes to reduce the possibility of not serving some ICUs, when executing the planned routes, due to demand uncertainty. However, after a certain value (too conservative) of safety stock, expected costs tend to increase due to excessive deterministic costs. The idea, then, is to find the most convenient safety stock policy providing the best trade-off among costs and reliability.

Our proposed method is composed of three main phases: *(i)* location phase, *(ii)* customer allocation and, *(iii)* vehicle routing. To deal with the location decisions, we randomly open depots until the total available capacity is enough to serve total demands. Then, in the allocation phase, ICUs are randomly selected and assigned to the nearest open depot with the available capacity to serve it. Besides, the available capacity of the corresponding depot is updated and the ICU is marked as assigned. This process is repeated until all nodes have been allocated to an open depot. In case that a given ICU can not be assigned to any open depot due to capacity constraints, a new depot is opened and the customer is allocated to it. Here, solution failure due to depot capacity it is not considered. In the routing phase, the starting node for each route is randomly selected. Next, a modified version of the nearest neighbor heuristic is applied. A route finishes when the next customer to be added to the route cannot be served due to vehicle capacity constraints, so the vehicle is sent back to the depot.

This three-phase process is executed during a given number of iterations and we keep the best solution found among them. The aforementioned solution is sent to a short simulation process to estimate stochastic costs and reliabilities. Stochastic costs are generated when a planned route can not serve a certain ICU and, as a consequence, a corrective round-trip from such ICU to the depot is executed to re-load the vehicle and resume the planned route. After the short simulation, the ILS framework is executed. To do so, we propose as a perturbation operator the interchange of an open depot with a closed depot, i.e., a previously open depot is closed and a previously closed depot is opened. The ICUs assigned to the closing depot are allocated to the opening one. A graphical representation of this operator can be seen in Fig. 2. Moreover, two local search operators have been designed. The first one is the exchange of two customers among different routes from the same depot [23] (see Fig. 3 top) while the second one is the exchange of two customers from routes belonging to different depots (see Fig. 3 bottom).

Promising solutions obtained by the ILS framework are passed through the short simulation process. Next they are stored in a pool of solutions which is

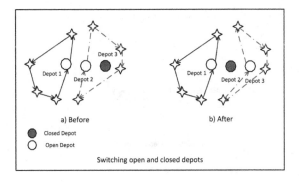

Fig. 2. Diversification operator

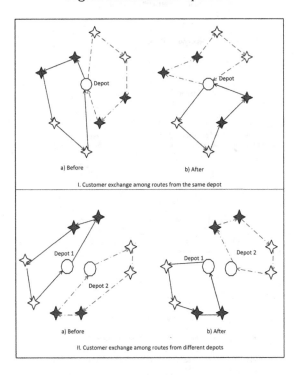

Fig. 3. Local-search operators

sorted by increasing costs. Finally, the top-10 solutions stored within the pool are passed through a more intensive simulation process to refine both stochastic costs and reliabilities. The process is depicted in Fig. 4.

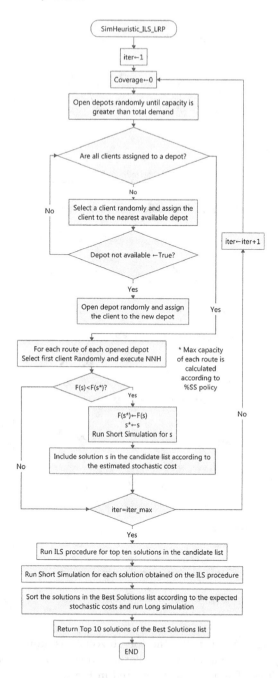

Fig. 4. Flowchart of the proposed method for the LRPSD

The reliability $reliab_r$ for each route r in solution S is computed as follows:

$$reliab_r = (1 - \frac{\sum_{n=0}^{TotalSimulationRuns} RouteFailures}{TotalSimulationRuns}) * 100\% \qquad (1)$$

It is important to note that each route within a solution could be seen as an independent component of a series system, i.e., the proposed solution will fail if, and only if, a failure occurs in any of its routes ($RouteFailures$). Therefore, the reliability index of a solution S with R routes can be calculated as $\prod_{r=1}^{R} reliab_R$.

5   Results and Analysis

The proposed algorithm was coded as an Excel macro using the Visual Basic for Applications (VBA) language. The version used for the spreadsheet was MS Excel 2013. As stated by [24], using spreadsheet-based solutions have several advantages such as interface familiarity, ease of use, flexibility, and accessibility. Moreover, since MS Excel is largely known around the world, using it as the engine for the spreadsheet provides additional benefits, such as integration with software packages that offer built-in functionalities to obtain/send data from/to MS Excel, and the possibility of customizing code in Visual Basic for Applications [24]. It is also important to mention that using spreadsheet-based solutions may result in low-cost solutions that may yield significant savings for enterprises, especially in non-developed countries.

We have tested our proposed method using different instances with real locations in Colombia. Instances were generated by retrieving -from Google Maps- latitude and longitude coordinates belonging to retailing and warehousing points located in Bogotá (Colombia's main city and its capital). The expected value for demands corresponds to the PPE required for each ICU assuming that each patient is served by a team consisting of one physician, one nurse, and one therapist. The team visits each patient once per hour, so 24 visits are required during a complete day. The capacity of DCs was generated to guarantee the satisfaction of total demands. Opening costs, in US$, for each facility corresponds to real construction costs of warehouses in Bogotá. The vehicle capacity corresponds to the real-load capacity of the Renault Kangoo, which is a vehicle broadly used in Colombia to execute urban distribution tasks. Instances considers 53 ICUs, 10 distribution centers, the distance matrix and variations on demands. The instances are named MQS-BOG#, where # identifies the number of the instance. All instances are available in https://cutt.ly/SpreadsheetSimheuristicILS.

All deterministic instances were modeled using the GAMS modeling language. However, due to the complexity of the problem, the Cplex solver was unable to find a solution after eight hours of execution time. Thus, we generate a set of reduced instances to have a fair comparison between the exact method and our proposed algorithm. To generate the reduced instances, we run our heuristic method and we get two open depots and their allocated customers from our best reported solution. Additional depots were randomly selected while ensuring the capacity to serve all ICU demand. All reduced instances are available in https:// cutt.ly/ReducedInstances. The experiments were carried out using a standard windows PC with Intel® Core™ i7 – 6th generation and 8 Gb RAM. Each instance was solved using five different random seeds. The obtained results are summarized in Table 1. GAMS/Cplex column shows the best solution reported by this software for each instance after 27,000 secs of computational time. It is worth mentioning that none of the solutions was proven as optimal. OBDS is the best deterministic solution reported by our algorithm among all executions, while OADS is the average value of the obtained solutions. Besides, GAP shows the percentual gap of OADS concerning GAMS/Cplex. We can see that, on average, our algorithm has an average gap of 1.64% compared to the results provided by GAMS/Cplex.

Table 1. Results - Deterministic case

INSTANCES	SUB-PROBLEM	GAMS/cplex(1)	OBDS(2)	GAP% (2)-(1)
MQS-BOG1	1	2,558,514.72	2,558,516.65	0.00008%
MQS-BOG1	2	2,389,528.66	2,389,529.89	0.00005%
MQS-BOG1	3	2,401,725.86	2,401,728.02	0.00009%
MQS-BOG1	4	2,806,031.04	2,806,034.95	0.00014%
MQS-BOG1	5	2,818,227.57	2,818,233.07	0.00020%
MQS-BOG1	6	2,649,244.88	2,649,246.32	0.00005%
MQS-BOG2	1	1,972,433.46	2,231,267.37	13.12257%
MQS-BOG2	2	2,558,516.41	2,558,519.50	0.00012%
AVERAGE				1.64041%

As this work concerns the stochastic version of the LRP, we have transformed the deterministic instances by assuming their demands as the expected value (EV) of the stochastic case. Besides, stochastic demands are revealed once the vehicle arrives at the UCIs by using the uniform probability distribution $\sim U[EV - 10\%, EV + 10\%]$. It is worth to mention that any other probability distribution according to the real demand's behavior could be used.

Our proposed method was tested using five different random seeds and different safety stock policies as a security buffer to handle demand uncertainty i.e., 0%, 1%, 3%, 5% and 7%. The results are presented in Table 2. The setup time for the simheuristic was 7 min without including the simulation time. For each safety stock policy, the best stochastic solution (OBSS), the average of our top 10 stochastic solutions (OASS-10), the expected reliability of the OBSS (Reliab.), and the Gap among OBSS and OASS-10 are reported. It is worth to mention that the top 10 solutions for each of the instances tested with each safety stock policy show the same location decisions, clients allocation are similar and the major differences among them are due to the configuration of each route. Furthermore, Fig. 5 shows the behavior of expected stochastic costs and reliabilities for a given instance when using different safety stock policies. As expected, when no protection is considered there are many route failures due to demand uncertainty. Therefore, we can see higher costs and lower reliability. On the other hand, when the value of safety stock policy increases, costs tend to decrease while reliability increases; however, when the safety stock is greater than 5%, costs start to increase again. Similar situation occurs for the different instances, i.e., costs start to decrease when the percentage of safety stock is increased and, at a certain point (ideal safety stock policy), the costs reach its minimum value. After that value, costs become higher. This situation occurs when decision-makers for protecting against uncertainty, tend to greatly increase the % of safety stock for augmenting reliability and, therefore, costs are increased due to a higher number of planned routes.

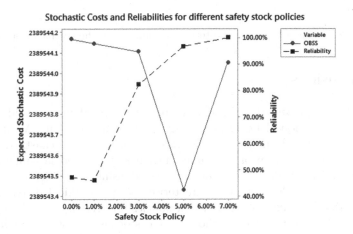

Fig. 5. Average values of expected costs and reliabilities of our best stochastic ssolutions for different safety stock policies

Table 2. Results - stochastic case

Safety stock policy	0%				1%				3%			
Instance name	OBSS	OASS-10	Reliab.	Gap	OBSS	OASS-10	Reliab.	Gap	OBSS	OASS-10	Reliab	Gap
MQS-BOG1	8,602,359,004.19	8,602,362,316.87	33%	0.0000039%	8,602,358,922.17	8,602,360,060.80	52%	0.000013%	8,602,358,781.84	8,602,360,594.29	75%	0.0000021%
MQS-BOG2	11,888,198,383.13	11,888,200,758.20	17%	0.000020%	11,888,199,113.51	11,888,200,274.41	49%	0.000010%	11,888,197,027.05	11,888,201,121.57	78%	0.0000034%
MQS-BOG3	18,747,950,727.91	18,747,951,361.18	47%	0.0000003%	18,747,943,157.57	18,747,946,203.56	46%	0.000016%	18,747,947,581.35	18,747,948,413.29	82%	0.0000004%

Safety stock policy	5%				7%							
Instance name	OBSS	OASS-10	Reliab.	Gap	OBSS	OASS-10	Reliab.	Gap				
MQS-BOG1	8,602,356,349.06	8,602,359,573.59	94%	0.0000037%	8,602,358,584.88	8,602,360,081.83	100%	0.000017 %				
MQS-BOG2	11,888,197,767.40	11,888,200,352.42	97%	0.000022%	11,888,195,544.36	11,888,198,946.79	100%	0.0000029%				
MQS-BOG3	18,747,943,818.55	18,747,946,987.14	97%	0.0000017%	18,747,950,202.88	18,747,952,654.84	100%	0.0000013%				

Moreover, in Fig. 6 we present the Best Deterministic Solution (OBDS) for a given instance against the two best performing stochastic solutions with a safety stock policy of 5%. Besides generating lower expected stochastic costs, the stochastic solutions, even if they are not the optimal ones, show less variability than the deterministic one in the stochastic case.

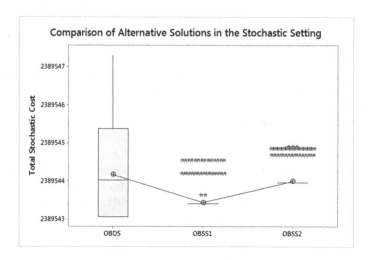

Fig. 6. Example of behavior of alternative solutions in the stochastic setting

6   Conclusions

This article has presented a decision support tool based in a hybrid method consisting of an Iterated Local Search algorithm combined with Monte Carlo simulation to deal with the Location Routing Problem with stochastic demands. This version finds applications in humanitarian logistics and last-mile deliveries problems. Our proposed method was tested using five different instances generated with Google Maps to characterise the geographical distribution of customers. Besides, different safety stock policies are tested.

Results show the behavior of the expected stochastic costs and reliabilities when using different safety stock policies per each tested scenario. Furthermore, results also show that this version of the LRP is a hard problem and its complexity increases according to the level of uncertainty. As expected, when no protection is considered there are many route failures due to demand uncertainty and higher costs and lower reliability are obtained. On the other hand, when the value of safety stock policy reach the ideal value, costs tend to decrease while reliability increases.

Regarding future research, there is a room for including other representations for uncertainties and to design other approaches for handling them. Moreover, adaptation for large scale problems will be also considered.

Acknowledgments. This work has been partially supported by the Master Program in Operations Management and the General Direction of Research from Universidad de La Sabana, grant EICEA-112-2018.

References

1. Gorbalenya, A.E., Baker, S.C., Baric, R.S.: The species severe acute respiratory syndrome-related coronavirus: classifying 2019-nCoV and naming it SARS-CoV-2. Nature Microbiol. **5**, 536–544 (2020)

2. Torrealba-Rodriguez, O., Conde-Gutiérrez, R.A., Hernández-Javier, A.L.: Modeling and prediction of covid-19 in Mexico applying mathematical and computational models. Chaos, Solitons Fractals **138**, 582–589 (2020). https://doi.org/10.1016/j.chaos.2020.109946

3. Wang, P., Zheng, X., Li, J., Zhu, B.: Prediction of epidemic trends in covid-19 with logistic model and machine learning technics. Chaos, Solitons Fractals, 110058 (2020). https://doi.org/10.1016/j.chaos.2020.110058

4. Saluddata Secretaria Distrital de Salud. Casos confirmados de covid-19 (Aug 2020). http://saludata.saludcapital.gov.co/osb/index.php/datos-de-salud/enfermedades-trasmisibles/covid19/

5. Cook, T.M.: Personal protective equipment during the covid-19 pandemic - a narrative review. Anaesthesia **75**(75), 920–927 (2020)

6. Chirico, F., Nucera, G., Magnavita, N.: Covid-19: protecting healthcare workers is a priority. Infect. Control Hosp. Epidemiol. **2020**(1), 1–4 (2020)

7. Juan, A.A., Faulin, J., Grasman, S.E., Rabe, M., Figueira, G.: A review of simheuristics: extending metaheuristics to deal with stochastic combinatorial optimization problems. Oper. Res. Perspect. **2**, 62–72 (2015)

8. Maranzana, F.E.: On the location of supply points to minimize transport costs. J. Oper. Res. Soc. **15**(3), 261–270 (1964)

9. Salhi, S., Rand, G.K.: The effect of ignoring routes when locating depots. Euro. J. Oper. Res. **39**(2), 150–156 (1989)

10. Nataraj, S., Ferone, D., Quintero-Araujo, C., Juan, A., Festa, P.: Consolidation centers in city logistics: a cooperative approach based on the location routing problem. Int. J. Ind. Eng. Comput. **10**(3), 393–404 (2019)

11. Almouhanna, A., Quintero-Araujo, C.L., Panadero, J., Juan, A.A., Khosravi, B., Ouelhadj, D.: The location routing problem using electric vehicles with constrained distance. Comput. Opera. Res **115**, 104864 (2020)

12. Quintero-Araujo, C.L., Caballero-Villalobos, J.P., Juan, A.A., Montoya-Torres, J.R.: A biased-randomized metaheuristic for the capacitated location routing problem. Int. Trans. Oper. Res. **24**(5), 1079–1098 (2017)

13. Prodhon, C., Prins, C.: A survey of recent research on location-routing problems. Euro. J. Oper. Res. **238**, 1–17 (2014)

14. Quintero-Araujo, C.L., Guimarans, D., Juan, A.A.:. A simheuristic algorithm for the capacitated location routing problem with stochastic demands. J. Simul. 1–18 (2019)

15. Shen, C.Y.: Logistic growth modelling of covid-19 proliferation in china and its international implications. Int. J. Infect. Dis. **96**, 582–589 (2020)

16. Loske, D.: The impact of covid-19 on transport volume and freight capacity dynamics: an empirical analysis in German food retail logistics. Trans. Res. Interdiscip. Perspectiv. **6**, 100165 (2020). https://doi.org/10.1016/j.trip.2020.100165

17. Zhang, M.-X., Yan, H.-F., Jia-Yu, W., Zheng, Y.-J.: Quarantine vehicle scheduling for transferring high-risk individuals in epidemic areas. Int. J. Environ. Res. Pub. Health **17**(7), 2275 (2020)

18. Yu, H., Sun, X., Solvang, W.D., Zhao, X.: Reverse logistics network design for effective management of medical waste in epidemic outbreaks: insights from the coronavirus disease 2019 (covid-19) outbreak in Wuhan (China). Int. J. Environ. Res. Pub. Health **17**(5), 1770 (2020)
19. Kaplan, E.H.: Covid-19 scratch models to support local decisions. Manufact. Serv. Oper. Manage. (2020). https://doi.org/10.2139/ssrn.3577867
20. Prins, C., Prodhon, C., Calvo, R.W.: Solving the capacitated location-routing problem by a grasp complemented by a learning process and a path relinking. 4OR **4**, 221–238 (2006)
21. Quintero-Araújo, C.L., Juan, A.A., Montoya-Torres, J.R., Muñoz-Villamizar, A.: A simheuristic algorithm for horizontal cooperation in urban distribution: Application to a case study in Colombia. In: 2016 Winter Simulation Conference (WSC), pp. 2193–2204 (2016)
22. Lourenço, H.R., Martin, O.C., Stützle, T.: Iterated local search: framework and applications, pp. 363–397. Springer, Boston (2010). ISBN 978-1-4419-1665-5. https://doi.org/10.1007/978-1-4419-1665-5_12
23. Osman, I.H.: Metastrategy simulated annealing and tabu search algorithms for the vehicle routing problem. Ann. Oper. Res. **41**(4), 421–451 (1993)
24. Erdoğan, G.: An open source spreadsheet solver for vehicle routing problems. Comput. oper. Res. **84**, 62–72 (2017)

Simulated Annealing Metaheuristic Approach for Generating Alternative Corridor Locations

Pedro Moreno-Bernal[1(✉)] and Sergio Nesmachnow[2]

[1] Universidad Autónoma del Estado de Morelos, Cuernavaca, Morelos, Mexico
pmoreno@uaem.mx
[2] Universidad de la República, Montevideo, Uruguay
sergion@fing.edu.uy

Abstract. This article presents a metaheuristic approach applying simulated annealing for the problem of planning corridors for oil pipelines facilities. The corridor location problem requires finding an optimal or a set of sub-optimal alternative routes between two locations, accounting for cost optimization and topographical constraints. The proposed simulated annealing applies a path-based search for exploring the space of possible routes. The experimental evaluation solves realistic problem instances considering real information of Veracruz, Mexico. The proposed approach is able to compute a set of near-optimal alternative routes that improve up to 34.1% of the terrain impact cost over a greedy pathfinding reference method, such as the ones included in traditional geographical information systems.

Keywords: Corridor location · Simulated annealing · Path analysis · Shortest path optimization

1 Introduction

The right-of-way (ROW) or *corridor* is a crucial requirement for the construction of linear installations such as pipelines, electric transmission lines, roads, among others [19]. The planning process of real corridors must consider legal, social, political, environmental, engineering, and economic factors, which make the design process complex [17]. Proper planning of ROW explores different paths based on design specifications, environmental laws, and good practices to determine the best option, which is the goal of the corridor location planning [4].

The corridor location problem (CLP) can be modeled as an optimization problem, subject to topographic constraints such as distance, obstacles, environmental impact, and safety [19]. Locating a corridor is analogous to identifying the shortest or least-cost path for the facility to be designed on a landscape [26]. Traditional approaches to solve the corridor location problem are based on generating a set of alternate routes through enumerative methods [24], using penalty techniques [14], or by specifying one or multiple intermediate points where the

© Springer Nature Switzerland AG 2021
D. A. Rossit et al. (Eds.): ICPR-Americas 2020, CCIS 1408, pp. 47–62, 2021.
https://doi.org/10.1007/978-3-030-76310-7_4

route must pass [22] depending on the size of the area across the landscape where the corridor could pass [8]. The problem of generating alternate routes has led researchers to develop solution methods that use different types of representation of the domain, through graphs; they use deterministic shortest path methods, such as Dijkstra and its variants to get only one path at a time (optimal route). However, the Dijkstra algorithm and its variants are costly, since the computing time to find the shortest path using exact methods increase based on the network size. For this reason, additional mechanisms generating more than one route at a time are useful contribution in this area.

Geographical Information Systems (GIS) facilitate the management of geographic information using raster (2D grid of measurable numerical characteristics of the projected area) and vector (representation of the projected area through points, lines, and polygons) data models. The efficient handling of the data models affects map analysis speed. Distance measurement permits identifying the optimal path by an accumulated surface cost raster layer (ASC) performed on spatial analysis [2,7]. ASC raster layer represents a weighted graph associating a value (weight) on every edge on an appropriate scale that corresponds to the goals and objectives for every raster cell involved in the corridor design. Generally, GIS tools use the Dijkstra algorithm and its variants for the spatial analysis, without considering the generation of alternate routes [10]. Unlike exact techniques, metaheuristics cannot guarantee to obtain the optimal solution to a problem. However, metaheuristic techniques are more efficient than exact techniques in solving complex optimization problems [20]. Simulated Annealing (SA) is a metaheuristic method that finds reasonable solutions for combinatorial problems over large solution spaces [12]. SA implements a local search guided through a stochastic process with a given probability. The quality solution can be improved using a neighborhood search in the SA algorithm. The neighborhood structure allows searching for approximate solutions, adjusted to the complexity and nature of the problem [6].

In this line of work, this article presents a SA metaheuristic approach for the CLP. The space exploration allows computing different alternatives routes between origin and destination points, generating competitive and spatially different routes alignments. Realistic problem instances are solved in the experimental evaluation, using real information of the Veracruz state, a well-know oil production area in Mexico. Accurate results are reported: the proposed SA allows computing a spatially diverse set of near-optimal alternatives that improves up to 31.4% the results of traditional methods included in GIS software.

The rest of the article is structured as follows. Section 2 describes the corridor location problem and the optimization model of the shortest route. Section 3 describes the proposed SA metaheuristic approach, including the method for generating the initial solution. Section 4 provides details of the experimental evaluation of the proposed approach, including the description of the real scenarios and the discussion of the results. Additionally, the corridor paths obtained by the proposed SA are displayed on GIS software. Finally, Sect. 5 presents the conclusions and formulate the main lines for future work.

2   The Corridor Location Problem

This section describes the corridor location problem, presents its mathematical formulation, and reviews related works about traditional approaches to solve the problem based on enumerative methods, penalty techniques, and by specifying intermediate points where the path must pass.

2.1   Problem Description

In the construction of linear facilities such as pipelines, communication networks, transmission lines, etc., it is essential to permit access to public land that will house the facilities. Therefore, one of the factors to consider is the use of existing or new ROWs [15,19]. Specific laws demand considering alternative routes in a ROW project to determine the optimal route for the corridor across public and private terrains. In the context of line-based facilities, two important concepts are distinguished, depending on the location perspective: in a path, the facility is placed on an existing network composed of street/road segments and intersections; instead, a corridor is open in terms of being a route that lies anywhere on the landscape [5]. The linear facility design process is split into two stages. In the first stage, the best compromise route is determined, based on the objectives of the interested parties. In the second stage, the engineering and construction design and the facility accessories must be included in the design process [17].

   This work focuses on the first stage of the facility design process. The goal is to generate alternative routes, to be analyzed in a posterior decision-making process to determine a proper route for corridor location. Relevant attributes are considered for route generation, including a cost function that considers topographical information and the height of the candidate locations for the route.

2.2   Mathematical Formulation

The CLP is formulated as the problem of finding the shortest path between source and destination in an undirected weighted graph $G = (N, A, h, c)$ [26].

   The graph G represents a Cartesian grid used to discretize the area where the corridor is to be built. $N = \{u_1, u_2, ..., u_n\}$ is the set of graph nodes. Each node represents the center of a cell in the rectangular grid specified by a point (x, y, z), where z is a non-negative integer value based on the digital elevation model (DEM), and x, y are in terms of latitude and longitude coordinates from the Universal Transversal Mercator (UTM) reference system. Each grid cell is linked to its eight neighboring cells. $A = \{(u_{i1}, u_{j1}), (u_{i2}, u_{j2}), ..., (u_{im}, u_{jm}\}, u_{ik}, u_{jk} \in N$ is the set of edges; $n = |N|$ and $m = |A|$. An edge $(u_i, u_j) \in A$ connects nodes u_i, and u_j. Function $h : N \rightarrow \mathbb{Z}^+$ assigns a non-negative height (the *elevation*) to each node. In turn, function $c : A \rightarrow \mathbb{R}^+$ defined by Eq. 1, assigns a non-negative cost to each edge $(u_i, u_j) \in A$.

$$c(u_i, u_j) = \frac{\sum_{k}^{|L|} L^k(i, j)}{|L|} \tag{1}$$

In Eq. 1, $L^k : \{1, ..., n\} \times \{1, ..., n\} \rightarrow \mathbb{R}$ is the k-th *topographic layer* considered for the area where the corridor is to be built. Each topographic layer represents specific information about roads, buildings, land use, and other relevant features about the terrain, and values for each cell (i, j) are classified on a zero to ten scale, according to a passing penalty between node u_i and node u_j. Higher passing penalty are associated to higher costs.

The problem proposes finding the minimum-cost path from a given source node s to a destination node t. The objective function to minimize (Eq. 2a) represents the total viability cost of edges chosen for the path. In the proposed formulation, each edge (u_i, u_j) has associated a binary decision variable x_{ij} that is equal to 1 if the edge is included in the optimal shortest path and 0 otherwise.

$$\min f(x) \quad = \sum_{(u_i, u_j) \in A} c(u_i, u_j) \cdot x_{ij} \tag{2a}$$

subject to

$$\sum_{(u_i, u_j) \in A} x_{ij} - \sum_{(u_j, u_i) \in A} x_{vu} = \begin{cases} 1 \text{ if } u_i = s \\ -1 \text{ if } u_i = t, \\ 0 \text{ if } u_i \neq s, t. \end{cases} \tag{2b}$$

$$x_{ij} \in \{0, 1\} \quad \forall (u_i, u_j) \in A \tag{2c}$$

$$h(u_i) \leq DEM_{max} \tag{2d}$$

Equation 2b express the flow conservation constraint, ensuring that the computed path starts on the origin node $s \in N$ and ends on the destination node $t \in N$. Equation 2c is the integrity constraint for the binary decision variable x_{ij}. Finally, Eq. 2d is the elevation penalty constraint, which states that node u_i avoid steep slopes in the path through an established maximum height (DEM_{max}) for the corridor route location.

2.3 Related Work

The iterative penalty method assigns penalty parameters to the edges of the graph once one of them has been selected to be part of a solution. The efficiency of the algorithm depends on the penalty strategy implemented. Regardless of the penalty strategy, many alternatives generated are parallel lines to each other. The method was initially proposed for road design [24] and used for routing hazardous materials delivery [11], among other applications. Lombard and Church [14] introduced the Gateway Shortest Path method (GPS), based on a shorter route problem with restrictions. It generates a short route called "Gateway Shortest Path" between a source and a destination, restricted to pass through a specific node (the *gateway*). The method allows identifying different acceptable alternate routes, with less computational effort than the iterative penalty method.

A different approach for generating alternate routes is the kth-shorter routes (KSR) method [25], which generates a classified list of the k lowest cost routes. After generating the first route, KSR excludes one edge from the network included in the solution; the network is modified iteratively to generate different alternatives. A similar method is Near-Shortest Path (NSP) [3], which generates k routes within a specific costs range, without classifying them.

Other proposed methods for generating alternate routes for corridors are Minimax [13] and p-dispersion [1]. The Minimax method is a greedy approach that begins by generating the shortest k routes between an origin and a destination, to build a subset of different routes. The p-dispersion method generates a broad set of candidate routes and selects a subset of p routes using a dispersion model that maximizes the minimum difference between any pair of selected routes. In contrast, Scaparra and Church [22] extended the GPS method by considering multiple gateway points to generate more complex paths, but with a higher computational cost.

Other approaches make use of parallel computing to run parallel methods such as KSR and NSP [16]. Exact algorithms guarantee to find the optimal global solution; however, in the worst case, the computation times can be exponential depending on the size of the problem [9]. The computation time and effort scale to the extent that solving the problem is intractable for realistic problem instances [8]. An alternative solution to deal with this type of problem is through non-deterministic heuristics, capable of generating optimal solutions in reasonable computation time. Zhang and Armstrong [26] proposed Multi-Objective Genetic Algorithm for the Corridor Selection Problem (MOGADOR), which generates a wide set of Pareto optimal and near-optimal solutions. The objective functions that it minimizes calculate the suitability of the accumulated corridor aligned to engineering costs, environmental cost, and socioeconomic cost. Experimental results showed that MOGADOR, through the designed and implementation of specific genetic operators for corridor alternatives outperforms conventional approaches based on shortest path algorithms, both in terms of computational complexity and the quality of generated route alternatives, contributing to GIScience-based spatial analysis.

3 The Proposed SA for the CLP

This section describes the proposed SA metaheuristic approach to solve the CLP.

3.1 Simulated Annealing Method

Simulated annealing is a stochastic search algorithm based on a heat treatment process by altering the structure of a solid in order to change its properties [12]. The annealing process requires heating and then slowly cooling the solid to obtain a robust structure [23]. In the annealing process, the free energy of the solid minimized; that is, the atoms of the solid moves freely (i.e., randomly) on high temperatures, and then the atoms are rearranged themselves through lower temperatures, producing a minimum energy state of the system. The cooling schedule permits gradually to decrease the temperature until it reached an equilibrium state. However, if the solid is cooled too quickly or the initial temperature of the annealing is too low, a higher energy state would reach arriving into an amorphous state. An analogy is formulated between the annealing of solids and an optimization problem to solve. The system state corresponds to a solution of the optimization problem; the free energy of the system is analogous to the objective function cost of the problem to be optimized; the slight system state change (perturb the system state) is analogous to a local search of solutions into a solution neighborhood (correspond to the mutation operator used on Evolutionary Algorithms); the cooling schedule corresponds to the control mechanism of the parameters (initial temperature, equilibrium state, cooling function, stopping criteria or final temperature) in SA algorithm; the final system state (frozen state) corresponds to the final solution by the SA algorithm (global optimum).

SA consists of empowering an iterated local search method through a procedure that allows accepting lower quality solutions than the current solution to escape from local optimum. In this way, SA explores and exploits the search space to find better solutions [20]. The SA process begins with high temperatures where random changes are accepted, altering the initial state continuously, with a slight decrease in temperature. In this way, the number of accepted changes decreases until none change is accepted. Finally, the process converges and stops with a profoundly modified initial state. SA is an iterative local search guided through a stochastic process, where a j state is accepted with a determined probability given by $\Pi(\Delta E, T) = exp^{-\Delta E/kT}$, where k is the Boltzmann constant. The Metropolis algorithm [18] simulates the change of energy in the cooling process of a physical system. SA works over a representation of candidate solutions, defined according to the specific features of the problem to solve.

The iterative method requires an initial solution, which is generated by applying a random initialization procedure or using a heuristic method incorporating knowledge of the problem. The exploration is performed changing the system state (candidate solution), which corresponds to generate local movements in the neighborhood of the current state. This state change is applied in order to increase diversity in the search. Once the state is slightly perturbed, if the energy change produced by the perturbation is negative, the new system state is accepted; otherwise, if it is positive, it is probabilistically accepted according to the Boltzmann distribution function. The starting temperature must be balanced between high and low values to conduct the local search in enough time to allow movements in the neighborhood. The initial temperature decreases gradually (cooling), if the decrease in temperature is slowly, better solutions obtained but with more significant computation time. The equilibrium state at each temperature must permit to apply sufficient movements proportional to the neighborhood size. Finally, the stopping criterion is related to the final temperature parameter, which is regularly a low value, near to zero.

3.2   SA for the CLP

The main features of the proposed SA algorithm for the CLP are described next.

Solution representation. A two-dimensional representation is used to encode candidate solutions for the CLP in the proposed SA. Two vectors of integer numbers ($\vec{x}$ and $\vec{y}$) are used to encode a path $p = \{u_0, u_1, \cdots, u_k\}$ between a source node s and a destination node t ($u_0 = s$, $u_k = t$). Value (x_i, y_i) in the representation correspond to the indexes of the i-th node in the path, in the Cartesian grid used to discretize the area where the corridor is to be built. Figure 1 presents a sample representation for a CLP solution from origin $s = (0,0)$ to destination $t = (9,9)$ on a 10×10 grid discretizing a given terrain.

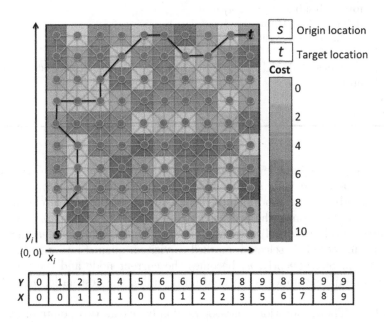

Y	0	1	2	3	4	5	6	6	6	7	8	9	8	8	9	9
X	0	0	1	1	1	0	0	1	2	2	3	5	6	7	8	9

Fig. 1. Sample representation for a CLP solution from origin $s = (0,0)$ to destination $t = (9,9)$ on a 10×10 terrain

Objective function. The proposed SA for CLP minimizes the cost function defined in the mathematical formulation (Eq. 2a).

Initial solution. The initial solution of SA for CLP is generated using an uninformed search, according to the pseudocode presented in Algorithm 1. Starting from the origin node, the neighboring node with the lower cost is selected in each iteration step, applying the Queen's movement (orthogonal plus diagonal), a standard method to reduce infeasible solutions in continuous terrain and increase cell connectivity that allows improving the accurate distance measurement by up to 30% over other movements [26].

Algorithm 1: Uninformed search for SA for CLP initialization

Data: $G(N, A, h, c)$, source node $s \in N$, target node $t \in N$
Result: Vector path $p = \{u_0, u_1, \cdots, u_k\}$

  1   Create a queue $q = \varnothing$ ▷ `queue of nodes`
  2   Create a vector $p = \varnothing$ ▷ `path to return`
  3   q.push(s)
  4   p.push(s)
  5   Label s as visited
  6   **while** ! q.empty() **do**
  7       $u_i = q$.pop()
  8   **if** $u_i == t$ **then**
  9   return vector p ▷ `destination node reached`
10   **else**
11   **for each** adjacent node u_j of u_i
12   **if** !visited(u_j) **then**
13   Label u_j as visited
14   q.push(u_j)
15   p.push(u_j)   ▷ `include node in path`
16   **end**
17   **end**
18 **end**

Algorithm 1 uses a queue structure q to store nodes of the graph. The process starts from the origin node s. This node is enqueued into q. Then, the process explores the adjacent neighbor node u_i before moving to the next level neighbors. If u_i is the destination t, the process finishes and returns vector p; otherwise, the procedure selects an adjacent node u_j of u_i and evaluates if has not been visited; in this case it is selected, marked as visited, queued into q and into the path vector p. Subsequently, u_j becomes the current node and the process of selecting its adjacent nodes is repeated until finding the destination node t.

Exploration. The exploration operator randomly selects two points of the current solution and explores alternate routes through prohibited and guided movements. Both random points must be different from the origin and destination, and different from each other. The random points are labeled as new starting point s_{new} (the nearest point to the origin s) and new destination point t_{new}. A pseudo-random path is created between s_{new} and t_{new}. Two criteria are considered to avoid infeasible solutions: if there is any common point between the current solution and the pseudo-random path, the solutions are merged in that point to create a new solution; otherwise, an A^* pathfinding algorithm is used to merge both paths and create a new feasible solution.

Annealing schedule. A geometric update scheme is applied for the temperature, i.e., $T^{k+1} = \alpha T^k$. The value of α is set by empirical analysis (see Sect. 4).

Stopping criterion. A fixed effort stopping criterion is applied: SA-CLP stops when the temperature reaches a fixed minimum value (determined empirically).

Algorithm 2 presents a pseudocode of the proposed SA for the CLP. The main difference with a canonical SA method is the information stored during the search. SA is a memoryless search algorithm; i.e., it only use information from the last step of the search to compute a new solution. However, the CLP requires a set of efficient and spatially different route alternatives. Thus, up to n_S local optima solutions are stored in a vector P during the local search process.

Algorithm 2: Proposed SA for the CLP

---

1 $k = 0$
2 initialize(s_k)   ▷ generate initial solution
3 $T_k = T_{max}$ ▷ initialize temperature
4 $P(n_S) \leftarrow \varnothing$ ▷ initialize the vector of solutions
5 **while** $T_k < T_f$ **do**
6     $i = 0$ ▷ Markov chain iterator
7   **while** $i <$ maxL **do**
8        $s^{k+1} =$ perturbation(s^k)   ▷ exploration
9   **if** $f(s^{k+1}) < f(s^k)$ **then**
10        $s^k \leftarrow s^{k+1}$ ▷ acceptance criterion
11   **else**
12        $\Delta f = f(s^{k+1}) - f(s^k)$
13        $\Pi(\Delta f) = \exp^{-(\Delta f/T_0)}$
14        $\beta =$ random(0,1)
15   **if** $\beta \leq \Pi(\Delta f)$ **then**
16         $s^k \leftarrow s^{k+1}$ ▷ accept sub-optimal solution
17   **end**
18   **end**
19   **end**
20   insert(P, s^k)   ▷ store in solutions vector
21    $T_k \leftarrow \alpha T_k$ ▷ temperature decay
22    $i \leftarrow i + 1$
23 **end**
24 **return** s^k

SA starts from the initial solution s^0 and proceeds to search for feasible solutions through an iterative local search in the neighborhood of size $N(s)$. At each iteration, a perturbed neighbor s^{k+1} is generated. Feasible perturbed solutions that improve the cost function (Eq. 2a) are always accepted. Perturbed solutions that do not improve cost are accepted according to a probability that depends on the current temperature T_k, and the Boltzmann probability function $\Pi(\Delta f)$). A maximum number of iterations (maxL) are performed to analyze different perturbed solutions in a Markovian search with the same value of the temperature parameter. The value of maxLenght is set by empirical analysis, accounting for the computational cost and the number of improved solutions in the Markovian search. The SA algorithm stops when the temperature reaches the final value T_f determined by empirical analysis (see Sect. 4).

This way, the proposed metaheuristic approach determines the optimal route between two points across a geographic terrain for the CLP. Likewise, space exploration allows generating different alternatives routes between origin and destination points for competitive and spatially different route alignments in the corridor location. The next section analyzes the experimental results of the proposed approach.

4   Experimental Results and Discussion

This section describes the experimental evaluation of the proposed SA for the CLP. Details about the platform used for development and execution are presented. Likewise, the studied scenario and problem instances are described. Experimental results of the parameter calibration are reported. Finally, results for the addressed real-world problem instances are reported and discussed.

4.1   Development and Execution Platform

The proposed SA method was developed in C++11 and evaluated on a Quad-core Xeon E5430(2.66 GHz), 8 GB RAM, from National Supercomputing Center (Cluster-UY), Uruguay [21]. QGIS 3 was used for data visualization.

4.2   Studied Area and Problem Instances

The studied area is located in Veracruz State, on the Gulf of México. Veracruz State has a territorial surface of $71\,699\,km^2$. The Veracruz Basin is an essential depot of gas and petroleum with an area of $34\,824\,km^2$, containing a potential number of oil extraction wells that distribute to other regions of Mexico.

Four problem instances were built using GIS data models of the studied area. Tested instances include remarkable spatial objects considered in a real-world pipeline facility construction. The assigned penalty impact cost to the spatial objects are: industrial facilities, airports, port lighthouse, runway aviation and bridges (10), electric substations, railways, port facilities, sports facilities and communication lines (9), water channel, water bodies, water ponds (human consumption), aqueducts, sand mining zones, cemeteries, urban areas and buildings (8), watershed, communication facilities (7), highways and rural locations (6), streets, racetracks, flood zones, rural paths and oil ducts/pipelines (5), green areas (not protected areas), sandy soils and growing areas (0). The DEM data model used has a raster BIL format, and each spatial object has a vector shapefile format. The UTM coordinates information of the spatial objects was converted from BIL raster to ASCII-TXT format using QGIS3, and from shapefile format to ASCII-TXT format using OGR/GDAL and Phyton 3. The methodology to build each problem instance considers creating an ASC raster through weighting the penalty impact costs of the spatial objects used from each topographic chart, using Eq. 1.

Problem instances #0 and #1 include the municipality of Veracruz (topographic chart E14B49); problem instance #2 includes the municipalities of Naranjos (E14B69) and Tierra Blanca (E14B79); and problem instance #3 includes part of the territories of Morelos, Puebla, and Veracruz states (E14B41-49, E14B51-59, E14B61-69, E14B71-79, and E14B81-89). Figure 2 shows the topographic charts of the tested problem instances at map scale 1:50,000.

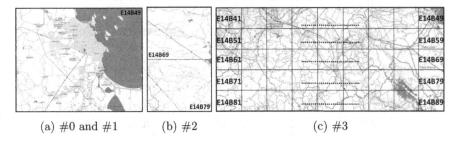

(a) #0 and #1　　　　(b) #2　　　　　　　　(c) #3

Fig. 2. Problem instances in the studied area, at map scale 1:50,000

4.3 Parameters Configuration

The SA parameters must be adjusted to determine the configuration that allows computing the best results. Parameters tuning is made by an empirical sensitivity analysis performed over problem instance #0. The studied parameters and candidate values included: $T_0 \in \{2.0, 1.3, 0.9, 0.3, 0.1\}$, $T_f \in \{10^{-2}, 10^{-3}, 10^{-4}, 10^{-5}, 10^{-6}\}$, $\alpha \in \{0.99, 0.98, 0.97, 0.96, 0.95\}$ and maxL $\in \{100, 50, 20, 10, 5\}$. Due to the stochastic quality of the simulated annealing, statistical analysis was applied to select the best configuration. 30 independent executions of SA were performed. Then, the Shapiro-Wilk test was applied to check normality. Results discarded the normality hypothesis with p-value $< 10^{-2}$. Thus, the Kruskal-Wallis (KW) non-parametric test was applied to analyze the differences between the results distributions. Table 1 reports the comparative analysis for each parameter value studied. The parameter values that allowed computing the best results on the tested instance (marked in bold in Table 1), with statistical significance according to the KW test were $T_0 = 2.0$ (KW p-value 0.017), $\alpha = 0.98$ (KW p-value 0.006), maxL $= 50$ (KW p-value 0.0016), and $T_f = 10^{-6}$ (KW p-value $> 10^{-2}$). Figure 3 graphically shows the behavior of the studied cooling schedule parameters values for the parameter setting instance.

4.4 Experimental Results

A greedy strategy was implemented as a baseline for evaluating the results of the proposed SA-CLP. The greedy algorithm iteratively builds solutions taking

Table 1. Results of the sensitivity analysis of SA-CLP parameters

T_0	$f(s^k)$			α	$f(s^k)$		
	Worst	Best	Median		Worst	Best	Median
2.0	3024.40	2049.51	**2096.15**	0.99	2915.40	2046.41	2127.45
1.3	2915.40	2046.41	2127.45	0.98	2856.50	2046.40	**2073.05**
0.9	2566.40	2047.51	2133.80	0.97	3356.40	2057.30	2138.91
0.3	3215.30	2047.81	2120.95	0.96	2829.00	2046.01	2117.05
0.1	2306.90	2106.00	2211.35	0.95	2269.10	2080.30	2161.75
maxL	$f(s^k)$			T_f	$f(s^k)$		
	Worst	Best	Median		Worst	Best	Median
100	3630.70	2054.21	2112.06	10^{-2}	3636.70	2048.80	2136.45
50	2543.20	2047.41	**2099.15**	10^{-3}	3195.50	2047.61	2128.40
20	2916.10	2047.81	2118.15	10^{-4}	2916.10	2047.81	2118.15
10	2688.10	2083.71	2132.55	10^{-5}	2745.00	2061.51	2156.80
5	2306.90	2106.00	2211.35	10^{-6}	2853.60	2051.40	**2108.95**

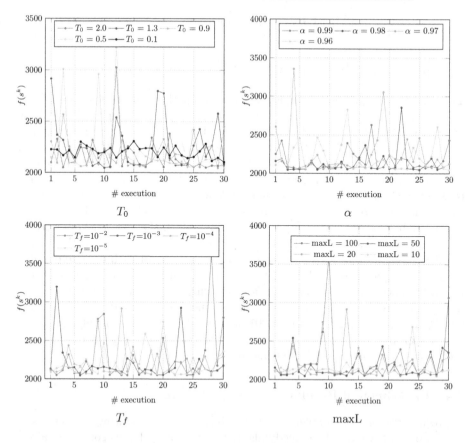

Fig. 3. Sample results of the sensitivity analysis for problem instance #0

local optimal decisions at each step (i.e., the adjacent node with the lowest cost is selected to be included in the path). When no feasible node is available to be selected, the method returns and chooses the second-best adjacent node. The procedure is repeated until reaching the destination node.

Table 2 reports the median of the best objective function values $(\overline{f(SA)})$ computed by the SA-CLP and the comparison with the reference algorithm $(f(greedy))$. The relative improvement on objective function values (Δ_f) over the reference algorithm are also reported.

Table 2. Cost solutions for the studied problem instances

Instance	Size	Origin, destination	$f(greedy)$	$f(SA)$	$\overline{f(SA)}$	Δ_f
#0	586×732	(44,365) (512,50)	3202.2	**2048.6**	2109.2	34.1%
#1	586×732	(292,551) (537,8)	2413.6	**1835.0**	2010.2	16.7%
#2	585×1467	(24,264) (113,1071)	9350.7	**7785.2**	8267.1	11.5%
#3	5341×6727	(221,47) (2525,6073)	63622.4	**55164.9**	57340.7	9.8%

Results in Table 2 indicate that the proposed SA-CLP is able to improve significantly over the greedy algorithm, regarding the cost function values. Considering the baseline results computed by the greedy algorithm, improvements of up to 34.1% were obtained in problem instance #0. Also, the solutions computed by the SA-CLP were between 9% and 16% better than the greedy algorithm in problem instances #1, #2, and #3. SA-CLP computed significantly competitive and slightly spatially different routes alignments on every problem instance. The SA-CLP execution with the parameter values reported in Subsect. 4.3 allows computing 492 different alternative routes between origin and destination points for the problem instances #0 and #1. For problem instances #2 and #3, SA-CLP computed 452 routes alignments each one. The competitive and spatially different route alignments for corridor location are displayed using a GIS tool. Figure 4 shows two samples of alternative routes for the corridor location in problem instances #1 and #2.

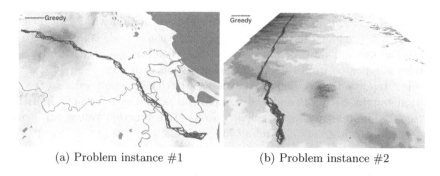

(a) Problem instance #1   (b) Problem instance #2

Fig. 4. Alternative routes for corridor location

5   Conclusions and Future Work

This article presented a metaheuristic approach to solve the CLP, a relevant problem related to the construction of linear facilities such as pipelines, electric transmission lines, and roads. A SA metaheuristic was applied to generate different and competitive alternative routes between origin and destination points.

The proposed exploration pattern applies an uninformed search for initialization and a specific exploration operator that modifies randomly selected subsegments of the current solution. Alternate routes are explored through prohibited and guided movements, and the A* pathfinding algorithm is applied for merging. Furthermore, the standard memoryless search pattern in SA is extended to store a vector of local optima solutions to be considered as alternative routes.

The experimental evaluation focused on four realistic problem instances defined over real information of the Veracruz state, a well-know oil production area on the Gulf of Mexico. Parameters configuration was performed via a sensitivity analysis that studied four relevant SA parameters: initial and final temperature, the decay factor of the annealing schedule and the length of the Markov chain used for exploration. Statistical analysis were applied to study the results distribution by checking normality and determining the significance of the median values for each parameter configuration. Using the best configuration from the previous analysis, the proposed SA allowed computing a spatially diverse set of near-optimal alternative routes that improves up to 34.1% the results of a greedy algorithm, such as the ones included in traditional GIS software.

The main lines for future work are related to extend and improve the neighborhood search to address the reduction of improvements over the traditional greedy approach when solving large problem instances, e.g., by applying hybrid approaches combining exact and metaheuristic methods. The experimental evaluation of the proposed approach should also be extended to consider a larger number of synthetic and real scenarios. Multiobjective versions of the problem and multiobjective metaheuristics should be studied to compute and analyze distance and penalty impact cost simultaneously.

References

1. Akgün, V., Erkut, E., Batta, R.: On finding dissimilar paths. Euro. J. Oper. Res. **121**(2), 232–246 (2000)
2. Berry, J.: Beyond Mapping: Concepts, Algorithms, and Issues in GIS. Wiley, Hoboken (1996)
3. Byers, T., Waterman, M.: Technical note-determining all optimal and near-optimal solutions when solving shortest path problems by dynamic programming. Oper. Res. **32**(6), 1381–1384 (1984)
4. Church, R., Loban, S., Lombard, K.: An interface for exploring spatial alternatives for a corridor location problem. Comput. Geosci. **18**(8), 1095–1105 (1992)
5. Church, R., Murray, A.: Line-based location. In: Church, R., Murray, A. (eds.) Business Site Selection, Location Analysis and GIS, chap. 7. Wiley, Hoboken (2008)
6. Cruz-Chavez, M., Martinez-Oropeza, A., Barquera, S.: Neighborhood hybrid structure for discrete optimization problems. In: IEEE Electronics, Robotics and Automotive Mechanics Conference, pp. 108–113. IEEE (2010)
7. Douglas, D.: Least-cost path in GIS using an accumulated cost surface and slopelines. Cartographica: Int. J. Geogr. Inf. Geovisualization **31**(3), 37–51 (1994)
8. Fournier, E.: Mogador revisited: improving a genetic approach to multi-objective corridor search. Environ. Plann. B Plann. Des. **43**(4), 663–680 (2016)
9. Hallam, C., Harrison, K., Ward, J.: A multiobjective optimal path algorithm. Digit. Sig. Process. **11**(2), 133–143 (2001)
10. Huber, D., Church, R.: Transmission corridor location modeling. J. Transp. Eng. **111**(2), 114–130 (1985)
11. Johnson, P., Joy, D., Clarke, D., Jacobi, J.: Highway 3.1: An enhanced highway routing model: Program description, methodology, and revised user's manual (1993)
12. Kirkpatrick, S., Gelatt, C., Vecchi, M.: Optimization by simulated annealing. Science **220**(4598), 671–680 (1983)
13. Kuby, M., Zhongyi, X., Xiaodong, X.: A minimax method for finding the k best "differentiated" paths. Geogr. Anal. **29**(4), 298–313 (1997)
14. Lombard, K., Church, R.: The gateway shortest path problem: generating alternative routes for a corridor location problem. Geogr. Syst. **1**, 25–45 (1993)
15. Medrano, A., Church, R.: Corridor location for infrastructure development: a fast bi-objective shortest path method for approximating the pareto frontier. Int. Reg. Sci. Rev. **37**(2), 129–148 (2013)
16. Medrano, A., Church, R.: A parallel algorithm to solve near-shortest path problems on raster graphs. In: Xi, X., Kindratenko, V., Yang, C. (eds.) Modern Accelerator Technologies for Geographic Information Science, pp. 83–94. Springer, Boston (2013). https://doi.org/10.1007/978-1-4614-8745-6_7
17. Medrano, A., Church, R.: Corridor location for infrastructure development: a fast bi-objective shortest path method for approximating the pareto frontier. Int. Reg. Sci. Rev. **37**(2), 129–148 (2014)
18. Metropolis, N., Rosenbluth, A., Rosenbluth, M., Teller, A., Teller, E.: Equation of state calculations by fast computing machines. J. Chem. Phys. **21**(6), 1087–1092 (1953)
19. Miller, H., Shaw, S.: Geographic information systems for transportation in the 21st century. Geogr. Compass **9**, 180–189 (2015)
20. Nesmachnow, S.: An overview of metaheuristics: accurate and efficient methods for optimisation. Int. J. Metaheuristics **3**(4), 320–347 (2014)

21. Nesmachnow, S., Iturriaga, S.: Cluster-UY: collaborative scientific high performance computing in Uruguay. In: Torres, M., Klapp, J. (eds.) ISUM 2019. CCIS, vol. 1151, pp. 188–202. Springer, Cham (2019). https://doi.org/10.1007/978-3-030-38043-4_16
22. Scaparra, M., Church, R., Medrano, A.: Corridor location: the multi-gateway shortest path model. Geogr. Syst. **16**, 287–309 (2014)
23. Talbi, E.G.: Metaheuristics: From Design to Implementation. Wiley, Hoboken (2009)
24. Turner, K.: Computer-assisted procedures to generate and evaluate regional highway alternatives. Purdue University, Technical report (1968)
25. Yen, J.: Finding the k shortest loopless paths in a network. Manage. Sci. **17**(11), 712–716 (1971)
26. Zhang, X., Armstrong, M.: Genetic algorithms and the corridor location problem: multiple objectives and alternative solutions. Environ. Plann. B Plann. Des. **35**(1), 148–168 (2008)

A Simulated Annealing Algorithm for Solving a Routing Problem in the Context of Municipal Solid Waste Collection

Matías Fermani[1], Diego Gabriel Rossit[1,2(✉)] , and Adrián Toncovich[1]

[1] Departmento de Ingeniería, Universidad Nacional del Sur, Bahía Blanca, Argentina
{diego.rossit,atoncovi}@uns.edu.ar
[2] INMABB, Universidad Nacional del Sur (UNS)-CONICET, Bahía Blanca, Argentina

Abstract. The management of the collection of Municipal Solid Waste is a complex task for local governments since it consumes a large portion of their budgets. Thus, the use of computer-aided tools to support decision-making can contribute to improve the efficiency of the system and reduce the associated costs. In the present work, a simulated annealing algorithm is proposed to address the problem of designing the routes of waste collection vehicles. The proposed algorithm is compared against two other metaheuristic algorithms: a Large Neighborhood Search (LNS) algorithm from the literature and a standard genetic algorithm. The evaluation is carried out on real instances of the city of Bahía Blanca and on benchmarks from the literature. The proposed algorithm was able to solve all the instances, having an average performance similar to the LNS, while the standard genetic algorithm showed less promising results.

Keywords: Municipal solid waste · Vehicle routing · Simulated annealing

1 Introduction

The generation of urban solid waste (MSW) is an inalienable consequence of the development of modern cities. Likewise, its correct management is one of the key elements for the sustainable development of a city [1]. The stages in the reverse logistics chain of MSW are diverse. According to Tchobanoglous et al. [2], they can be classified into waste generation; handling, separation, accumulation and processing of waste at source; collection, transfer and transportation; separation, processing and transformation of solid waste; and finally disposal. In these stages, there are many decisions to be made. For this reason, proposing computer-aided tools that allow assisting decision-making agents can contribute to a more efficient use of resources.

This work focuses on the collection, transfer and transportation stage of waste. In particular, a simulated annealing based metaheuristic algorithm is proposed to solve this problem. The proposed algorithm is compared against other metaheuristic algorithms in the literature in order to study its performance. The experimentation is carried out on real instances of the city of Bahía Blanca, Argentina, as well as in well-known benchmarks of the literature.

© Springer Nature Switzerland AG 2021
D. A. Rossit et al. (Eds.): ICPR-Americas 2020, CCIS 1408, pp. 63–76, 2021.
https://doi.org/10.1007/978-3-030-76310-7_5

This article is structured as follows. Section 2 presents the problem to be addressed and a bibliographic review of the main related works, Sect. 3 outlines the used resolution algorithms, Sect. 4 describes the computational experimentation and, finally, Sect. 5 presents the main conclusions of this work.

2 Problem Statement

The management of municipal solid waste in the city of Bahía Blanca is a crucial activity for local authorities, not only because of the broad environmental and social impact associated with its operation but also because it consumes a large portion of the municipality's budgetary resources [33]. Therefore, approaches that allow optimization of collection logistics may be relevant to achieve a more efficient service provision [17, 34]. It is within this framework that proposals have been made to migrate from the current door-to-door collection system to a containerized system [17, 33]. Continuing with these studies, the next section presents a mathematical model to plan the next stage in the reverse logistics chain of urban solid waste: the design of the collection routes for the waste accumulated in the containers. In addition, a review of the main related works in the literature is carried out.

2.1 Mathematical Model

The problem of collecting the waste accumulated in containers can be represented as a Capacitated Vehicle Routing Problem (CVRP). Thus, it can be modeled as follows. Given a set of containers C and a superset $\underline{C} = C \cup c_0$, where c_0 is the warehouse where the vehicles start and end their trips, the following variables and parameters are defined: binary variable x_{ij} that adopts value 1 if a vehicle uses the path from container $i \in C$ to container $j \in C$ and 0 otherwise; the continuous positive variable u_i that indicates the load of the vehicle before visiting container $i \in C$; parameter Q is defined as the capacity of the vehicle; parameter d_{ij} as the distance between containers $i \in C$ and container $j \in C$; and parameter q_i as the amount of waste to be collected at container $i \in C$. In this way, the following model is proposed in Eqs. (1)–(6), using the two-index formulation developed by Miller et al. [3].

$$\text{Minimize } FO = \sum_{i,j \in \underline{C}} x_{ij} d_{ij} \tag{1}$$

Subject to:

$$\sum_{j \in \underline{C}} x_{ij} = 1, \forall i \in C \tag{2}$$

$$\sum_{j \in \underline{C}} x_{ji} = 1, \forall i \in C, \tag{3}$$

$$u_j - u_i \leq Q(1 - x_{ij}) - q_j, \forall i, j \in \underline{C} \tag{4}$$

$$q_i \leq u_i \leq Q, \forall i \in C \tag{5}$$

$$x_{ij} \in \{0, 1\}, \forall i, j \in \underline{C} \tag{6}$$

The proposed objective is to minimize the total distance traveled expressed in Eq. (1). Equations (2) and (3) guarantee that each container is visited only once, having a single successor container and a single predecessor container on the route. Equation (4) prevents the formation of subtours. Equation (5) ensures that the vehicle's capacity is not exceeded. Equation (6) establishes the binary nature of the variable x_{ij}.

2.2 Literature Review

The problem of collection of containerized waste in a city has been frequently addressed in the literature. In Beliën et al. [4] and Han and Ponce Cueto [5], extensive reviews of these works are presented.

In the case of Argentina, some studies can be found on applications of VRP models to solve the problem of collection route design. For example, in Bonomo et al. [6] and Larrumbe [7] mathematical programming methods are implemented to schedule the collection routes for waste containers in the Southern area of the City of Buenos Aires. On the other hand, in Bonomo et al. [8] a different model is presented for this city, which aims at minimizing the travel distances simultaneously with minimizing wear and tear on vehicles. In the city of Concordia, Bertero [9] presents an application to design the city's collection routes, making an effort to minimize the number of turning maneuvers to facilitate the implementation of the routes by the authorities. On the other hand, Bianchetti [10] exhibits an algorithm to solve the zoning of the city of San Miguel de Tucumán in order to optimize the use of resources, reassigning trucks to the downtown area of the city. In Braier et al. [11, 12], through mathematical programming models, the collection of recyclable waste is planned for the city of Morón. In Rossit et al. [13], a comprehensive approach is presented that determines the location of the containers, as well as the design of the routes that collect their content in simulated instances of Bahía Blanca. For the same city, Cavallin et al. [14, 15] present vehicle routing models that balance the distances among the different routes to design the trips of the informal collectors of recyclable waste.

3   Proposed Solution Method

As aforementioned, the collection of MSW in containers constitutes an application case of the VRP (Vehicle Routing Problem) problem. In computational complexity theory, this type of problem is classified as NP-hard [16], that is, it is at most as difficult as problems for which efficient solving algorithms that run in polynomial time have not yet been developed, regarding the size of the problem [17]. In this type of problems, metaheuristic tools allow obtaining good solutions in reasonable computational times [35]. The proposed solution algorithm is based on a simulated annealing (SA) strategy and was adapted from the previous algorithm developed in Toncovich et al. [18, 19]. SA is a local search-based method that was developed from an analogy with the physical phenomenon of annealing to solve complex optimization problems [20]. Local search methods search for the solution with the best value of the chosen criteria in the current solution neighborhood, accept it as the current solution, and repeat this procedure until it is not possible to improve the solution in the explored neighborhood. By systematically applying this procedure, a local optimum for the problem is generally obtained. To avoid being trapped in a local optimum, a diversification mechanism must be incorporated in order to adequately explore the solution space. In the simulated annealing metaheuristic, the diversification strategy allows moves, with some probability, toward solutions that worsen the current value of the objective function.

To get a good approximation to the optimal solution of the problem during the search process, it is necessary to restart the search regularly from one of the solutions accepted during the search process selected at random. The proposed implementation of the algorithm incorporates the classic parameters and variables of simulated annealing which are indicated below.

- t: current iteration.
- S_0: initial solution.
- S_A: current solution.
- S_c: candidate solution.
- $V(S)$: neighborhood for solution S, given by the set of solutions that can be obtained from solution S through a basic perturbation or change.
- T: control parameter that simulates the temperature of the real annealing process. It is a positive value that varies within the interval $T \in [T_f, T_0]$ during the execution of the algorithm, where T_0 is the initial temperature and $T_0 > T_f$.
- N_T: number of iterations performed by the algorithm for a certain temperature value T at iteration t.
- $T(t)$: function that determines the cooling schedule, i.e., how T evolves during the process. In this case, a geometric progression of the form $T(t) = \alpha T(t-1)$ is used, with $\alpha \in [0, 8; 0, 99]$.
- N_{cont}: the number of iterations without improvement in the objective function at a given moment in iteration t.
- N_{stop}: maximum allowable number of iterations without improvement.

The following pseudo-code is applied to determine a potentially optimal solution.

Simulated annealing algorithm

Initialization
An initial solution S_0 is created randomly.
The solution is evaluated, $FO(S_0)$, using Equation (1).
S_0 is accepted as current solution, S_A.
Set $N_{cont} = 0, t = 1, T = T_0$.
While stop criteria do not met ($N_{cont} = N_{stop}$ or $T \leq T_f$) **do:**
   Set $n = 1$.
   While $n < N_T$ **do:**
      A random perturbation is performed on S_A to create a new solution S_c such that $S_c \in V(S_A)$.
      S_c is evaluated, $FO(S_c)$.
      The value ΔFO is calculated, $\Delta FO = FO(S_c) - FO(S_A)$.
      If $\Delta FO \leq 0, S_A = S_c, N_{cont} = 0$.   *Obs:* S_c is accepted as the new incumbent solution.
      Else, $p_A = e^{-\frac{\Delta FO}{T}}$. *Obs:* S_c is accepted as the new incumbent solution with probability p_A.
      $\beta = random[0,1]$:
      If $\beta \leq p_A$, **then** $S_A = S_c, N_{cont} = 0$;
      Else, $N_{cont} = N_{cont} + 1$.
      $n = n + 1$
   End While
   $t = t + 1$
   $T(t) = \alpha T(t - 1)$
End While
Return best S_A found.

3.1 Baseline Metaheuristics

The metaheuristics for comparison are the Large Neighborhood Search (LNS) algorithm developed by Erdoğan [21] and a standard genetic algorithm. Erdoğan's metaheuristic was adapted from the adaptive LNS heuristic developed by Pisinger and Ropke [22], which extends Shaw's original heuristic [23]. A simplified explanation of its operation and pseudo-code (see Fig. 1) can be found in Cavallin et al. [24].

The operation of the heuristic consists of two main stages. The first is the construction of an initial solution that is obtained by inserting clients into the available routes in accordance with the objective to be minimized (operator *Sol-Inicial*). Then an enhancement is made using the *Local-Search* operator that applies four local search techniques. The first three are known as *Exchange* (two clients are exchanged), *1-OPT* (a client is extracted from one route and reinserted into another) and *2-OPT* (two route segments are exchanged). These operators are detailed in Groër et al. [25]. The fourth local search operator is *Vehicle-Exchange*. This operator tries to exchange the vehicles assigned to two routes, which it is useful when working with a heterogeneous fleet.

The second stage aims at improving the solution from the previous stage. The *Destroy-And-Rebuild* operator is applied, which considerably alters the current solution to explore another region of the search space and, thus, allow to escape from local optimum. In particular, large portions of the routes are extracted from the current solution and the extracted clients are reinserted in those positions that minimize the objective function (*Cost*). The solution is accepted if it has a smaller cost than the best solution found so far or, in case the solution does not have a smaller cost than the previous solution, it can still be accepted with a probability *p*-value. This process is repeated iteratively up to the time limit imposed by the user.

```
1: procedure LNS(input)            ▷ Como input se ingresa la información
    correspondiente a los clientes, vehículos y lugares de descarga (depots)
2:     S ← SolInicial()
3:     S' ← LocalSearch(S)
4:     BestS ← S'
5:     repeat
6:         S' ← DestroyAndRebuild(S')
7:         S' ← LocalSearch(S')
8:         if (Cost(S') <= Cost(S')) then
9:             BestS ← S'
10:        else
11:            p ← Random(0, 1)
12:            if (p <= pvalue) then
13:                S' ← BestS
14:            end if
15:        end if
16:    until Time elapsed > Time limit
17:    return BestS
18: end procedure
```

Fig. 1. LNS algorithm implemented in [21]. Image Source: [24].

The standard genetic algorithm was developed using the Distributed Evolutionary Algorithms in Python (DEAP) framework [26]. It has the following characteristics:

Representation of Solutions. Solutions are encoded as a permutation of integers of length equal to the number of containers n. Each index in the vector represents the visit order in the tour and the corresponding integer value represents one of the containers.

Initialization. The population of size #P is initialized by applying a random procedure to generate the permutations with a uniform distribution.

Genetic Operators. The recombination operator is the Partially Mapped Crossover (PMX), which is applied on two selected individuals with p_c probability. The mutation operator is based on Swap Mutation and swaps two elements of the permutation. The mutation applies to an individual with probability p_m. The proposed operators guarantee the feasibility of the solution.

Selection. The selection is made through a binary tournament, from which the individual with the best fitness is selected.

Replacement. In each iteration, the new population is made up of the best 10% individuals (with better fitness) of the previous population and the rest of new individuals generated through genetic operators.

Fitness Assessment. The fitness function is decoded by reading alleles from left to right and inserting a depot visit when necessary.

Parametric Analysis. The parameters that were analyzed through a statistical analysis were the size of the population $\#P$ - the values considered were 100, 150 and 200 - the probability of mutation p_m – the values considered were 0.05, 0.1, 0.15, 0.20 and 0.25- and p_c crossover - the values considered were 0.5, 0.6, 0.7, 0.8, 0.9 and 1-. The parametric analysis was carried out on the instances P-n16-k8, P-n19-k2 and P-n20-k2 proposed in Augerat [27]. For each instance and for each parametric configuration, 50 independent executions were carried out. The Shapiro-Wilk test was applied to study whether the fitness distribution had a normal distribution or not. Since many executions did not follow a normal distribution, the medians were evaluated through the Friedman test, selecting the following configuration: $\#P = 200$, $p_c = 0.9$ and $p_m = 0.25$.

Cut Due to Stagnation. A deadlock cut-off mechanism was incorporated. The algorithm stops when no improvement in the fitness of the best individual is obtained during a time interval equal to 50% of the maximum execution time (Eq. (7)). Otherwise, it continues until the first generation - cycle of evaluation, selection, crossing, mutation and replacement - that exceeds the maximum execution time.

4 Results

The results obtained correspond to the resolution of four real scenarios of the MSW collection logistics problem of Bahía Blanca constructed in Herran Symonds [28] and Signorelli Nuñez [29], and different benchmark problems of the literature.

4.1 Tests Carried Out on Real Scenarios

The instances presented in Herran Symonds [28] and Signorelli Nuñez [29] constitute real cases based on the existing MSW background in the city of Bahía Blanca.

Two different variations for the disposal of the containers contents are studied in these real instances. These variations consider the classification of waste in containers of humid and dry fractions while considering different collection frequencies for the dry waste. Thus, the following four scenarios are developed:

Scenario 1:

- Waste collection frequency: 6 times per week.
- Type of waste collected: humid waste.

Scenario 2:

- Waste collection frequency: 4 times per week.
- Type of waste collected: dry waste.

Scenario 3:

- Waste collection frequency: 6 times per week.
- Type of waste collected: humid waste.

Scenario 4:

- Waste collection frequency: 3 times per week.
- Type of waste collected: dry waste.

For the purpose of comparison, the scenarios were also solved by an exact method. This was implemented using CPLEX software version 12.6.0.0 in a GAMS environment. For this, the mathematical formulation presented in Sect. 2.1 was used with the addition of the valid inequality of the minimum number of trips to accelerate convergence. A time limit of 15,000 s was set for each run.

Both the SA and LNS algorithms were programmed using the Visual Basic for Applications (VBA) interface of MS Excel. While, to apply the GA algorithm, the DEAP in Python was used.

Each algorithm was run on a personal computer, performing 50 runs for each instance in order to obtain a more representative sample of the performance of the algorithms.

The results obtained are shown below in Tables 1 and 2. In the first one, the best distance - FO according to Eq. (1) - obtained for each MSW scenario through each of the proposed solution methods can be observed. For the CPLEX solution, the gap calculated by the software is reported, which indicates that the optimal value was not obtained in scenarios 2 and 4. In the second table, three indicators of the sets of solutions obtained for each metaheuristic are contrasted, a measure of centralization, such as the average value of the results obtained, a measure of dispersion, that is, the standard deviation, and the average gap. The gap is calculated as the difference between the average value and the value obtained by CPLEX, divided by the latter.

Table 1. Results of the MSW problem: solutions of CPLEX and metaheuristic approaches.

Instance	Exact solver CPLEX		Best result of FO for each metaheuristic		
	FO	CPLEX *gap* %	SA	LNS	GA
Scenario 1	22613,90	0,00	22814,00	22877,00	36712,00
Scenario 2	41533,28*	26,35	43763,00	43491,00	48561,00
Scenario 3	27150,00	0,00	27707,00	27648,00	49463,00
Scenario 4	62800,63*	18,33	65995,00	65938,00	71185,00

* Since the optimal solution for these scenarios was not found, the values correspond to the best value obtained by CPLEX within the execution time (Upper Bound).

Table 2. Results of the RSU problem: average performance of each metaheuristic.

Scenario	SA			LNS			GA		
	Avg. FO	Std. dev.	gap %	Avg. FO	Std. dev.	gap %	Avg. FO	Std. dev.	gap %
Scenario 1	23310,12	216,82	3,08	23542,56	246,68	4,11	39041,70	1164,59	72,65
Scenario 2	44294,64	291,85	6,65	44229,92	303,77	6,49	51047,52	1247,72	22,91
Scenario 3	28143,00	296,63	3,66	28032,34	223,30	3,25	53408,00	1396,83	96,99
Scenario 4	66366,31	202,49	5,68	66356,19	180,61	5,66	73290,82	1096,94	16,70

The maximum execution time of all metaheuristics was established proportionally to the number of nodes in the instance - excluding the depot - and is given by Eq. (7).

$$Maximum\,Execution\,Time\,[sec] = \frac{(n° \text{ of nodes} * 15 * 60)}{50} \tag{7}$$

It should be taken into account that both the SA and the GA described here have cut-off mechanisms due to stagnation that can lead to a real execution time that is less than the maximum. This is not the case for the LNS whose execution time will be equal to the maximum.

Although the execution time of the real scenarios in CPLEX was limited to 15,000 s, in scenarios 1 and 3, in which the optimal value was reached, the optimal solution was found at 1892 s and 4367 s, respectively.

In order to evaluate its performance, Table 3 records the number of iterations required by each metaheuristic in the time established by the previously mentioned equation.

Table 3. Average time and iterations required for MSW scenarios.

Scenario	Maximum execution time (sec)	Average number of iterations		
		SA	LNS	GA
Scenario 1	1242	76860,12	33700,22	30739,22
Scenario 2	1242	211455,16	16724,16	50439,62
Scenario 3	1422	265653,78	11563,00	51479,40
Scenario 4	1422	1516847,40	6129,72	63967,20

4.2   Tests Carried Out on Benchmark Problems of the Literature

To extend the metaheuristic comparison, some instances of the benchmark proposed in Christofides and Eilon [30], called Set E, were used. Both the instances and the optimal solutions were downloaded from the VRP-REP digital repository [32].

Table 4 shows the best result found in each instance by the algorithms under study, while Table 5 shows the performance of the metaheuristics through the same parameters used to evaluate the performance in real scenarios of the previous section.

Table 4. Results of set E: best known solutions of each metaheuristic.

Instance	Optimal solution	Best result of each metaheuristic		
		SA	LNS	GA
E-n13-k4	247.00	247.00	247.00	247.00
E-n22-k4	375.00	375.00	375.00	375.00
E-n23-k3	569.00	569.00	619.00	569.00
E-n30-k3	534.00	503.00*	534.00	529.00*
E-n31-k7	379.00	379.00	379.00	427.00
E-n33-k4	835.00	835.00	835.00	859.00
E-n51-k5	521.00	521.00	521.00	582.00

*The solutions found have a lower value of FO, but a greater number of routes (k = 4) than the optimal solution, a fact that is also observed in Table 5.

Table 5. Results of set E: Average performance of each metaheuristic.

Instance	SA			LNS			GA		
	Avg. FO	Std. dev.	*gap %*	Avg. FO	Std. dev.	*gap %*	Avg. FO	Std. dev.	*gap %*
E-n13-k4	247.08	0.27	0.03	247.00	0.00	0.00	247.72	1.09	0.29
E-n22-k4	376.14	3.24	0.30	375.00	0.00	0.00	393.94	14.90	5.05
E-n23-k3	569.04	0.20	0.01	619.00	0.00	8.79	625.086	41.06	9.85
E-n30-k3	503.04	0.28	-5.80	535.08	1.45	0.20	583.86	36.37	9.34
E-n31-k7	394.34	13.11	4.05	379.64	2.16	0.17	485.44	31.18	28.08
E-n33-k4	836.44	2.21	0.17	835.00	0.00	0.00	903.42	23.11	8.19
E-n51-k5	522.48	3.81	0.28	521.84	2.76	0.16	645.56	29.05	23.91

*The gap was calculated as the difference between the average and the optimal value, divided by the latter.

As in the case of Table 3, the average number of iterations reached for each metaheuristic in the time given by Eq. (7) were recorded (Table 6).

Table 6. Average time and number of iterations performed per instance.

Instance	Maximum Execution Time (sec)	Average number of iterations		
		SA	LNS	GA
E-n13-k4	216	76860.12	33700.22	30739.22
E-n22-k4	378	211455.16	16724.16	50439.62
E-n23-k3	396	265653.78	11563.16	51479.40
E-n30-k3	522	1516847.40	6129.72	63967.2
E-n31-k7	540	2390054.58	9992.40	6800.86
E-n33-k4	576	3614994.88	4502.92	68233.54
E-n51-k5	900	6460189.50	2788.66	91230.70

4.3 Analysis of Results

From the analysis of the results, it is verified that the three metaheuristic procedures are valid for use in solving CVRP-type problems. It is necessary to clarify that both SA and GA find a value lower than the optimal value reported in the bibliography for the E-n30-k3 instance given that, since they do not have a restriction of the maximum number of routes, they present a solution with an additional route. It is also because of this reason that a negative value of the gap is observed in this instance in Table 5 for SA.

The experience carried out on the benchmark and the real scenarios of Bahía Blanca, showed a good performance for both the proposed SA algorithm and the LNS of the literature by yielding considerable small values of the average gap. On the other hand, GA showed the worst performance compared to the SA and LNS, which is again reflected in the average gap. This might be due to the use of a standard genetic algorithm with a simple decoding function taken from applications in unrestricted problems (Traveling Salesman Problem).

Regarding the average iterations carried out, it is observed that SA and GA reach a greater number of iterations as the number of nodes grows, contrary to the case of LNS for which the number of iterations decreases with the increment in the number of nodes.

5   Conclusions

Providing new computer-aided tools to optimize municipal solid waste (MSW) logistics is of paramount importance in current societies. This work focuses on solving waste collection problems for the Bahía Blanca area. A comparative study is presented among the exact solution approach and three metaheuristic solution tools for the MSW collection problem in Bahía Blanca. The exact solution approach is based on a mathematical programming formulation of the CVRP model solved using the CPLEX software. On the other hand, as regards the metaheuristic solution strategies, a simulated annealing algorithm is proposed, which is compared against an algorithm taken from the literature based on Large Neighborhood Search (LNS) and a standard genetic algorithm. In addition, the comparison of metaheuristics is extended with instances from a well-known benchmark from the literature. The proposed simulated annealing algorithm has a similar performance to the LNS algorithm from the literature - obtaining values close to the optimal solution on several occasions - and remarkably exceeds the performance of the standard genetic algorithm. Finally, it can be considered that the application of these metaheuristic algorithms allowed to obtain competitive results with small computational effort.

The lines for future work include to experiment with larger scenarios of the city of Bahía Blanca, using the proposed metaheuristics, which were validated in the present work. On the other hand, it is proposed to continue improving the mathematical programming formulation, implemented through CPLEX, by including additional valid inequalities with the aim of reducing solution times. Finally, it is proposed to develop a problem-specific coding function of the genetic algorithm with an approach that is better adapted to the type of CVRP problem addressed in this paper.

Acknowledgements. The authors of this work wish to acknowledge the funding received from the Universidad Nacional del Sur for the research projects PGI 24/J084 and PGI 24/ZJ35. In addition, the first author of this work is grateful for the funding received from the *Consejo Interuniversitario Nacional* of Argentina (CIN) through a scholarship *Becas de Estímulo a las Vocaciones Científicas (Becas EVC – CIN)*.

References

1. Hoornweg, D., Bhada-Tata, P.: What a waste: a global review of solid waste management. World Bank 15. Washington DC, United States of America (2012)
2. Tchobanoglous, G., Kreith, F., Williams, M.E.: Introduction. In: Tchobanoglous, G., Kreith, F. (eds.) Handbook of Solid Waste Management, vol. 1, 2nd ed. McGraw-Hill, EUA (2002)
3. Miller, C.E., Tucker, A.W., Zemlin, R.A.: Integer programming formulation of traveling salesman problems. J. ACM (JACM) **7**(4), 326–329 (1960)
4. Beliën, J., De Boeck, L., Van Ackere, J.: Municipal solid waste collection and management problems: a literature review. Transp. Sci. **48**(1), 78–102 (2012)
5. Han, H., Ponce Cueto, E.: Waste collection vehicle routing problem: literature review. PROMET Traffic Transp. **7**(4), 345–358 (2015)
6. Bonomo, F., Durán, G., Larumbe, F., Marenco, J.: Optimización de la Recolección de Residuos en la Zona Sur de la Ciudad de Buenos Aires. Revista de Ingeniería de Sistemas **23**, 71–88 (2009)
7. Larrumbe, F.: Optimización de la Recolección de Residuos en la Zona Sur de la Ciudad de Buenos Aires. Master Thesis. Universidad de Buenos Aires. Buenos Aires, Argentina (2009)
8. Bonomo, F., Durán, G., Larumbe, F., Marenco, J.: A method for optimizing waste collection using mathematical programming: a buenos aires case study. Waste Manag. Res. **30**(3), 311–324 (2012)
9. Bertero, F.: Optimización de recorridos en ciudades. Una aplicación al sistema de recolección de residuos sólidos urbanos en el Municipio de Concordia. Master Thesis. Universidad Nacional de Rosario. Rosario, Argentina (2015)
10. Bianchetti, M.L.: Algoritmos de zonificación para recolección de residuos. Master Thesis. Universidad de Buenos Aires. Buenos Aires, Argentina (2015)
11. Braier, G., Durán, G., Marenco, J., Wesner, F.: Una aplicación del problema del cartero rural a la recolección de residuos reciclables en Argentina. Revista de Ingeniería de Sistemas **29**, 49–65 (2015)

12. Braier, G., Durán, G., Marenco, J., Wesner, F.: An integer programming approach to a real-world recyclable waste collection problem in Argentina. Waste Manag. Res. **35**(5), 525–533 (2017)
13. Rossit, D.G., Broz, D., Rossit, D.A., Frutos, M., Tohmé, F.: Modelado de una red urbana de recolección de residuos plásticos en base a optimización multi-objetivo. In: Proceedings of the XXVI EPIO and VIII RED-M, Bahía Blanca, Argentina (2015)
14. Cavallin, A., Vigier, H.P., Frutos, M.: Aplicación de un modelo CVRP-RB a un caso de logística inversa. In: Proceedings of the XXVI EPIO and VIII RED-M, Bahía Blanca, Argentina (2015a)
15. Cavallin, A., Vigier, H.P., Frutos, M.: Logística inversa y ruteo en el sector de recolección informal de residuos sólidos urbanos. In: Avances en Gestión Integral de Residuos Sólidos Urbanos 2014–15. Instituto Nacional de Tecnología Industrial, Buenos Aires, Argentina (2015b)
16. Lenstra, J.K., Kan, A.R.: Complexity of vehicle routing and scheduling problems. Networks **11**(2), 221–227 (1981)
17. Rossit, D.G., Toutouh, J., Nesmachnow, S.: Exact and heuristic approaches for multi-objective garbage accumulation points location in real scenarios. Waste Manag. **105**, 467–481 (2020)
18. Toncovich, A., Burgos, T., Jalif., M.: Planificación de la logística de recolección de miel en una empresa apícola. In: Proceedings of the X Congreso Argentino de Ingeniería Industrial. Ciudad Autónoma de Buenos Aires, Argentina (2017)
19. Toncovich, A., Rossit, D.A., Frutos, M., Rossit, D.G.: Solving a multi-objective manufacturing cell scheduling problem with the consideration of warehouses using a Simulated Annealing based procedure. Int. J. Ind. Eng. Comput. **10**(1), 1–16 (2019)
20. Kirkpatrick, S., Gelatt, C.D., Vecchi, M.P.: Optimization by simulated annealing. Science **220**(4598), 671–680 (1983)
21. Erdoğan, G.: An open source spreadsheet solver for vehicle routing problems. Comput. Oper. Res. **84**, 62–72 (2017)
22. Pisinger, D., Ropke, S.: A general heuristic for vehicle routing problems. Comput. Oper. Res. **34**(8), 2403–2435 (2007)
23. Shaw, P.: Using constraint programming and local search methods to solve vehicle routing problems. In: Maher, M., Puget, J.-F. (eds.) Principles and Practice of Constraint Programming—CP98, pp. 417–431. Springer, Heidelberg (1998). https://doi.org/10.1007/3-540-49481-2_30
24. Cavallin, A., Rossit, D.G., Savoretti, A.A., Sorichetti, A.E., Frutos, M.: Logística inversa de residuos agroquímicos en Argentina: resolución heurística y exacta. In: Proceedings of the XV Simposio en Investigación Operativa - 46 Jornadas Argentinas de Informática e Investigación Operativa. Córdoba, Argentina (2017)
25. Groër, C., Golden, B., Wasil, E.: A library of local search heuristics for the vehicle routing problem. Math. Program. Comput. **2**(2), 79–101 (2010)
26. Fortin, F.A., De Rainville, F.M., Gardner, M.A.G., Parizeau, M., Gagné, C.: DEAP: evolutionary algorithms made easy. J. Mach. Learn. Res. **13**(1), 2171–2175 (2012)
27. Augerat, P.: Approche polyèdrale du problème de tournées de véhicules. PhD Thesis. Institut polytechnique de Grenoble. Grenoble, Francia (1995)
28. Herran Symonds, V.: Ubicación de contenedores diferenciados de RSU. Master Thesis. Universidad Nacional del Sur. Bahía Blanca, Argentina (2019)
29. Signorelli Nuñez, M.: Análisis del sistema actual de recolección de RSU en el Barrio Universitario de la ciudad de Bahía Blanca. Master Thesis. Universidad Nacional del Sur. Bahía Blanca, Argentina (2019)
30. Christofides, N., Eilon, S.: An algorithm for the vehicle-dispatching problem. J. Oper. Res. Soc. **20**(3), 309–318 (1969)

31. Christofides, N., Mingozzi, A., Toth P.: The vehicle routing problem. In: Christofides, M., Mingozzi, A., Toth, P., Sandi, C. (eds.) Combinatorial Optimization, vol. 14. John Wiley, Chichester (1979)
32. Mendoza, J., Hoskins, M., Guéret, C., Pillac, V., Vigo, D.: VRP-REP: a vehicle routing community repository. In: Proceedings of the Third meeting of the EURO Working Group on Vehicle Routing and Logistics Optimization (VeRoLog'14), Oslo, Norway (2014)
33. Cavallin, A., Rossit, D.G., Herrán Symonds, V., Rossit, D.A., Frutos, M.: Application of a methodology to design a municipal waste pre-collection network in real scenarios. Waste Manag. Res. 38(1_suppl), 117–129 (2020)
34. Rossit, D.G., Nesmachnow, S., Toutouh, J.: A bi-objective integer programming model for locating garbage accumulation points: a case study. Revista Facultad de Ingeniería Universidad de Antioquia 93, 70–81 (2019)
35. Nesmachnow, S., Rossit, D.G., Toutouh, J.: Comparison of multiobjective evolutionary algorithms for prioritized urban waste collection in Montevideo, Uruguay. Electron. Notes Disc. Math. 69, 93–100 (2018)

Optimal Dispatch of Electric Buses Under Uncertain Travel Times

Víctor Manuel Albornoz, Karam Anibal Kharrat$^{(\boxtimes)}$, and Linco Jose Ñanco

Departamento de Industrias, Campus Santiago Vitacura,
Universidad Técnica Federico Santa María, Av. Santa María, 6400 Santiago, Chile
{victor.albornoz,linco.nanco}@usm.cl, karam.kharrat.14@sansano.usm.cl

Abstract. Santiago de Chile has one of the largest electric bus fleet in the world. One of the main problems that the public transport system must address is optimal dispatch of these buses, taking into account a given schedule of departures and travel times that are considered stochastic. In this work, we plan the dispatches of two depots in a full day of operation, divided into certain time windows, in which there are certain behaviors of travel times and availability of buses. A multistage stochastic optimization model is formulated and a particular instance is solved in order to determine the respective sequence of dispatches and times when the battery of each bus will be charged in order to meet the initially proposed departure plan.

Keywords: Smart cities · Electric buses · Multistage stochastic programming · MDVSP

1 Introduction

Important aspects such as sustainability and reduction of greenhouse gas emissions are present in the design and operation of current public transport systems. In this sense, both in Europe and in Asia, electric buses have already been operating for some years and, in the specific case of Santiago de Chile, this is true since March 2019 [1]. In fact, at the end of 2019, Santiago had a total of 386 electric buses [2] and in June, 2020, the Ministry of Transport and Telecommunications of Chile announced that it will soon double the current fleet, reaching 776 electric vehicles in the short term [3]. These buses not only seek to reduce the system's CO_2 emissions, but also to improve the quality of the service; given that they have air conditioning, device charging stations, among other functions. However, their use also poses operational challenges when meeting scheduled departure compliance, as dispatch must take battery level into account and schedule the charging of the buses. The latter, added to the uncertainty of traffic conditions during the day, constitute variables to be taken into account when scheduling bus dispatch.

In [4] the problem of conventional vehicle programming (VSP) is detailed, where the different antecedents associated with this problem, types of vehicle

© Springer Nature Switzerland AG 2021
D. A. Rossit et al. (Eds.): ICPR-Americas 2020, CCIS 1408, pp. 77–86, 2021.
https://doi.org/10.1007/978-3-030-76310-7_6

programming and their variations are presented. However, neither electric buses nor stochastic aspects are covered in these models, and that is the novelty presented by our contribution.

Regarding the variations in the base model in conventional vehicles, [5] studies the problem in large cities that have multiple bus depots. To address this problem, it is broken down into scheduling problems for each depot treated individually. The proposed model is used in a full day of operation subdivided into morning, afternoon and night and resolved through an ad hoc heuristic.

In [6] the dispatch of electric buses is addressed without considering battery change, contemplating a discretization of battery levels to decide how many trips the bus will make before having to return to the depot to charge. The authors solve four instances of their model, where the first two are small and medium in size and were solved through an integer linear programming model. The other two were resolved by column generation. It is worth mentioning that these authors do not consider stochastic behaviors in their model.

On the other hand, in [7] a fleet with electric and hybrid buses is considered, where given an initial schedule of departures, the optimal dispatch of buses is sought, and battery charging problems were achieved with this fleet mix. It is worth mentioning that in this problem several service lines are served, but all buses start their journey from a common depot. The model presented corresponds to an MIP and does not consider aspects under uncertainty.

In [8], a dynamic model with robust strategies is proposed that considers randomness in travel conditions for the dispatch scheduling of electric buses. In particular, the representation of the network, both in the deterministic model and in its stochastic version, is done through *time span* nodes. This means that each trip, charging station and bus depot is represented by its respective node and is assigned a time slot. It is worth mentioning that the problem studied considers a single bus depot, and both deterministic and stochastic models that are linear in mixed–integer variables and resolved through strategies such as: path–based formulation, Branch–and–price and initial path set generation.

This article addresses a problem with multiple electric bus depots and stochastic traffic conditions, in which the representation of the network is *time–expanded* (through nodes *time span*) and a multistage stochastic optimization model is formulated. To the best of our knowledge, there are no works that cover these topics in the same model. The rest of the article is structured as follows: Sect. 2 describes the main background of the problem and the proposed stochastic optimization model. Section 3 describes the background of the case study addressed and our preliminary results. Finally, Sect. 4 gives an account of the conclusions and implications for future research.

2   Methodology

The problem addressed consists of scheduling dispatches of electric buses under stochastic traffic conditions and considers detailed scheduling of the outputs and battery charge of each bus in each depot. An important aspect of the problem is

the variability observed in travel times, making necessary the explicit incorporation of this variable under uncertainty in a decision–making model for the problem. An important part of the literature on optimization models under uncertainty is consolidated under the named stochastic programming models, which
extend the deterministic ones by explicitly including random variables in the
model parameters. In particular, a two–stage stochastic program with recourse
provides an optimal solution in which there are decision variables, called here–
and–now, whose value is prior to the realization of the random parameter, and
decision variables called recourse variables, whose value is adopted in response
to the decision of the here–and–now variables and of a specific realization of the
random parameter (scenario), which allows the necessary flexibility for the proposed model. In this work, as in many others, decisions are made over multiple
time intervals and the realizations of the random parameters change with an
intertemporal dependence, giving rise to a multistage stochastic program with
recourse that goes further than previous ones, see for example [9].

Thus, the model proposed in this paper addresses the operation of a time
interval, which is subdivided into multiple stages (of time), according to the
behavior of the stochastic travel time. To better comprehend the proposed multistage stochastic programming model, we define what we understand by stage,
state, and scenario.

a) **Stages**. A time frame is taken from one day of operation, and it is subdivided
   into different time intervals that define the stages of the model. In each of
   these stages, the decision should be made as to which bus is going to perform
   the scheduled departure, taking into consideration the information from the
   previous stages and the realization of the parameter associated with the
   travel time at that stage.
b) **States**. As it has been previously pointed out, in this model the parameter
   under uncertainty is the travel time and therefore this parameter is modeled
   as a finite and discrete number of states, where each one has an associated
   a conditional probability of occurrence, defined for each stage in which the
   planning horizon has been subdivided.
c) **Scenarios**: A scenario corresponds to the realization of a series of states
   from the first to the last stage. This can be represented as a "tree", where
   each state is the branch of a node, and a scenario would correspond to a
   certain path from the first node in the "tree" to a node in the last stage.
   The probability of occurrence of each scenario corresponds to the conditional
   probability of all the states belonging to said scenario to occur.

To exemplify the above, a scenario tree is presented in Fig. 1, which has three
stages and four scenarios. Using this example for the problem, each stage can
belong to a finite number of trips. In the first node, the buses that will serve
the scheduled departures must be adopted independently of the travel times,
while the decisions in the second stage present two types of states, for example
one optimistic and one pessimistic travel time for buses available for dispatch
and the same occurs in the third stage according to the resulting state in the

previous stage. Each one of these branches has an assigned probability, where at each stage the sum of the probabilities of occurrence of all the states in said stage must equal 1. Regarding the scenarios, in this example, four scenarios can be seen, which correspond to the four possible paths from node 1 to nodes 4, 5, 6 and 7. It should be mentioned that the sum of the probability of occurrence of all scenarios must be equal to 1.

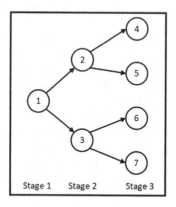

Fig. 1. Scenario tree with 3 stages.

Motivated by the problem faced by a public transport company in Santiago, Chile, the main assumptions considered in the formulation of the model are as follows:

– Deadhead travels are not allowed.
– Only one type of charger exists.
– Battery change is not allowed.
– The fleet consists only of electric buses.
– If a bus visits a charge station, it will be there for a fixed time, independent of charge level.

The problem, in turn, is represented by a graph $G = (N, A)$, where N represents the set of nodes and A the set of links. N is made up by the node of "origin" o_k; the "end" node d_k from each electric bus depot k (the two nodes are located in the same physical place: the warehouse k); the set of trips I, in which all the trips of k warehouses are considered. Each trip has an already programmed scheduled exit and stochastic travel times that must be catered only once and for a single bus; and the set of virtual nodes V^k, which are found in depots, but are distinguished from o_k and d_k in the sense of facilitate modeling. These virtual nodes are, in turn, divided into V_1^k and V_2^k, where the former represents an electric bus that is heads to the depots, charges its battery and then continues with the travel sequence. V_2^k represents when an electric bus goes to the depot, does not charge its battery and simply waits to continue with the travel sequence. It is worth mentioning that these nodes are called *time span nodes*, which means that they are not only located in a spatial point, but also temporal, so each node is assigned a time interval. Therefore, the entire network turns out to be a *time–expanded network*.

The sets, parameters and decision variables of the model are described below:

Sets

S : Set of scenarios.
K : Set of electric bus depots.
A_s^k : Set of arcs from warehouse k under scenario s.
N^k : Set of nodes from depot k.
I : Set of trips that must be attended.
V^k : Set of virtual nodes from terminal k, which are divided into sets V_1^k and V_2^k.

V_1^k : Buses that arrive at depot k, charge their batteries, and then continue attending the sequence of trips.

V_2^k : Buses that arrive at depot k, do not charge batteries, and then at some point go back to attending the trip sequence.

W_j : Set of nodes that if, when finalizing a trip, decides to charge its battery (V_1^k), coinciding with part of the bus charging time, that, after completing trip j decided to charge its battery.

Parameters

$e_{i,j}$: Battery consumption from node i to node j.
W : percentage of battery that charges in a station.
r^k : Quantity of vehicles in depot k.
E_{max} : Maximum battery level allowed.
E_{min} : Minimum battery level allowed.
Cr^k : Quantity of charging stations available in depot k.
M : Penalty for not attending a trip.
p_s : Probability of scenario s occurring.
c_{ij} : Operational costs for making the trip of the edge (i, j).

Decision Variables

$x_{i,j,s}^k$: Binary variable that takes a value equal to 1 if the bus from depot k, under scenario s, makes a trip to node j after having made a trip to node i, and 0 otherwise.
$z_{i,j,s}^k$: Binary variable that takes a value equal to 1 if the bus from depot k, under scenario s, **does not** make a trip to node j after having made a trip to node i, and 0 otherwise.
$g_{j,s}$: Real variable that indicates the consumption of battery accumulated by the bus at the end of its trip to node j under scenario s.
$Aux1_{i,j,s}^k$: Real auxiliary variable.
$Aux2_{i,j,s}^k$: Real auxiliary variable.

The first decision variable has the purpose of choosing the sequence of trips that each electric bus will make. The second decision variable corresponds to whether or not to attend a particular trip because it is not possible. Here, it is important to note that the latter helps understand which trips will need the support of another additional bus in order to attend all trips. The variable $g_{j,s}$ is used to keep count of the battery consumed by each bus during its travel sequence. Finally, two auxiliary variables will be used for linear representation of the restrictions related to the accumulated battery consumption.

$$\min \quad \sum_{s \in S} p_s \left[\sum_{k \in K} \sum_{(i,j) \in A_s^k} c_{i,j} x_{i,j,s}^k + \sum_{k \in K} \sum_{(i,j) \in A_s^k} M z_{i,j,s}^k \right] \tag{1}$$

s.t.

$$\sum_{i:(i,j) \in A_s^k} (x_{i,j,s}^k + z_{i,j,s}^k) = 1, \quad \forall j \in I, \forall k \in K, \forall s \in S \tag{2}$$

$$\sum_{j:(o_k,j) \in A_s^k} x_{o_k,j,s}^k \le r^k, \quad \forall k \in K, \forall s \in S \tag{3}$$

$$\sum_{i:(i,j) \in A_s^k} x_{i,j,s}^k - \sum_{p:(j,p) \in A_s^k} x_{j,p,s}^k = 0, \quad \forall j \in I \cup V_1^k \cup V_2^k, \forall k \in K, \forall s \in S \tag{4}$$

$$g_{j,s} = \sum_{i:(i,j) \in A_s^k} (Aux1_{i,j,s}^k + e_{i,j} \cdot x_{i,j,s}^k), \quad \forall j \in I \cup V_2^k, \forall k \in K, \forall s \in S \tag{5}$$

$$Aux1_{i,j,s}^k \le E_{max} \cdot x_{i,j,s}^k, \quad \forall i \in N^k, \forall j \in N^k, \forall k \in K, \forall s \in S \tag{6}$$

$$Aux1_{i,j,s}^k \le g_{i,s}, \quad \forall i \in N^k, \forall j \in N^k, \forall k \in K, \forall s \in S \tag{7}$$

$$Aux1_{i,j,s}^k \ge g_{i,s} - (1 - x_{i,j,s}^k) \cdot E_{max}, \forall i \in N^k, \forall j \in N^k, \forall k \in K, \forall s \in S \tag{8}$$

$$g_{j,s} = \sum_{i:(i,j) \in A_s^k} (Aux2_{i,j,s}^k - W \cdot x_{i,j,s}^k), \quad \forall j \in V_1^k, \forall k \in K, \forall s \in S \tag{9}$$

$$Aux2_{i,j,s}^k \le E_{max} \cdot x_{i,j,s}^k, \quad \forall i \in N^k, \forall j \in N^k, \forall k \in K, \forall s \in S \tag{10}$$

$$Aux2_{i,j,s}^k \le g_{i,s}, \quad \forall i \in N^k, \forall j \in N^k, \forall k \in K, \forall s \in S \tag{11}$$

$$Aux2_{i,j,s}^k \ge g_{i,s} - (1 - x_{i,j,s}^k) \cdot E_{max}, \forall i \in N^k, \forall j \in N^k, \forall k \in K, \forall s \in S \tag{12}$$

$$g_{i,s} + \sum_{j:(i,j) \in A_s^k} e_{i,j} \cdot x_{i,j,s}^k \le (1 - E_{min}), \quad \forall i \in I, \forall k \in K, \forall s \in S \tag{13}$$

$$\sum_{j:(j,r) \in A_s^k} x_{j,r,s}^k + \sum_{w \in W_j} \sum_{i \in I} x_{i,w,s}^k \le Cr^k, \quad \forall r \in V_1^k, \forall k \in K, \forall s \in S \tag{14}$$

$$x_{i,j,s}^k \in \{0,1\}, \ z_{i,j,s}^k \in \{0,1\}, \quad \forall i \in N^k, \forall j \in N^k, \forall k \in K, \forall s \in S \tag{15}$$

$$Aux1_{i,j,s}^k \ge 0, \ Aux2_{i,j,s}^k \ge 0 \quad \forall i \in N^k, \forall j \in N^k, \forall k \in K, \forall s \in S \tag{16}$$

$$g_{j,s} \ge 0, \quad j \in I \cup V_1^1 \cup V_1^2, \forall s \in S \tag{17}$$

$$Nonanticipativity \ constraints \tag{18}$$

The objective function (1) minimizes the total costs of the scheduling performed in the use of vehicles and battery charging, including a term that penalizes unattended trips. Regarding the model's restrictions, constraint (2) ensures that each trip must be attended once, either by the available bus fleet or by some support bus in circulation $(z_{i,j,s}^k)$. Constraint (3) ensures that no more electric buses can be used than are available in each bus depot. Constraint (4) ensures the conservation of the bus fleet. Constraint (5) allows to calculate the accumulated energy consumption of the bus when making a trip from node i to node j under the scenario s. Constraints (6)–(8) allow a linearization of the nonlinear equation $g_{j,s} = \sum_{i:(i,j)\in A_s^k} (g_i + e_{i,j}) \cdot x_{i,j,s}^k$, whose linear representation is the previous equation. In the same way, linear constraint (9) updates the accumulated energy consumption of the electric bus when it charges its battery, using constraints (10)–(12) together with an auxiliary variable $(Aux2_{i,j,s}^k)$. Constraint (13) does not allow an electric bus to attend a trip if its battery percentage is less than the minimum percentage necessary to leave the depot. Constraint (14) ensures that the number of buses visiting the depot k charging station does not exceed its capacity. Constraints (15)–(17) show the nature of the decision variables. Finally, constraint (18) is essential for stochastic models with scenarios, since it ensures that if two different scenarios are identical up to a given stage the decisions must also be identical up to this stage. In Fig. 1, for example, there is a case of 4 scenarios and 3 stages. Therefore, if one takes all the trips corresponding to node 1, the same decisions would be made for that node in all scenarios. Similarly, if we take node 2 from the second stage, the decisions made in that node should be the same for scenario 1-2-4 and for scenario 1-2-5.

The resulting model corresponds to a multistage stochastic mixed–integer linear program with recourse. The following section describes a preliminary case study based on the problem that defines the dispatch of an electric bus route from a public transport company in Santiago.

3 Case Study

3.1 Statistic Analysis

In this section an instance is made using real information of the operation during some months of 2019 of an electric bus route of the public transport of the city of Santiago, Chile. In the first place, the Kolmogórov–Smirnov test was used to determine whether travel times follows a specific probability distribution for different days and time intervals during the week. Figure 2 shows a normal fitted distribution with data from a given day of the week.

Then the days of the week are grouped according to similar behaviors. In this way, the week is divided into 3 groups: Monday–Tuesday–Wednesday, Thursday–Friday and Saturday–Sunday. Given the observed variability, Table 1 shows the expected times and the standard deviation of different time windows on Saturday–Sunday. It is worth mentioning that on the weekend there is no registration between 00:01 and 08:00 because at that time the company uses conventional buses.

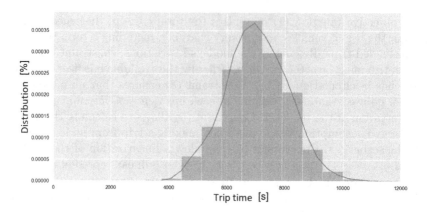

Fig. 2. Probability distribution fit to trip times.

The case study is carried out on Saturday, specifically in the time window 12:00–17:15, in which the mean times and standard deviations are recalculated within the same time windows with intervals of times every 15 min, in order to be able to separate the problem in multiple stages within this time window.

In this preliminary instance, 2 bus depots are used, where each of the buses must attend 22 trips (44 trips in total), which have a departure frequency every 15 min. In addition, there are 11 buses and 7 charging stations in each depot. It is considered as an assumption that all buses to be used start with 100% of their battery charged and that it is not possible for the bus to attend a trip if its battery level is less than 30%. The problem is divided into 3 stages, where the first 3 trips from each depot belong to the first stage, the next 6 trips to the second stage and the rest to the third stage. An optimistic and a pessimistic state is considered for each stage according to the standard deviation of each stage, thus generating 4 scenarios that form a tree of scenarios like the one shown in Fig. 1. The probability of occurrence of each state is 50%, and the trips that are in the first stage have deterministic trip times.

Table 1. Average times and standard deviation of different time windows.

Time window	Mean [HH:MM:SS]	Std. dev.[HH:MM:SS]
08:00 – 12:00	01:46:33	00:14:37
12:00 – 17:15	01:53:41	00:14:25
17:15 – 22:15	01:44:08	00:10:58
22:15 – 00:00	01:27:21	00:17:43

3.2 Computational Results

A given spreadsheet provides the list of trips per depot with the time each trip should begin. Using Python 3.7, this information is processed to form the *time-expanded network* and deliver all this information to the model formulated in Pyomo. The model was solved with Gurobi 9.0.2 on a computer with a 1.60 [GHz] Intel Core i5 processor and 8 [Gb] of RAM. The execution time of the resolved instance was 15,3 s.

For each scenario the model provides the sequence that each bus must follow, and the trips that cannot be attended due to lack of fleet, where the penalty of the latter amounts to a value of the order of CLP\$ 10^6. On the other hand, operational costs are of the order of CLP\$ 10^3.

The value of the objective function in this instance turns out to be CLP\$ 17.620.764. To measure the quality and importance of the stochastic solution, we computed the EVPI value (*Expected Value of Perfect Information*). This value represents the maximum amount that the decision maker is willing to pay for having all the information about the future, and is calculated as the difference between the optimal value of the proposed *multistage stochastic program with recourse* and the average of the optimal value of the same model solved separately for each scenario (known as *wait–and–see* value).

For this case study, the value of *wait–and–see* turned out to be CLP\$ 12.916.522 so that the EVPI value turned out to be CLP 4.704.242. It can be seen that the EVPI represents a value close to the 30% of the value of the model with recourse. Now, results from the EVPI, means that the decision maker is willing to pay a maximum of CLP\$ 4.704.242 for having information about the future, which gives relevance to the use of the proposed methodology.

4 Conclusions and Future Research

In this paper, we proposed a multistage stochastic optimization model for scheduling departures of an electric bus fleet with multiple depots, where travel time under traffic conditions is considered as an uncertain parameter. Also, the model uses a representation of the network through *time–expanded* nodes, which is a novel approach, when compared to the current literature for this type of problem. To the best of our knowledge, there is no evidence of studies that cover all the aspects considered.

Additionally, we analyzed the operating data in some months of 2019 of an electric bus line from the public transportation system in the city of Santiago, Chile, and it was determined that the distribution that best adjusted to the travel times is the normal distribution. Then, the days of the week were divided into three groups that shared similar behaviors within certain time windows, so that during full day of operation for days belonging to these groups could be divided into several time intervals.

An instance was resolved using the operation information from Saturday. This instance was solved for a time window with a three–stage stochastic optimization model with recourse. The selected solver manages to resolve this instance in about 15 s. In order to measure the quality of the optimal solution, the EVPI value was calculated, reached a high value in relative terms, so it is expected that the proposed stochastic optimization model would have a great impact when used in this problem.

As future research, we expect to add some flexibility in the dispatch, which will allow eventual delays in departures, with their corresponding penalty per minute of delay. Similarly, we will add more than one electric bus route, and perform this multistage modeling with higher time windows. Regarding the resolution times, so far they do not seem to be very high, but if by adding these new aspects the resolution times increase considerably, it will be necessary to evaluate an algorithmic resolution method.

Acknowledgements. This research was partially supported by the Postgraduate and Programs Directorate from Universidad Técnica Federico Santa María (USM Master Scholarship).

References

1. Revistaei. Llegan a chile 100 nuevos buses eléctricos que se sumarán a red metropolitana de movilidad (2019). https://n9.cl/kafq. Accessed Aug 2020
2. Sostenible, R.: Santiago se consolida como la segunda capital a nivel mundial con mayor cantidad de buses eléctricos (2020). https://n9.cl/wcn8a. Accessed Aug 2020
3. Ministerio de Transportes y Telecomunicaciones. Duplicaremos la flota de buses red en los próximos dos meses y reasignamos el 41% de los recorridos de express (2020). https://www.mtt.gob.cl/archivos/25575. Accessed Aug 2020
4. Ball, M.O., Magnanti, T.L., Monma, C.L., Nemhauser, G.L.: Handbooks in Operations Research and Management Science: Network Routing. Elsevier, Hoboken (1995)
5. Banihashemi, M., Haghani, A.: Optimization model for large-scale bus transit scheduling problems. Transp. Res. Rec. **1733**(1), 23–30 (2000). https://doi.org/10.3141/1733-04
6. van Kooten Niekerk, M.E., van den Akker, J.M., Hoogeveen, J.A.: Scheduling electric vehicles. Public Transp. **9**, 155–176 (2017). https://doi.org/10.1007/s12469-017-0164-0
7. Rinaldi, M., Parisi, F., Laskaris, G., D'Ariano, A., Viti, F.: Optimal dispatching of electric and hybrid buses subject to scheduling and charging constraints. In: 2018 21st International Conference on Intelligent Transportation Systems (ITSC), pp. 41–46. IEEE (2018). https://doi.org/10.1109/ITSC.2018.8569706
8. Tang, X., Lin, X., He, F.: Robust scheduling strategies of electric buses under stochastic traffic conditions. Transp. Res. Part C Emerg. Technol. **105**, 163–182 (2019). https://doi.org/10.1016/j.trc.2019.05.032
9. Birge, J.R., Louveaux, F.: Introduction to Stochastic Programming. Springer Science & Business Media, Berlin (2011)

Intelligent Systems and Decision Sciences

Production Planning with Remanufacturing and Environmental Costs

Paula Martínez, Diego Molina, Luciana Vidal$^{(\boxtimes)}$, Pedro Piñeyro⬤, and Omar Viera

Department of Operations Research, Institute of Computer Science, Faculty of Engineering, Universidad de la República, Julio Herrera y Reissig 565, Montevideo, Uruguay
{paula.martinez.vizoso,diego.molina,luciana.vidal.jaureguy,
ppineyro,viera}@fing.edu.uy

Abstract. In this paper we address an extension of the economic lot-sizing problem with remanufacturing, in which environmental costs are considered in the objective function in addition to the traditional economic costs. Separate costs for emissions, water and energy consumptions are assumed for the manufacturing and remanufacturing activities as well as to carry on positive inventories of both used and serviceable products. We provide a mixed-integer linear programming formulation for the problem and suggest several solving procedures based on the Tabu Search metaheuristic. In order to evaluate the suggested procedures, we compare them against an optimization solver over a benchmark set of instances of the literature extended for the problem. From the results obtained of the numerical experimentation, we can conclude that the suggested procedures are able to achieve near optimal solutions in a time-effective way. This effect is particularly significant for large instances of the problem. In addition, we consider an objective function with weighted costs, in order to analyze the effect of including environmental costs in the hybrid production-remanufacturing systems.

Keywords: Lot-sizing problems · Remanufacturing · Environmental costs · Optimization · Heuristics · Tabu Search

1 Introduction

Production planning is crucial for industrial companies [1]. The basic production problem involves to determine how much and when to produce of a certain product in order to fulfill the demand requirements on time, minimizing the sum of the involved costs of production and storage. This problem is commonly known in the literature as the Economic Lot Sizing Problem (ELSP) [2].

Manufacturing has an impact on the planet's ecology. The emissions coming from the processing and use of fossil fuels that reach the atmosphere and the industrial garbage impact all forms of life around the world, causing irreversible damage to the ecosystem. These are just some of the environmental problems caused by pollution from factories [1, 3].

Circular Economy is an alternative that promotes to change the traditional linear production model consisting on producing, using and discarding, for one where used

© Springer Nature Switzerland AG 2021
D. A. Rossit et al. (Eds.): ICPR-Americas 2020, CCIS 1408, pp. 89–101, 2021.
https://doi.org/10.1007/978-3-030-76310-7_7

products are reincorporated at the first phase of the production cycle. It proposes to recover and restore used products, components and materials applying strategies such as reuse, repair and remanufacturing [4].

Remanufacturing is an industrial recovery process of used product which guarantees that a recovered product offers at least the same functionalities as a new one [5]. To achieve this goal, it goes through different stages that involves dismantling the product, restoring and replacing components and testing the individual parts and the whole product to ensure that it is within its original design specifications.

Taking into account the aforementioned, it is important to note that the focus should not only be on reducing economic costs of the production process but also on minimizing the negative impact on the environmental of their activities. This can be achieved by redesigning the production cycle of factories to focus on more environmentally friendly strategies such as collecting and remanufacturing used products.

Regarding production planning, we note that Wagner and Whitin [2] were the first to suggest an efficient algorithm for the ELSP. It is based on a dynamic programming approach of computational complexity $O(n^2)$, with n the number of periods. Zangwill [6] extended the algorithm for the case of general concave costs and Hanafizadeh et al. [7] extended with uncertain costs. In 1993, Aggarwal and Park [8] developed a more efficient algorithm of $O(n \log n)$ time.

A relative recently extension of the ELSP is to consider the remanufacturing of used products returned to the origin in order to satisfy the demand requirements. This problem extension is known in the literature as the Economic Lot Sizing Problem with Remanufacturing (ELSR). Teunter et al. [9] analyze this problem, suggesting a Mixed Integer-Linear Programming (MILP) formulation for the cases of joint and separate set-up scheme. In addition, the Silver-Meal (SM), Minimum Unit Cost (LUC) and Fragmented Period Balancing (PPB) heuristics are adapted for the problem. Piñeyro and Viera [10] tackle the ELSR with disposal options. A Basic Tabu Search (BTS) based-on procedure is proposed and evaluated. In Piñeyro and Viera [11] the performance of the BTS procedure is improved by means of applying a property on the form of the ELSR optimal solutions.

Absi et al. [12] and Retel Helmrich et al. [13], describe that in recent years, a significant effort has been made to mitigate the environmental impact of industrial processes. For instance, the United Nations Kyoto Protocol in 1998 [14], and the European Commission's emissions trading legislation in 2010 [15]. In Retel Helmrich et al. [13], the emission in both the set-up and the inventory holding costs are studied for the ELSP. To solve the problem, they suggest a Lagrangean heuristic and a Polynomial-Time Approximation Scheme (FTPAS) based-on algorithm.

The Cap-and-Trade emerged as a public policy to counteract CO_2 emissions, offering economic incentives for those who manage to reduce them. Benjaafar et al. [16] and Akbalik and Rapine [17] suggest and study models for exploring this concept. In Benjaafar et al. [16] and Yuan et al. [18], the concept of Certified Emission Reduction (CER) is also incorporated in order to reduce the carbon emissions.

The problem of lot-sizing that incorporates remanufacturing options under a carbon emission constraint is considered by Zouadi et al. [19, 20]. Piñeyro and Viera [21] consider the ELSR with recovery targets established as a percentage of the total returns that must be remanufactured. Establishing a recovery target may be a mechanism to reduce the negative impact of the new item manufacturing activity. Rapine et al. [22] also tackle a production system with capacity constraints and environmental considerations. The objective is to minimize the sum of all the costs involved and the use of energy established as a non-linear function. Giglio et al. [23] address an integrated lot sizing and energy-efficient job shop scheduling problem, minimizing the sum of the economic costs and the energy costs related to the utilization of machines and processing times. Finally, Heck and Schmidt [1] identify three relevant factors to minimize the environmental impact in a production system: energy use, carbon emissions and water consumption. A mathematical model is defined with an objective function that includes the cost of energy, emissions and water usage in addition to the economic costs of production.

In this paper we propose a MILP formulation for the ELSR considering environmental costs in addition to the traditional economic costs. The main difference with the problem addressed in Zouadi et al. [19, 20] is that we consider not only emissions, but also water and energy consumption. In addition, we relax the upper bounds on the emissions, since it is considered as part of the environmental costs.

For this ELSR extension, three heuristic procedures based on the Tabu Search metaheuristic are suggested and evaluated by means of a benchmark set of ELSR instances adapted to this problem. The suggested procedures are based on the BTS procedure introduced in Piñeyro and Viera [10] for the ELSR, extending it to consider environmental costs as well as other strategies for the exploration phase.

The rest of the paper is organized as follows. Section 2 provides the problem definition and the mathematical model for the ELSR extension under consideration. Section 3 addresses the description of the solution procedures suggested for the problem. The numerical experimentation and the analysis of the results obtained are provided in Sect. 4. Conclusions and possible directions for future research are given in Sect. 5.

2 Problem Definition

We address a hybrid production-remanufacturing system of a single product in which the demand must be satisfied on time by either the production of new items or remanufacturing used ones returned to the origin, i.e., remanufactured products look as-good-as new. Henceforth, the terms return and used product, as well as production and manufacturing, are used interchangeably. Demand and returns values are assumed non-negative integers known in advance for each one of the periods within the finite planning horizon. Economic and environmental costs are assumed for manufacturing and remanufacturing, as well as for the inventory of serviceable and used products. Three types of environmental costs are considered: water consumption costs, energy consumption costs and emissions costs. Set-up costs are assumed for both manufacturing and remanufacturing, and linear costs for holding inventory of both used and serviceable products. The objective is to determine the quantities to produce and remanufacture at each period in order to minimize the sum of the economic and environmental costs.

The notation used for the problem is listed below.

- Parameters:

- n: number of periods, with $n > 0$.
- $T = \{1,..., n\}$: set of periods within the planning horizon.
- D_t: serviceable items demanded in period $t \in T$.
- R_t: used items returned in period $t \in T$.
- h_t^r: economic unit cost of holding used inventory in period $t \in T$.
- h_{at}^r: water consumption unit cost of holding used inventory in period $t \in T$.
- h_{ct}^r: emissions unit cost of holding used inventory in period $t \in T$.
- h_{et}^r: energy unit cost of holding used inventory in period $t \in T$.
- h_t^s: economic unit cost of holding serviceable inventory in period $t \in T$.
- h_{at}^s: water consumption unit cost of holding serviceable inventory in period $t \in T$.
- h_{ct}^s: emissions unit cost of holding serviceable inventory in period $t \in T$.
- h_{et}^s: energy unit cost of holding serviceable inventory in period $t \in T$.
- K_t^m: economic set-up cost for manufacturing in period $t \in T$.
- K_{at}^m: water set-up cost for manufacturing in period $t \in T$.
- K_{ct}^m: emissions set-up cost for manufacturing in period $t \in T$.
- K_{et}^m: energy set-up cost for manufacturing in period $t \in T$.
- K_t^r: economic set-up cost for remanufacturing in period $t \in T$.
- K_{at}^r: water set-up cost for remanufacturing in period $t \in T$.
- K_{ct}^r: emissions set-up cost for remanufacturing in period $t \in T$.
- K_{et}^r: energy set-up cost for remanufacturing in period $t \in T$.
- M: big number.

- Decision variables:

- I_t^s: inventory level of serviceable products in period $t \in T \cup \{0\}$.
- I_t^r: inventory level of returns in period $t \in T \cup \{0\}$.
- x_t^m: number of items to manufacture in period $t \in T$.
- x_t^r: number of items to remanufacture in period $t \in T$.
- y_t^m: equal to 1 if there is manufacturing in period $t \in T$, 0 otherwise.
- y_t^r: equal to 1 if there is remanufacturing in period $t \in T$, 0 otherwise.

We refer to the hybrid system described above as the Economic Lot-Sizing Problem with Remanufacturing and Environmental Costs (ELSR-EC). It can be formulated as the following MILP, extending those suggested for the ELSR in [9, 11]:

$$\min \sum_{t \in T} \left\{ \left(K_t^m y_t^m + K_t^r y_t^r + h_t^s I_t^s + h_t^r I_t^r\right) + \left(K_{et}^m y_t^m + K_{et}^r y_t^r + h_{et}^s I_t^s + h_{et}^r I_t^r\right) + \left(K_{ct}^m y_t^m + K_{ct}^r y_t^r + h_{ct}^s I_t^s + h_{ct}^r I_t^r\right) + \left(K_{at}^m y_t^m + K_{at}^r y_t^r + h_{at}^s I_t^s + h_{at}^r I_t^r\right) \right\} \quad (1)$$

subject to:

$$I_t^s = I_{(t-1)}^s + x_t^r + x_t^m - D_t, \quad \forall t \in T \quad (2)$$

$$I_t^r = I_{(t-1)}^r + R_t - x_t^r, \quad \forall t \in T \tag{3}$$

$$x_t^m \leq My_t^m, \quad \forall t \in T \tag{4}$$

$$x_t^r \leq My_t^r, \quad \forall t \in T \tag{5}$$

$$I_0^s = I_0^r = 0 \tag{6}$$

$$y_t^m, y_t^r \in \{0, 1\}, \quad \forall t \in T \tag{7}$$

$$x_t^m, x_t^r, I_t^s, I_t^r \geq 0, \quad \forall t \in T \tag{8}$$

Objective function (1) is composed by four components: economic costs, energy consumption costs, emissions costs and water consumption costs. Equations (2) and (3) are the inventory balance equations for serviceable and used products, respectively. Constraints (4) and (5) state that the binary variables related to the set-up costs for production and remanufacturing must be equal to 1 if a positive amount is produced or remanufactured, respectively. Constraint (6) establishes that the initial inventory level for both serviceable and used products is assumed equal to zero. Finally, constraints (7) and (8) define the domain for the decision variables of the problem.

3 Heuristic Procedures for the ELSR-EC

In this section we present the main solving procedure and its variants, suggested for the ELSR-EC. The procedure TS-EC (Tabu Search with Environmental Costs) is based on the Tabu Search metaheuristic (Glover et al. [24]). This metaheuristic is an iterative exploration process based on information stored in memory (tabu list) to allow to escape from local optimal of optimization problems. In the following subsections the different components of TS-EC will be explained. This procedure can be considered an extension of the BTS procedure proposed by Piñeyro and Viera [10] for the ELSR.

3.1 TS-EC Procedure

A feasible solution of the ELSR-EC is represented as a pair of (0,1) n-tuples, one for indicating the positive periods of remanufacturing and the other for the positive periods of manufacturing. A value 1 in position t of the remanufacturing (manufacturing) in the n-tuple means remanufacturing (manufacturing) can be positive in period t, with $1 \leq t \leq n$. On the other hand, a value 0 in position t means remanufacturing (manufacturing) must be zero.

The TS-EC procedure consists of the following phases:

- **Initial phase**: the initial n-tuple of the remanufacturing plan is generated.

- **Generation phase**: given the n-tuple of the remanufacturing periods, the remanufacturing amounts are determined and then, the optimal production plan (periods and quantities) is obtained.
- **Exploration phase**: given the feasible ELSR-EC solution from the previous phase, a new solution is sought among its neighbors, with lower ecological and economic costs than the given (or current) solution.

For the initial phase, we use a value of zero for every position of the n-tuple of the remanufacturing plan, i.e. zero remanufacturing. Given the n-tuple of the remanufacturing plan, the generation phase is performed in order to determine a feasible solution for the ELSR-EC. The remanufacturing plan is determined according to the following rule also used in [10]:

$$x_i^r = min\{I_{i-1}^r + R_i, D_{i,(j-1)}\} \quad \forall 1 \le i < j \le n+1 \tag{9}$$

$$D_{i,(j-1)} = \sum_i^{j-1} D_k \tag{10}$$

Equation (9) means that the remanufacturing quantity of a period i is determined as the minimum between the number of available returns in period i and the accumulated demand, described in (10), until the preceding period of the period j fixed as the next period of positive remanufacturing. We note that the optimal production plan can be obtained from a given remanufacturing plan [10]. The algorithm of Wagner and Whitin [2] was adapted to consider the initial inventory of serviceable products and the environmental costs of the production activity.

Finally, the exploration phase is performed with the aim to improve the feasible solution of the ELSR-EC obtained from the generation phase. First the neighbors of the remanufacturing n-tuple are determined by means of a Hamming distance of 1 (swap moves). It is checked that each neighbor does not belong to the tabu list. We consider a limited size tabu list managed according to the FIFO rule (First in First Out). A new candidate solution for the problem is determined using the procedure of the generation phase. The exploration phase runs until the stop condition is reached: the number of iterations is greater than a certain maximum value. The solution with the lowest economic and environmental cost of all the evaluated solutions, is returned.

3.2 TS-EC Variants

In order to make the exploration phase of the TS-EC more diverse, we develop two variants of the procedure by means of the jump transformation defined as follows. Given a feasible solution of the ELSR-EC, a jump is defined as a shift of certain number of components of the n-tuple of its remanufacturing plan, from 0 to 1 or vice versa. The following jump transformations are used in this paper:

- Negative: all components from 1 to n of the remanufacturing n-tuple are shifted.
- Random: a random number of the components of the remanufacturing n-tuple are shifted.

We refer as TS-ECNO and TS-ECRO to the TS-EC variants with the negative and the random jump transformations, respectively. The following pseudocode presents an outline of the new procedures.

```
jumpCount = 0;
remanufacturingTuple = zeroTuple;
solution = TS-EC(remanufacturingTuple);
while (jumpCount < jumpCountMax) {
  remanufacturingTuple = solution.remanufacturingTuple()
  if (variant == TS-ECNO) {
    transformedTuple = Negative(remanufacturingTuple);
  }
  if (variant == TS-ECRO) {
    transformedTuple = Random(k, remanufacturingTuple);
  }
  solution = TS-EC(transformedTuple);
  jumpCount ++
}
```

We note that the Random () function in the pseudocode of above receives the number k of components of the remanufacturing n-tuple to be shifted. The position of each component is determined randomly.

4 Numerical Experimentation

In this section we provide the results of the numerical experimentation carried out to evaluate the performance of the heuristic procedures presented in the previous section. The cost gap between the solution obtained from the heuristics and that from AMPL/CPLEX, is calculated as follows:

$$cost\ gap = \frac{heuristic\ \cos t - exact\ method\ cost}{exact\ method\ cost} \times 100 \tag{11}$$

The parameter values were generated according to a uniform distribution (UD), considering the ranges suggested by Retel Helmrich et al. [13]. We consider two different scenarios for returns and set-up costs of remanufacturing, "Low" and "High". In order to facilitate the comparison of the results, the unit production and remanufacturing costs are assumed zero.

- Demand (D_t): UD(100, 200).
- Returns (R_t): UD(0, 100) and UD(100, 200).
- Set-up cost for production (K_t^m, K_{at}^m, K_{ct}^m, K_{et}^m): UD(2500, 7500).
- Set-up cost for remanufacturing (K_t^r, K_{at}^r, K_{ct}^r, K_{et}^r): UD(500, 1500) and UD(2500, 7500).
- Inventory cost for final products (h_t^s, h_{at}^s, h_{ct}^s, h_{et}^s): UD(10, 20).
- Inventory cost for returns (h_t^r, h_{at}^r, h_{ct}^r, h_{et}^r): UD(0, 10) y DU(10, 20).

We note that there are 512 different configurations of parameter values. For each configuration we generate 10 different instances, which results in 5,120 instances in total. The number of periods (n) is equal to 20 for all the instances.

It is important to note that TS-EC and TS-ECNO are deterministic, whereas TS-ECRO can give different results depending on the executions since the periods to be shifted are chosen randomly. Therefore, for the TS-ECRO we take the average value of three runs for each one of the instances.

There are also certain parameters that are specific to the heuristics, which are listed below:

- MAX_ITER: number of iterations for the exploration phase.
- MAX_JUMPS: number of jumps transformation to be performed.
- LIST_SIZE: number of n-tuples to be stored in the tabu list.
- RANDOM_NUM: number of periods to be shifted at the remanufacturing n-tuple.

In Table 1 we provide the values chosen for the parameters of the heuristics. To achieve this, a preliminary study was carried out in 512 instances, comparing the results against CPLEX solver.

Table 1. Heuristic parameter values for $n = 20$.

	TS-EC	TS-ECNO	TS-ECRO
MAX_ITER	50	50	50
MAX_JUMPS	-	2	4
LIST_SIZE	1,000,000	1,000,000	1,000,000
RANDOM_NUM	-	-	16

All the numerical experiments were carried out on a 2.81 Hz Windows 10 i7-7700HQ laptop, with 16 GB RAM.

4.1 Performance Analysis for $n = 20$

Tables 2 and 3 show the results for the costs gap and runtimes (in milliseconds) of the numerical experimentation. The first three columns in Table 1 describe the scenario (for example: Setup – Economic – High scenario). Then, there are columns for reporting the average, median, standard deviation and maximum values for the heuristics. Table 3 also includes the runtimes of CPLEX.

From Table 2 we can observe that the solutions found by the heuristics are similar to those found by CPLEX, with an average gap less than 1.5%. Then, we can conclude that the three heuristics are competitive with CPLEX.

Regarding the standard deviation in Table 3, we can note that the solutions found by CPLEX show a greater dispersion in terms of running time than the solutions from the heuristics.

Table 2. Cost gaps for $n = 20$.

			Average			Median			Standard deviation			Maximum		
			TS-EC	TS-ECNO	TS-ECRO	TS-EC	TS-ECNO	TS-ECRO	TS-EC	TS-ECNO	TS-ECRO	TS-EC	TS-ECNO	TS-ECRO
All instances			1.2%	0.7%	0.8%	0.6%	0.4%	0.4%	0.4%	0.2%	0.3%	36.9%	12.9%	12.9%
Returns		Low	0.8%	0.5%	0.5%	0.5%	0.3%	0.3%	0.8%	0.6%	0.5%	5.1%	4.3%	4.3%
		High	1.6%	1.0%	0.9%	0.8%	0.6%	0.5%	2.7%	1.3%	1.2%	36.9%	12.9%	12.9%
Setup	Economic	Low	1.3%	0.8%	0.8%	0.7%	0.5%	0.5%	2.3%	1.2%	1.1%	31.6%	12.9%	12.9%
		High	1.1%	0.6%	0.6%	0.6%	0.4%	0.4%	1.8%	0.8%	0.7%	36.9%	12.5%	7.1%
	Water	Low	1.3%	0.9%	0.8%	0.7%	0.5%	0.5%	2.3%	1.2%	1.1%	36.9%	12.9%	12.9%
		High	1.1%	0.6%	0.5%	0.6%	0.4%	0.3%	1.7%	0.8%	0.7%	27.9%	12.5%	7.9%
	Energy	Low	1.3%	0.9%	0.8%	0.7%	0.5%	0.5%	2.3%	1.2%	1.2%	36.9%	12.9%	12.9%
		High	1.1%	0.6%	0.5%	0.6%	0.4%	0.3%	1.8%	0.7%	0.6%	29.1%	6.7%	4.5%
	Emitions	Low	1.3%	0.9%	0.8%	0.7%	0.5%	0.5%	2.3%	1.2%	1.1%	36.9%	12.9%	12.9%
		High	1.1%	0.6%	0.5%	0.6%	0.4%	0.3%	1.8%	0.8%	0.6%	31.6%	7.1%	6.5%
Inv.	Economic	Low	1.3%	0.8%	0.8%	0.7%	0.5%	0.5%	2.0%	1.1%	1.0%	29.1%	12.9%	12.9%
		High	1.0%	0.6%	0.6%	0.5%	0.4%	0.3%	2.1%	0.9%	0.8%	36.9%	9.1%	9.1%
	Water	Low	1.3%	0.9%	0.8%	0.7%	0.5%	0.5%	2.2%	1.2%	1.1%	28.4%	12.9%	12.9%
		High	1.0%	0.6%	0.6%	0.6%	0.4%	0.3%	1.9%	0.8%	0.7%	36.9%	12.5%	5.9%
	Energy	Low	1.4%	0.9%	2.4%	0.7%	0.6%	2.4%	2.4%	1.1%	0.0%	36.9%	12.9%	2.4%
		High	1.0%	0.6%	0.6%	0.5%	0.4%	0.3%	1.7%	0.8%	0.8%	29.1%	11.4%	11.4%
	Emitions	Low	1.3%	0.9%	0.8%	0.8%	0.5%	0.5%	2.1%	1.2%	1.1%	36.9%	12.9%	12.9%
		High	1.0%	0.6%	0.6%	0.5%	0.4%	0.3%	2.0%	0.8%	0.7%	31.6%	10.6%	10.6%

Table 3. Running times (ms) for $n = 20$.

			Average				Median				Standard deviation				Maximum			
			CPLEX	TS-EC	TS-ECNO	TS-ECRO	CPLEX	TS-EC	TS-ECNO	TS-ECRO	CPLEX	TS-EC	TS-ECNO	TS-ECRO	CPLEX	TS-EC	TS-ECNO	TS-ECRO
All instances			339	81	245	359	290	73	218	349	32	11	44	9	2500	716	1219	1473
Returns		Low	452	80	250	342	422	81	246	336	205	13	21	26	2000	435	616	653
		High	305	68	199	361	266	67	198	350	193	9	13	38	1000	315	426	665
Setup	Economic	Low	300	86	232	341	266	82	237	334	186	21	25	30	1000	459	615	922
		High	374	72	209	366	344	70	208	353	208	13	19	49	1125	406	553	798
	Water	Low	275	76	218	380	250	73	212	304	170	15	25	46	1063	398	560	886
		High	361	84	238	359	328	81	241	339	206	20	28	52	1000	432	786	898
	Energy	Low	258	77	212	354	219	72	202	346	177	19	29	35	1094	473	599	790
		High	336	84	240	336	297	82	239	328	196	24	50	38	1047	650	991	911
	Emitions	Low	259	72	202	366	218	68	199	355	183	16	23	45	1984	385	653	802
		High	334	69	207	338	282	67	205	333	197	13	22	40	1032	315	502	1473
Inv.	Economic	Low	269	86	245	357	219	83	244	347	175	15	29	39	1093	380	741	804
		High	278	77	225	345	234	74	220	340	179	14	27	28	1125	377	703	694
	Water	Low	317	76	258	392	266	73	215	401	204	14	107	55	1250	392	1011	939
		High	352	73	212	371	297	71	210	363	225	13	19	42	1407	335	611	811
	Energy	Low	421	101	422	373	359	81	446	358	279	49	134	50	2500	716	1175	1023
		High	431	72	217	358	375	69	212	346	277	14	31	40	1797	380	822	887
	Emitions	Low	353	116	339	375	297	104	243	359	232	43	157	45	1500	508	1219	811
		High	431	90	286	356	375	72	220	355	248	40	110	25	2093	711	854	861

From Tables 2 and 3, we conclude that TS-ECNO and TS-ECRO variants outperform the TS-EC original heuristic in terms of cost gaps. In addition, we note that in average, TS-ECNO achieves the best results in terms of both, cost gaps and running times. In general, we also can conclude that the three heuristics are competitive compared to the CPLEX solver.

4.2 Cost Weighting Analysis

We note that the objective function of ELSR-EC considers the sum of the economic and environmental costs. However, environmental costs are composed by set-up costs for emissions, energy and water consumptions. In this section we update the costs as follows: environmental costs are multiplied by α and the economic costs by β, with $0 \leq \alpha, \beta \leq 1$, in order to analyze the solutions under different weights on the cost components of the objective function. We consider 512 instances (one per configuration) and the TS-ECNO heuristic.

In Table 4 we summarize the results obtained. First column presents the different weights of the costs considered. The rest of the columns present the difference between the economic cost of the weighted function and the economic cost of the original function for all the measurements (average, median, standard deviation and maximum). The same is shown for the environmental costs.

Table 4. Weighted cost results for TS-ECNO compared to AMPL/CPLEX.

Weights	Average		Median		Standard deviation		Maximum	
	Gap cost Econ. (%)	Gap cost Env. (%)	Gap cost Econ. (%)	Gap cost Env. (%)	Gap cost Econ. (%)	Gap cost Env. (%)	Gap cost Econ. (%)	Gap Cost Env. (%)
$\alpha=0, \beta=1$	-7.759%	13.874%	-6.977%	9.585%	6.380%	18.980%	-42.854%	197.415%
$\alpha=1/3, \beta=1$	-4.604%	3.199%	-3.052%	1.810%	5.529%	4.659%	-32.299%	49.587%
$\alpha=1, \beta=0$	6.979%	-1.026%	4.252%	-0.475%	9.551%	1.749%	95.380%	-13.794%

For the case ($\alpha=0$, $\beta=1$), only the economic costs are minimized. Analyzing the results in the average case, we note that the environmental costs increase around 14% with a maximum increase of 197.4%. Therefore, we can conclude that not including environmental costs in the objective function can cause a significant ecological impact. On the other hand, the economic cost in average decreases around 8% compared to the problem with the original cost function, i.e., $\alpha=\beta=1$.

Since the environmental costs have three components (water, energy and emission), the case ($\alpha=1/3$, $\beta=1$) is intended to balance the weight of the three environmental aspects and of the economic costs. We note that the average of environmental cost increases only 3.2% with respect to the original objective function. In the opposite direction, is observed for the economic costs. However, the impact of reducing the weight of the environmental costs may be significant as it can be seen in the maximum gaps.

Finally, for the last scenario ($\alpha=1$, $\beta=0$) with only environmental costs in the objective function, we note the economic costs increase in average near to 7%, but the maximum gap is 95%. In addition, the benefits on the environmental side are only 1%.

4.3 Performance Analysis for $n=70$

Given the good performance obtained for the three heuristics for the case of 20 periods, we decided to evaluate them for large instances of 70 periods. Because of the size of the problem with 70 periods, we consider only 14 instances chosen arbitrary from the 512 configurations defined at the beginning of Sect. 4. The values of the parameters for the heuristic procedures are reported in Table 5 (the format is the same as Table 1).

In Table 6 we present the results obtained from the heuristics for the 14 large instances. First column indicates the name of the solving procedure and the second column the maximum number of iterations for the heuristic procedures. The rest of the columns are for the cost gaps and runtimes (average, median, deviation and maximum).

Considering the average values of Table 6 we can conclude that the suggested heuristics are competitive with CPLEX. In particular, for both the 35 and 45 iterations cases, we can observe that TS-ECRO achieves the lowest cost gaps (1.204% for 35 iterations and 1.121% for 45 iterations) in a less running time than CPLEX (1.439 and 1.766 versus 2.543 min, respectively).

Table 5. Heuristic parameter values for $n = 70$.

	TS-EC	TS-ECNO	TS-ECRO
	15	15	15
	25	25	25
MAX_ITER	35	35	35
	45	45	45
MAX_JUMPS	-	2	4
LIST_SIZE	1,000,000	1,000,000	1,000,000
RANDOM_NUM	-	-	16

Table 6. Cost gaps and runtimes for $n = 70$.

	Iterations	Average		Median		Standard deviation		Maximum	
		Cost gap (%)	Runtime (m)	Cost gap (%)	Runtime (m)	Cost gap (%)	Runtime (m)	Cost gap (%)	Runtime (m)
CPLEX	-	0.000%	2.543	0.000%	0.684	0.000%	4.443	0.000%	15.087
TS-EC	15	12.945%	0.107	8.186%	0.103	18.098%	0.007	79.969%	0.123
	25	3.544%	0.207	2.183%	0.206	5.037%	0.008	22.270%	0.227
	35	2.350%	0.254	1.876%	0.253	2.165%	0.009	10.129%	0.276
	45	2.248%	0.440	1.876%	0.439	1.828%	0.004	8.706%	0.452
TS-ECNO	15	8.365%	0.372	8.186%	0.373	2.698%	0.010	12.000%	0.386
	25	2.826%	0.536	2.183%	0.534	1.003%	0.007	5.529%	0.555
	35	2.130%	0.764	1.876%	0.757	1.394%	0.038	6.833%	0.900
	45	2.058%	1.104	1.876%	1.103	1.204%	0.005	5.941%	1.120
TS-ECRO	15	1.880%	0.619	1.751%	0.623	0.681%	0.021	3.850%	0.641
	25	1.251%	1.439	1.182%	0.936	0.624%	0.029	3.264%	1.003
	35	1.204%	1.439	1.206%	1.254	0.724%	0.669	3.474%	3.941
	45	1.121%	1.766	0.971%	1.766	0.718%	0.031	3.496%	1.833

5 Conclusions and Future Research

In this paper we address the economic lot-sizing problem with remanufacturing, considering environmental costs in addition to the traditional economic costs. We provide a MILP formulation for the problem and suggest three heuristic procedures based on the Tabu Search metaheuristic to solve it. To evaluate their performance, we compare them against AMPL/CPLEX, considering both costs and running times of the solutions.

Although the three heuristic procedures, TS-EC, TS-ECNO and TS-ECRO, show a good performance compared to the CPLEX solver, we note that for small and medium size instances, TS-ECNO achieves the best performance, considering the balance between runtime and cost gaps. On the other hand, for large instances, TS-ECRO achieves the best performance probably due to its random nature. In particular, we note that for the large instances the average runtimes for all procedures are significantly lower than those of CPLEX. In addition, the runtimes of the heuristic procedures are more stable, since their standard deviation is less than that obtained from the solver.

In order to improve the performance of heuristic procedures, we can modify the stop condition considering a maximum number of iterations without improvements instead of the total number of iterations as it is currently done. Another possible way to improve the procedures may be to analyze the form of the optimal solutions of the ELSR-EC as in Piñeyro and Viera [11] for the ELSR. Regarding the model, we observe that the objective function considered is the sum of the economic and environmental costs. A more realistic approach would be to consider a multi-objective function or even a multi-level model to better represent the relationship between economic and environmental costs.

References

1. Heck, M., Schmidt, G.: Lot-Size Planning with non-linear cost functions supporting environmental sustainability. In: Zavoral, F., Yaghob, J., Pichappan, P., El-Qawasmeh, E. (eds.) NDT 2010. CCIS, vol. 88, pp. 1–6. Springer, Heidelberg (2010). https://doi.org/10.1007/978-3-642-14306-9_1
2. Wagner, H., Whitin, T.: Dynamic version of the economic lot size model. Manag. Sci. 5(1), 89–96 (1958)
3. Green Peace Organization, https://www.greenpeace.org/, Accessed 24 June 2019
4. Ellen Macarthur Foundation, What is a circular economy?, https://www.ellenmacarthurfoundation.org, Accessed 21 Aug 2019
5. The European Remanufacturing Network, What is remanufacturing?, https://www.remanufacturing.eu, Accessed 21 May 2020
6. Zangwill, W.: Minimum concave cost flows in certain networks. Manag. Sci. 14(7), 429–450 (1968)
7. Hanafizadeh, P., Shahin, A., Sajadifar, M.: Robust Wagner-Whitin algorithm with uncertain costs. J. Ind. Eng. Int. 15, 435–447 (2019)
8. Aggarwal, A., Park, J.K.: Improved algorithm for economic lot size problems. Oper. Res. 41, 549–571 (1993)
9. Teunter, R., Bayindir, Z., Van Den Heuvel, W.: Dynamic lot sizing with products returns and remanufacturing. Int. J. Prod. Res. 44(20), 4377–4400 (2006)
10. Piñeyro, P., Viera, O.: Inventory policies for the economic lot-sizing problem with remanufacturing and final disposal options. J. Ind. Manag. Optim. 5(2), 217–238 (2009)
11. Piñeyro, P., Viera, O.: The economic lot-sizing problem with remanufacturing: analysis and an improved algorithm. J. Remanuf. 5(1), 1–13 (2015). https://doi.org/10.1186/s13243-015-0021-8
12. Absi, N., Dauzère-Pérès, S., Kedad-Sidhoum, S., Penz, B., Rapine, C.: Lot sizing with carbon emission constraints. Eur. J. Oper. Res. 227(1), 55–61 (2012)
13. Retel Hemlrich, M., Jans, R., Van Den Heuvel, W., Wagelmans, A.: The economic lot-sizing problem with an emission capacity constraint. Eur. J. Oper. Res. 241(1), 50–62 (2014)
14. United Nations: Kyoto Protocol to the United Nations Framework Convention on Climate Change (1998)
15. European Commission: Communication from the Commission to the European Parliament, the Council, the European Economic and Social Committee and the Committee of the Regions (2010)
16. Benjaafar, S., Li, Y., Daskin, M.: Carbon footprint and the management of supply chains: insights from simple models. IEEE Trans. Autom. Sci. Eng. 10(1), 99–116 (2013)
17. Akbalik, A., Rapine, C.: Single-item lot sizing problem with carbon emission under the cap-and-trade policy. In: Proceedings of the 2014 International Conference on Control, Decision and Information Technologies, pp. 30–35 (2014).
18. Yuan, B., Gu, B., Xu, C.: The multi-period dynamic optimization with carbon emissions reduction under cap-and-trade. Disc. Dyn. Nature Soc. 2019(1), 1–12 (2019)

19. Zouadi, T., Yalaoui, A., Reghioui, M., El Kadiri, K.: Hybrid manufacturing/remanufacturing lot-sizing problem with returns supplier's selection under, carbon emissions constraint. IFAC-PapersOnLine **49**(12), 1773–1778 (2016)
20. Zouadi, T., Yalaoui, A., Reghioui, M.: Hybrid manufacturing/remanufacturing lot-sizing and supplier selection with returns, under carbon emission constraint. Int. J. Prod. Res. **56**(3), 1233–1248 (2018)
21. Piñeyro, P., Viera, O.: Heuristic procedure for the economic lot-sizing problem with remanufacturing and recovery targets. J. Remanuf. **8**(1–2), 39–50 (2018). https://doi.org/10.1007/s13243-018-0044-z
22. Rapine, C., Goisque, G., Akbalik, A.: Energy-aware lot sizing problem: complexity analysis and exact algorithms. Int. J. Prod. Econ. **203**(1), 254–263 (2018)
23. Giglio, D., Paolucci, M., Roshani, A.: Integrated lot sizing and energy-efficient job shop scheduling problem in manufacturing/remanufacturing systems. J. Cleaner Prod. **148**, 624–641 (2017)
24. Glover, F., Laguna, M., Martí, R.: Principles and Strategies of Tabu Search. In: Handbook of Approximation Algorithms and Metaheuristics, 2nd edn. CRC Press, Boca Raton (2018)

Simultaneous Lot-Sizing and Scheduling with Recovery Options: Problem Formulation and Analysis of the Single-Product Case

Pedro Piñeyro[1]([🖂]) [iD] and Daniel Alejandro Rossit[2,3] [iD]

[1] Department of Operations Research, Institute of Computer Science,
Faculty of Engineering, Universidad de la República, Julio Herrera y Reissig 565,
11300 Montevideo, Uruguay
ppineyro@fing.edu.uy
[2] Department of Engineering, Universidad Nacional del Sur,
Av. Alem 1253, 8000 Bahía Blanca, Argentina
[3] INMABB-UNS-CONICET, Av. Alem 1253, 8000 Bahía Blanca, Argentina
daniel.rossit@uns.edu.ar

Abstract. We address an extension of the discrete lot-sizing and scheduling problem in which the demand of products can be also satisfied by remanufacturing used products returned to the origin. A mixed-integer linear programming formulation is provided for the problem, assuming dynamic demand and returns values, and time-invariant costs of setup and holding inventory. We then present a numerical experimentation carried out with the mathematical model for the case of a single product, in order to evaluate the efficiency in both costs and solving times compared to the traditional problem without returns. From the results obtained we conclude that the problem with recovery options can lead to economic benefits only under certain assumptions on the costs and amounts of returns. In addition, according to the runtimes obtained for large instances of the problem, it seems to be much more difficult to solve than its traditional version without returns.

Keywords: Lot-sizing · Scheduling · Remanufacturing · Mathematical programming · Optimization

1 Introduction

The discrete lot-sizing and scheduling problem (DLSP) refers to the problem of simultaneously determining the lots and sequence of production for a set of products, in order to meet the demand requirements on time, minimizing the sum of production and inventory holding costs. Unlike other lot-sizing problems, in the DLSP only one item can be produced at each period and the production process runs at full capacity, i.e., small bucket periods and all-or-nothing production policy are assumed. Therefore, the quantity of production in a period established as

© Springer Nature Switzerland AG 2021
D. A. Rossit et al. (Eds.): ICPR-Americas 2020, CCIS 1408, pp. 102–112, 2021.
https://doi.org/10.1007/978-3-030-76310-7_8

a positive production period is determined based on a certain production speed expressed in units per period. In addition, setup costs are only incurred when there is product changeover, i.e., configuration costs of the production line are only incurred at the beginning of the production process of a product.

A mixed-integer linear programming (MILP) and a heuristic solving approach for the DLSP in the case of a single machine is proposed in [3]. In [4] the DLSP with sequence-dependent setup costs is formulated and analyzed. Salomon et al. in [17] present a classification scheme and complexity results for the DLSP. In particular, they demonstrated that the DLSP with stationary costs, positive setup costs and single machine is NP-hard. However, they note that the DLSP for a single product can be efficiently solved as in [8] and [24]. Several extensions and solving approaches have been proposed in the literature for the DLSP, such as [23] for costs with non-speculative motives and [7] for start-up times and backlogging in the context of a tire curing scheduling problem. More recently, the DLSP with sequence-dependent changeover costs and time is addressed in in [5]. Bi-objective formulations for the DLSP considering carbon emissions and renewable energy are presented in [10]. We refer the readers to [1] and [25] for recent surveys about the DLSP and extensions.

In recent years, the circular economy (CE) paradigm has emerged as an alternative to face the negative impact on the environment of the traditional linear model of producing, consuming and disposing [21]. As a recovery option, remanufacturing can be considered a critical component of the CE as it attempts to maintain the value of used products (also called cores or returns) and to reduce the use of raw material and energy [9], thus achieving a win-win-win situation for producers, consumes and environment. Remanufacturing is an industrial process in which it is warrantied that a used product is returned to its original condition or even better. It often involves inspection, sorting, disassembling, cleaning, testing, reprocessing and reassembling tasks. Remanufacturable products include automotive parts, electric home appliances, personal computers, cellular phones, cameras, vending machines, among others [11]. In the field of production planning, we can find the seminal works of [15,16] and [26] that extend the traditional economic lot-sizing problem (ELSP) in order to include returns flow and remanufacturing process. Teunter et al. [22] introduce the economic lot-sizing problem with remanufacturing (ELSR) with either joint or separate setup scheme. They provide an efficient algorithm for the joint case and extend the Silver-Meal heuristic for the problem. In [12] consider the ELSR with disposal of returns and in [18] suggest several improvements for the extended Silver-Meal heuristic. In [2,13,19] are suggested and evaluated different solving procedures for the ELSR based on metaheuristic approaches, and [14] considers the ELSR with recovery targets. We refer to [21] for a recent survey about production planning with recovery options.

In this paper, we extend the DLSP in order to include the returns flow, and assuming that demand can be also satisfied by remanufacturing used products. We refer to this problem extension as the discrete lot-sizing and scheduling problem with remanufacturing (DLSR). We provide a MILP formulation for the problem and present and analyze the results of a numerical experimentation carried out in order to evaluate the efficiency in both costs and solving time of this hybrid production-remanufacturing system in the case of a single product. As far as we know, this DLSP extension has not been considered in the literature before. However, we note that there are some works that combine scheduling and remanufacturing, but for different problem assumptions [6,20,27].

The remainder of this paper is organized as follows. In Sect. 2 we present the description and the MILP formulation suggested for the problem under consideration. Section 3 provides the results and the analysis of the numerical experimentation carried out in order to evaluate the efficiency of the DLSR for the case of a single product. Finally, in Sect. 4 we present the conclusions and some possible directions for future research.

2   Problem Description and Formulation

The problem tackled in this paper can be stated as follows. We address a dynamic hybrid production-remanufacturing system with a single machine for processing $J > 0$ products under all-or-nothing production policy and over a finite planning horizon $T > 0$. Small-bucket periods are assumed. Therefore, only one product can be produced or remanufactured, but not both in the same period. The production and remanufacturing speeds for each product j, p_j and r_j respectively, are expressed as unit per period. Demand for a product j in a period t must be satisfied on time (backlogging is not allowed) from produced or remanufactured products generated in the same period (zero lead-time is assumed) or in a previous period and held in the inventory of serviceable products. Demand and returns values are assumed known in advance for each product and period over the planning horizon. Production and remanufacturing setup costs are incurred when there is product changeover. There are also linear costs for carrying on one unit of serviceable or used product from one period to the next one. Setup and inventory holding costs are assumed known and time-invariant. The objective is to determine the lots and sequence of products to be produced and remanufactured in order to satisfy the demand requirements on time, minimizing the sum of the costs involved.

Below we present the notation used for the problem from here and throughout the rest of the paper.

Sets and indices:

$$J : \text{Number of products, with index } j \in \{1, ..., J\}.$$
$$T : \text{Number of periods, with index } t \in \{0, 1, ..., T\}.$$

Parameters:

D_{jt} : Demand of product j in period $t > 0$.

R_{jt} : Returns of product j available in period $t > 0$.

K_j^p : Production setup cost of product j.

K_j^r : Remanufacturing setup cost of product j.

h_j^s : Holding cost for serviceable product j per unit and per period.

h_j^u : Holding cost for used product j per unit and per period.

p_j : Production speed for product j (units per period).

r_j : Remanufacturing speed for product j (units per period).

Variables:

I_{jt}^s : Inventory of serviceable product j at the end of period t.

I_{jt}^u : Inventory of used product j at the end of period t.

y_{jt}^p : $y_{jt}^p = 1$ indicates that product j is produced in period t;

$\quad y_{jt}^p = 0$, otherwise.

y_{jt}^r : $y_{jt}^r = 1$ indicates that used product j is remanufactured in period t;

$\quad y_{jt}^r = 0$, otherwise.

z_{jt}^p : $z_{jt}^p = 1$ indicates a production setup for product j in period $t > 0$;

$\quad z_{jt}^p = 0$, otherwise.

z_{jt}^r : $z_{jt}^r = 1$ indicates a remanufacturing setup for product j in period $t > 0$;

$\quad z_{jt}^r = 0$, otherwise.

Based on the notation of above, the DLSR can be formulated as the following MILP. We note that it can be considered an extension of that introduced in [8] and [24] for the DLSP.

$$\min \quad \sum_{j=1}^{J}\sum_{t=1}^{T}(K_j^p z_{jt}^p + K_j^r z_{jt}^r + h_j^s I_{jt}^s + h_j^u I_{jt}^u) \tag{1}$$

$$\text{s.t.} \quad I_{jt}^s = I_{j(t-1)}^s + p_j y_{jt}^p + r_j y_{jt}^r - D_{jt}, \quad \forall j, t > 0 \tag{2}$$

$$I_{jt}^u = I_{j(t-1)}^u + R_{jt} - r_j y_{jt}^r, \quad \forall j, t > 0 \tag{3}$$

$$\sum_{j=1}^{J}(y_{jt}^p + y_{jt}^r) \leq 1, \quad \forall t > 0 \tag{4}$$

$$z_{jt}^p = y_{jt}^p - y_{j(t-1)}^p, \quad \forall j, t > 0 \tag{5}$$

$$z_{jt}^r = y_{jt}^r - y_{j(t-1)}^r, \quad \forall j, t > 0 \tag{6}$$

$$I_{j0}^s = I_{j0}^u = y_{j0}^p = y_{j0}^r = 0, \quad \forall j \tag{7}$$

$$I_{jt}^s, I_{jt}^u \geq 0, \quad \forall j, t \tag{8}$$

$$y_{jt}^p, y_{jt}^r, z_{jt}^p, z_{jt}^r \in \{0, 1\}, \quad \forall j, t \tag{9}$$

The objective function (1) minimizes the sum of setup and inventory holding costs. Constraints (2) and (3) are the inventory balance equations for serviceable and used products, respectively. Constraints (4) establish that: i) only one product can be produced or remanufactured in certain period; ii) production and remanufacturing can not be carried out in the same period. Constraints (5) and (6) are for establishing the production and remanufacturing setup cost incurred when there is product changeover, respectively. The constraints (7) mean that the initial inventory level for all products is assumed zero and that the system is initially idle. Finally, (8) and (9) are the domain constraints for the decision variables.

3   Numerical Experimentation

In this section we provide the results and the analysis of the numerical experiments carried out to evaluate the mathematical model suggested for the DLSR. Specifically, we are interested in compare the efficiency in both costs and solving time for the problem with and without returns, i.e., the DLSR and the DLSP. We note that the MILP of Sect. 2 in which the parameters, variables and specific constraints related to returns are removed, corresponds to the formulation of the DLSP.

To facilitate the understanding of the results and to make a more effective comparison, we consider for the experiments the case of a single product ($j = 1$).

3.1   Experiments Design

The instances were generated based on the benchmark set introduced in [22] and [18] for the ELSR, also used in [2, 13, 19] and [14]. All them have a planning horizon T of 22 periods (workdays of a month). Demand values D_{1t} were generated according to a normal distribution with mean 100 and 20% of coefficient variation. Returns R_{1t} were also generated using a normal distribution with means 50, 70 and 100, and 20% of coefficient variation. The setup cost of production K_1^p is set to 2000 and the setup cost of remanufacturing K_1^r can be 200, 500 and 2000. The unit cost for holding inventory of serviceable products h_1^s is set to 1 and for used products h_1^u can be 0.2, 0.5 and 0.8. Production speed p_1 is set to 200 units per period, and the remanufacturing speed r_1 can take values on 200, 300 and 400 units per period. Note that, as is to be expected in practice, it is assumed that remanufacturing costs are at most equal to production costs and that remanufacturing speed is at least equal to production speed. We generate 10 instances for each on of the 3^4 configurations, that is 810 instances in total. In addition, to evaluate the solving time, we generate one instance for each configuration, with planning horizons of 22, 44, 66, 88, 110 and 132 periods, that is $81 \times 6 = 486$ instances in total.

The model was coded in AMPL and solved for all the instances of above with CPLEX 12.9.0.0 with a time limit of 1800 s, on a PC with 8 CPUs Intel Core i7-6700 3.40 GHz, 64-bit, 24 GB of RAM and CentOS Linux 7.

3.2   Analysis of Results for Instances with $T = 22$

Next, we present and analyze the results obtained for the instances with a planning horizon of $T = 22$ periods. The values shown in Table 1 are the average cost, the standard deviation and maximum cost of the optimal solutions obtained with CPLEX for the DLSR and DLSP. We decided not to report the runtimes, since they were very low for both problems, less than half a second for all the 810 instances. In the first row of the table, labeled as *"All instances"*, we show the average cost considering all the instances, which is 12480.1 for the DLSR and 10828.9 for the DLSP. This represents around of 15% increase in the costs for the DLSR in average. However, we can note that in the cases of low-cost returns ($K_1^r = 200$ or $h_1^u = 0.2$), the costs are in average a little more favorable for the DLSR. The average costs of these cases are highlighted in the table (underlined and in bold). In fact, for each one of the 90 instances with $K_1^r = 200$ and $h_1^u = 0.2$, we want to remark that the cost of the DLSR optimal solutions is lower than the cost of the DLSP optimal solutions, even reaching differences of more than 35% in favor of the DLSR. We also note that the total number of instances for which the cost is favorable for the DLSR is 255, that is 30% of the total instances. On the other hand, we note that the maximum gap between the average costs of the DLSR and DLSP is for the cases of high-cost returns ($K_1^r = 2000$ and/or $h_1^u = 0.8$).

In the cases of variations on the average quantity of returns (R_{1t}), we note that they do not have a significant impact on the cost gaps. However, we note a favorable trend to the DLSR as the amount of returns increase, as it can be noted for the case of large returns ($R_{1t} = 100$). The opposite behaviour can be observed for the remanufacturing speed (three last rows of Table 1): as the remanufacturing speed increases, also do the average cost gap between DLSR and DLSP. Therefore, it would be more economically convenient to consider similar speeds for production and remanufacturing, although in fact the remanufacturing speed could be higher, as it often requires less labor effort, materials, and energy.

In Table 1, the higher standard deviation and maximum values obtained for the DLSR compared to the DLSP, may be due to the fact that the variations are in the parameters related to the returns, which do not affect the DLSP. However, this appreciable dispersion in the costs of the DLSR optimal solutions observed for all the instances, would show that the variation in the parameter values has a significant impact on the efficient resolution of the problem.

The cases in which the DLSR obtained lower costs than DLSP, that is, $K_1^r = 200$ and $h_1^u = 0.2$, are presented in more detail in Tables 2 and 3, respectively.

Table 2 shows the average cost of the optimal solutions for the DLSR and DLSP for the instances with $K_1^r = 200$. The rest of the parameters (R_{1t}, h_1^u and r_1) vary as described in Sect. 3.1. In column "Gap" we report the difference in percentage between the average costs of the optimal solutions determined as: $100 \times (DLSR\_cost - DLSP\_cost)/DLSP\_cost$. From Table 2 we first note that the DLSR average costs are lower than DLSP in most cases (negative values in the Gap column). In terms of R_{1t}, it can be seen that as the value of returns increases, the cost of the DLSR optimal solutions decreases, reaching a percentage gap near to -14%. For the cases of h_1^u, we note that the sensitivity

Table 1. Summary of results for the 810 instances with $T = 22$ for DLSR and DLSP. The columns represent the average, standard deviation and maximum value of the optimal solutions obtained with CPLEX.

		Average		Stand. desv.		Maximum	
		DLSR	DLSP	DLSR	DLSP	DLSR	DLSP
All instances		12480.1	10828.9	2858.7	413.6	21086.2	12065.0
R_{1t}	**50**	12653.5	10837.5	2206.8	402.9	18098.0	12065.0
	70	12625.5	10841.8	2671.6	438.0	19351.6	11930.0
	100	12161.3	10807.3	3523.8	399.4	21086.2	12007.0
K_1^r	**200**	**10447.5**	**10834.4**	1890.4	407.0	14865.8	12065.0
	500	11553.2	10816.9	1712.9	413.3	15839.8	11868.0
	2000	15439.6	10835.3	2066.4	421.6	21086.2	12007.0
h_1^u	**0.2**	**10600.6**	**10804.5**	2116.3	393.0	15177.6	11794.0
	0.5	12582.6	10850.2	2483.1	442.9	17746.0	12065.0
	0.8	14257.1	10831.9	2684.7	403.4	21086.2	11930.0
r_1	**200**	11519.7	10846.1	2672.3	419.6	17711.6	12065.0
	300	12610.2	10819.1	2825.5	433.9	20403.2	12007.0
	400	13310.3	10821.5	2796.2	386.9	21086.2	11930.0

Table 2. Comparison of the average costs of the DLSR and DLSP optimal solutions for the case $K_1^r = 200$.

			DLSR	DLSP	Gap
$K_1^r = 200$	R_{1t}	**50**	11217.9	10789.1	4.0%
		70	10778.4	10877.5	−0.9%
		100	9346.2	10836.6	−13.8%
	h_1^u	**0.2**	8736.0	10831.8	−19.3%
		0.5	10520.5	10816.1	−2.7%
		0.8	12086.0	10855.3	11.3%
	r_1	**200**	9415.7	10881.8	−13.5%
		300	10527.7	10810.3	−2.6%
		400	11399.1	10811.1	5.4%

of the DLSR is significant with respect to this parameter, since the cost gaps vary from -19.3% ($h_1^u = 0.2$) to 11.3% ($h_1^u = 0.8$). Regarding the remanufacturing speed r_1, the analysis is similar to the cases of h_1^u: the cost of the DLSR is very sensitive and increases as the value of r_1 increases, as we mentioned at the beginning of this subsection for the analysis of the results presented in Table 1. In addition, we observe that the costs are significantly favorable for the DLSR in the case in which the production and remanufacturing speed are the same ($p_1 = r_1 = 200$).

Table 3. Comparison of the average costs of the DLSR and DLSP optimal solutions for the case $h_1^u = 0.2$.

			DLSR	DLSP	Gap
$h_1^u = 0.2$	R_{1t}	50	10938.7	10812.1	1.2%
		70	10636.5	10856.8	−2.0%
		100	10226.5	10744.7	−4.8%
	K_1^r	200	8736.0	10831.8	−19.3%
		500	9862.4	10808.2	−8.8%
		2000	13203.3	10773.7	22.6%
	r_1	200	9881.1	10833.5	−8.8%
		300	10740.5	10801.3	−0.6%
		400	11180.1	10778.9	3.7%

Table 3 shows the results for the instances with $h_1^u = 0.2$. The meaning of the columns is the same as in Table 2. From Table 3 we can note that there are more instances in which the costs of the DLSR is lower than the costs of the DLSP (negative values in column "Gap"). The effect observed for the returns cases R_{1t} is similar to that seen in Table 2, but less marked: as returns amounts increase, the cost of the DLSR decreases. A high sensitivity for the costs of the DLSR optimal solutions is observed against different values of K_1^r, with a percentage gap from −19.3% for $K_1^r = 200$ to 22.6% for $K_1^r = 2000$. For the cases of the remanufacturing speed (r_1), the results are also similar for that observed in Table 2, but, once again, not so marked.

3.3 Analysis of Results for Large Instances

We present in this subsection the results of the experiments carried out to evaluate the computational effort required to solve the problems DLSR and DLSP. In Table 4 we present the runtimes in seconds for each problem, along with the number of times that the time limit of 1800 s, established to CPLEX, is attained. In Fig. 1 we show the CPU times required by CPLEX for solving the problems for the different values of T under consideration, represented on a logarithmic scale. As it can be seen in Table 4 and Fig. 1, DLSR is considerably more computationally expensive than DLSP. While the CPU times for the DLSP are around 1 s (even for the larger case of $T = 132$), in the case of the DLSR, they grow suddenly (more than 500 s for $T = 132$). This is evidenced by the number of times that the time limit of execution is reached by CPLEX for the DLSR, which begins to increase in larger instances. On the other hand, for the DLSP the time limit is never attained.

Finally, we note that the cost gaps for the large instances show, on average, the same behaviour as that observed for the 810 instances with $T = 22$ periods, that is, around a 15% increase for the DLSR compared to the DLSP.

Table 4. Average runtimes (in seconds) and number of times that the execution time limit is reached by CPLEX to solve the problems DLSR and DLSP for the different values of T considered.

	DLSR		DLSP	
T	Runtime	#Tim. lim.	Runtime	#Tim. lim.
22	0.07	0	0.006	0
44	0.53	0	0.10	0
66	9.02	0	0.26	0
88	134.03	4	0.82	0
110	362.44	13	1.11	0
132	513.67	19	1.34	0

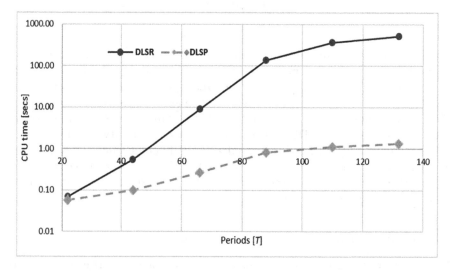

Fig. 1. CPU times in average required by CPLEX to solve the problems DLSR and DLSP for the different values of T. The CPU time is expressed in logarithmic scale in seconds.

4   Conclusions

In this paper we have tackled an extension of the discrete lot-sizing and scheduling problem in which the demand requirements of products can be also satisfied by remanufacturing used products returned to the origin. We refer to this problem as the discrete lot-sizing and scheduling problem with remanufacturing (DLSR). Remanufacturing can be considered as a key element of the Circular Economy, a paradigm that has emerged to counteract the negative effect on the environment of the traditional linear production of produce, use and dispose.

We provide a MILP formulation for the problem and analyze the results of a numerical experimentation carried out for the case of a single product, with the aim to compare the efficiency in both costs and solving time of the DLSR against the DLSP (the traditional problem without returns). From the numerical experiments conducted, we conclude that the problem extension with recovery options can offer economic benefits in the case of low-cost returns and, to a lesser extent, for those cases in which the number of returns is relatively high compared to the demand requirements. In addition, we note that similar production and remanufacturing speeds appear to be a more favorable configuration, even in those cases with large returns. In terms of solving times, we show that the DLSR is much harder to solve than the DLSP, in particular, as the size of the instances increases. Thus, we conjecture that the DLSR belongs to the complexity class of NP-hard problems.

Taking into account the results of the numerical experimentation summarized above, future research on DLSR should involve the development of efficient and effective solving procedures for the problem, possibly extending those suggested for the DLSP or ELSR. Another interesting topic to address is the analysis of the order of complexity of the problem. This analysis should involve not only the general case, but also particular cases such as the case of a single product as in [24] for the DLSP, or the case of sufficient returns at the first period as in [15] and [16] for ELSR. It would be also interesting to study the DLSR with more than one machine or processing line. In particular, the case in which production and remanufacturing are allowed in the same period. Considering the environmental side of the problem, it would be interesting to investigate the returns usage rate in the optimal or high quality solutions. Also including more realistic assumptions, such as the non-uniform condition of the returns (heterogeneous quality), may be an interesting extension to consider for the problem tackled here.

References

1. Copil, K., Wörbelauer, M., Meyr, H., Tempelmeier, H.: Simultaneous lotsizing and scheduling problems: a classification and review of models. OR Spectr. **39**(1), 1–64 (2016). https://doi.org/10.1007/s00291-015-0429-4
2. Fazle Baki, M., Chaouch, B.A., Abdul-Kader, W.: A heuristic solution procedure for the dynamic lot sizing problem with remanufacturing and product recovery. Comput. Oper. Res. **43**, 225–236 (2014)
3. Fleischmann, B.: The discrete lot-sizing and scheduling problem. Euro. J. Oper. Res. **44**(3), 337–348 (1990)
4. Fleischmann, B.: The discrete lot-sizing and scheduling problem with sequence-dependent setup costs. Euro. J. Oper. Res. **75**(2), 395–404 (1994)
5. Gicquel, C., Lisser, A., Minoux, M.: An evaluation of semidefinite programming based approaches for discrete lot-sizing problems. Euro. J. Oper. Res. **237**(2), 498–507 (2014)
6. Giglio, D., Paolucci, M., Roshani, A.: Integrated lot sizing and energy-efficient job shop scheduling problem in manufacturing/remanufacturing systems. J. Clean. Prod. **148**, 624–641 (2017)
7. Jans, R., Degraeve, Z.: An industrial extension of the discrete lot-sizing and scheduling problem. IIE Trans. **36**(1), 47–58 (2004)

8. Kuik, R., Salomon, M., Van Hoesel, S., Van Wassenhove, L.N.: The single item discrete lotsizing and scheduling problem: Linear description and optimization. Technical report, Erasmus University (1989)
9. Kurilova-Palisaitiene, J., Sundin, E., Poksinska, B.: Remanufacturing challenges and possible lean improvements. J. Clean. Prod. **172**, 3225–3236 (2018)
10. Liu, C.H.: Discrete lot-sizing and scheduling problems considering renewable energy and CO2 emissions. Prod. Eng. **10**(6), 607–614 (2016)
11. Matsumoto, M., Umeda, Y.: An analysis of remanufacturing practices in Japan. J. Remanuf. **1**(1), 2 (2011)
12. Piñeyro, P., Viera, O.: Inventory policies for the economic lot-sizing problem with remanufacturing and final disposal options. J. Ind. Manage. Optim **5**(2), 217–238 (2009)
13. Piñeyro, P., Viera, O.: The economic lot-sizing problem with remanufacturing: analysis and an improved algorithm. J. Remanuf. **5**(1), 1–13 (2015). https://doi.org/10.1186/s13243-015-0021-8
14. Piñeyro, P., Viera, O.: Heuristic procedure for the economic lot-sizing problem with remanufacturing and recovery targets. J. Remanuf. **8**(1), 39–50 (2018)
15. Richter, K., Sombrutzki, M.: Remanufacturing planning for the reverse Wagner/whitin models. Euro. J. Oper. Res. **121**(2), 304–315 (2000)
16. Richter, K., Weber, J.: The reverse Wagner/whitin model with variable manufacturing and remanufacturing cost. Int. J. Prod. Econ. **71**(1), 447–456 (2001)
17. Salomon, M., Kroon, L.G., Kuik, R., Van Wassenhove, L.N.: Some extensions of the discrete lotsizing and scheduling problem. Manage. Sci. **37**(7), 801–812 (1991)
18. Schulz, T.: A new silver-meal based heuristic for the single-item dynamic lot sizing problem with returns and remanufacturing. Int. J. Prod. Res. **49**(9), 2519–2533 (2011)
19. Sifaleras, A., Konstantaras, I., Mladenović, N.: Variable neighborhood search for the economic lot sizing problem with product returns and recovery. Int. J. Prod. Econ. **160**, 133–143 (2015)
20. Sun, H., Chen, W., Liu, B., Chen, X.: Economic lot scheduling problem in a remanufacturing system with returns at different quality grades. J. Clean. Prod. **170**, 559–569 (2018)
21. Suzanne, E., Absi, N., Borodin, V.: Towards circular economy in production planning: challenges and opportunities. Euro. J. Oper. Res. **287**(1), 168–190 (2020)
22. Teunter, R.H., Bayindir, Z.P., Den Heuvel, W.V.: Dynamic lot sizing with product returns and remanufacturing. Int. J. Prod. Res. **44**(20), 4377–4400 (2006)
23. Van Eijl, C., Van Hoesel, C.: On the discrete lot-sizing and scheduling problem with wagner-whitin costs. Oper. Res. Lett. **20**(1), 7–13 (1997)
24. Van Hoesel, S., Kuik, R., Salomon, M., Van Wassenhove, L.N.: The single-item discrete lotsizing and scheduling problem: optimization by linear and dynamic programming. Discrete Appl. Math. **48**(3), 289–303 (1994)
25. Wörbelauer, M., Meyr, H., Almada-Lobo, B.: Simultaneous lotsizing and scheduling considering secondary resources: a general model, literature review and classification. OR Spectr. **41**(1), 1–43 (2018). https://doi.org/10.1007/s00291-018-0536-0
26. Yang, J., Golany, B., Yu, G.: A concave-cost production planning problem with remanufacturing options. Naval Res. Logist. **52**(5), 443–458 (2005)
27. Zanoni, S., Segerstedt, A., Tang, O., Mazzoldi, L.: Multi-product economic lot scheduling problem with manufacturing and remanufacturing using a basic period policy. Comput. Ind. Eng. **62**(4), 1025–1033 (2012)

Study of the Location of a Second Fleet for the Brazilian Navy: Structuring and Mathematical Modeling Using SAPEVO-M and VIKOR Methods

Isaque David Pereira de Almeida[1] , José Victor de Pina Corriça[1] ,
Arthur Pinheiro de Araújo Costa[1] , Igor Pinheiro de Araújo Costa[2,3](✉) ,
Sérgio Mitihiro do Nascimento Maêda[2] , Carlos Francisco Simões Gomes[2] ,
and Marcos dos Santos[4]

[1] Brazilian Navy, Rio de Janeiro, RJ 20081-240, Brazil
[2] Fluminense Federal University (UFF), Niterói, RJ 24210-346, Brazil
costa_igor@id.uff.br
[3] Naval Systems Analysis Centre (CASNAV), Rio de Janeiro, RJ 20091-000, Brazil
[4] Military Institute of Engineering (IME), Urca, RJ 22290-270, Brazil

Abstract. The Brazilian Navy's main mission is to control maritime areas, deny the use of the sea to the enemy and design the Naval Power, thus defending the Brazilian Jurisdictional Waters (BJW). For this, a robust naval force is needed to enable the fulfillment of its mission. The Brazilian Navy Fleet is located in the interior of Guanabara Bay, where the Port of Rio de Janeiro is also located. This fact weakens the National Defense since there is the possibility of carrying out, by ships, submarines, or mines enemies, a dam or blockade of the only exit to the sea. Therefore, this paper aims to select the most suitable city to be the headquarters of the Second Fleet of the Brazilian Navy. Several localities distributed throughout practically all navigable areas of the Brazilian territory were analyzed as possible headquarters for the Second Squadron. For the structuring of the problem, the Soft Systems Methodology (SSM) was used. For the decision-making process, hybrid modeling was used, with the SAPEVO-M method to obtain the weights of the criteria and the VIKOR method for the evaluation of alternatives. For the composition of the model, the criteria Distance to the Sea (KM), Width of Departure (KM), Distance from Rio de Janeiro (MN), Infrastructure for Repairs, and State of the Sea (Scale) were selected. After the application of the methods, the city of Belém was selected as the most indicated to be the headquarters of the new Brazilian Fleet.

Keywords: Management · Multicriteria analysis · SAPEVO-M · VIKOR · Brazilian Navy

1 Introduction

The National Defense Strategy (END), enacted in 2008 and updated in 2012, in order to ensure the security of the country both in peacetime and in crises, establishes guidelines

© Springer Nature Switzerland AG 2021
D. A. Rossit et al. (Eds.): ICPR-Americas 2020, CCIS 1408, pp. 113–124, 2021.
https://doi.org/10.1007/978-3-030-76310-7_9

for the proper preparation and training of the Brazilian Armed Forces (FFAA). The END was also developed to meet the equipment needs of the Military Commands, reorganizing the defense industry so that the most advanced technologies are under national control and instituting medium and long-term strategic actions aimed at modernizing the national defense structure [1].

Brazil's natural maritime vocation is supported by its extensive coastline, waterways, the magnitude of its maritime trade and the undeniable strategic and economic importance of the South Atlantic, which hosts the "Blue Amazon", an area that incorporates the high potential for living and non-living resources, such as Brazil's largest oil and natural gas reserves. The oceans are also important climate conditioners, besides serving as a cradle for submarine cables, whose data traffic is responsible for virtually all of the country's communication with the world [2].

The naval power must possess sufficient capacity and credibility to deter any adverse forces from conducting hostile actions in the AJB. The action of naval units in the South Atlantic, where the "Blue Amazon" is inserted, and in the Amazon and Paraguay rivers basins will be an essential factor for strengthening this deterrence. Thus, the Force must be prepared, both to act in an interstate crisis, and to monitor and suppress the actions of adverse groups practicing illegal activities in the AJB [2].

Over the past few years, much has been discussed about the constitution of the so-called Second Fleet, supported by a Naval Base capable of meeting the logistical needs arising. Numerous reasons justify a greater presence of the Brazilian Naval Power in the Blue Amazon [3].

With the survey of some points in the history of wars around the globe, there is a need to deal with some difficulties that Brazil may face if there is an armed conflict. An important point is a possibility of being carried out, by enemy ships or submarines, a dam or blockade at the exit to the sea. As Brazil has only one fleet headquarters – Rio de Janeiro – the country could go through serious problems due to the lack of a Second Fleet. Another important point, of great value to society, which justifies the Second Squadron, is the need for patrols and naval inspections, ensuring national sovereignty throughout the Blue Amazon.

However, the budgetary constraints are known and the need to prioritize the construction of submarines and corvettes, more essential to the Naval Power, led to the postponement of the Second Squadron project and its Naval Base [3].

In the study proposed by this paper, localities distributed through all navigable areas of the Brazilian territory are analyzed as possible headquarters for the Second Fleet.

In the decision-making process, Production Engineering becomes a fundamental mechanism in advising managers [4]. Part of this area, Operational Research (OR) is the comprehensive and multidisciplinary field that employs mathematical and analytical models for the solution of complex problems [5].

One relevant feature for a method of supporting decision-making is the availability of software implementing the method, as well as its graphical representation and exploitation of results [6]. Considering the structure of the proposed problem and the availability of software, this article will use hybrid modeling composed of the SAPEVO-M method to obtain the weights of the criteria and VIKOR method for the evaluation of alternatives.

Taking into account that the country is undergoing a serious budgetary constraint, only cities that already have a previously established naval base structure will be evaluated, considering that the construction of a large maintenance complex requires a large amount to the public coffers.

2 Problem Structuring

Problems Structuring Methods (PSM) are one of the stages of the decision-making process that aims to organize issues, questions and/or dilemmas for which decision propositions are sought [7]. PSM is widely accessible in the OR and in the movement of systems for understanding and structuring complex problems [8].

Every model has validity only within a certain managerial context, which, in addition to taking into account the specificities of the organization studied, also has a "space-time" validity [9].

Among the most used PSM, this paper will use the Soft Systems Methodology (SSM) [10], which has been explored in a variety of research fields, as well as serves equally diverse practical interests.

SSM presents seven stages of application [10], which two will be addressed in this article for the problem structuring:

1 - explore an unstructured problematic situation; and.

2 - express it by making a rich picture.

In the first stage, the brainstorming technique was used by the authors to demonstrate the perceptions of the group about all possible information, without interference or judgments to define the problem. In the second stage, a rich picture was constructed (Fig. 1) to express all relevant aspects of the problem.

The rich picture is a simple SSM tool, extremely useful for opening the discussion around individual perceptions towards a broad view of the different issues affecting the situation. They are created freely and unstructured to capture the participants' interpretation of a real situation [10, 11].

The rich picture shows that the Brazilian Navy Squadron is located inside Guanabara Bay, where the Port of Rio de Janeiro is also located. This fact weakens the national defense since there is the possibility of being carried out a dam or blockade of the only exit to the sea. This is considered a Political and Strategic movement that can lead Brazil to experience severe problems.

After analyzing the rich picture and consulting three Brazilian Navy officers with more than 20 years of experience, five criteria were established: Distance to the Sea (DS, Exit Width (EW), Distance from Rio de Janeiro by sea (DRJ), Infrastructure for Maintenance (IM) and State of the sea (SS).

Also, the alternatives of cities to be established as the headquarters of the second fleet were established, all of them with previously established naval base structure: Salvador/BA, Natal/RN, Rio Grande/RS, Ladário/MS, Belém/PA and Manaus/AM.

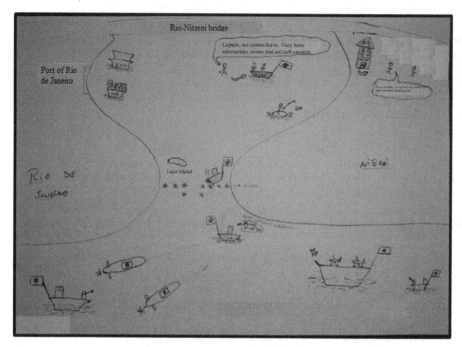

Fig. 1. Rich picture, entitled "Need for a second Fleet".

2.1 Presentation of Criteria

The five criteria were chosen to compose the model are as follows:

I – Distance to the Sea (DS): It was considered to be the approximate distance of the existing Military Organizations, which could be reformulated to become the probable Second Squadron, to the point where it is considered the largest free area for maneuvering the naval means.

II - Exit Width (EW): The exit width to the sea is essential, since the smaller the maneuvering space, the greater the probability of applying some type of barrier or blockade so that it is not possible to have access to the sea. Besides, a larger entry/exit width allows access to larger ships to the future second fleet headquarters.

III - Distance to Rio de Janeiro by sea (DRJ): Considering that the Brazilian coast is considerably extensive and taking into consideration the existence of a headquarter in Rio de Janeiro, the further away this Second Squadron is, the better it will be for a larger area of operation concerning the protection of the entire Brazilian coast.

IV - Infrastructure for Maintenance (IM): It was considered as being the supply of specialized labor and bases for repairs of ships that will be crowded in the second fleet.

V - State of the sea (SS): Considered as being the general condition of ocean waves in a given locality and time. The state of the sea is characterized by statistics, which include wave height, frequency and spectral density. In this article, the data referring to the state of the sea were recorded by the official website of the Brazilian Navy Hydrography Center [12] through the Douglas Scale of the State of the Sea.

2.2 Presentation of Alternatives

After the establishment of the criteria, and seeking to evaluate cities with previously established infrastructure, distributed along the Brazilian coast, six cities were selected to compose the model: Salvador - BA, Natal - RN, Rio Grande - RS, Ladário - MS, Belém - PA and Manaus - AM.

Table 1 presents the attributes of the cities in each criterion.

Table 1. Data from the cities evaluated.

Criterion	Feature	Salvador/Ba	Natal	Rio Grande	Ladário	Belém	Manaus
DS (Km)	1st exit	1,7	5,5	18	No access	100	No access
	2nd exit	25	–	–	–	130	–
EW (m)	1st exit	450	200	500	200	1200	1500
	2nd exit	9000	–	–	–	1500	–
DRJ (NM)	In Nautical Miles	745	1233	755	–	2218	–
IM	Naval Base + City Structure	4	3	1	2	2	2
SS	Douglas Sea State Scale	2	2	5	1	1	1

3 Background

In this section, the SAPEVO-M and VIKOR methods will be described, which will be used in hybrid modeling as tools to support decision-making.

3.1 SAPEVO-M Method

In general, the desired result in a given Multicriteria Decision Support problem can be identified among four types of reference problems [13]:

I - Problematic P.α - aims to clarify the decision by choosing a subset of alternatives as restricted as possible. Therefore, the desired result is a choice;

II - Problematic P.β - aims to clarify the decision by a screening resulting from the allocation of each alternative to a class (or category). Therefore, the desired result is a sort action;

III - Problematic P.γ - aims to clarify the decision by an organization obtained by the regrouping of part or all actions in equivalence classes, which are ordered partially or completely, according to the preferences of the decision-maker(s). Therefore, the desired result is a sort or ranking procedure;

IV - Problematic P.δ - aims to clarify the decision by a description of the actions and their consequences. Therefore, the desired result is a cognitive procedure or a description.

The Simple Aggregation of Preferences Expressed by Ordinal Vectors - Multi Decision Makers method (SAPEVO-M) represents a new version of the SAPEVO ordinal method [14], especially for problems of the P.γ type. This evolution of the original version extended the use of the method to multiple decision-makers [15].

The SAPEVO method unfolds the decision-making problem from three basic steps [15], which are:

1. Transforms the ordinal preferences of the criteria into a vector of criterion weights;
2. Transforms the ordinal preferences of alternatives for a given set of classification criteria into partial weights of alternatives; and.
3. Determines the global weights of the alternatives.

The method has been applied in several areas, such as the selection of equipment for a bakery [16] and the selection of a troop landing ship for the Argentine navy [17].

SapevoWeb Computer System. The SapevoWeb system was developed from a partnership between the technical staff of the Naval Systems Analysis Centre (CASNAV), a research group of the Graduate Program in Production Engineering of the Fluminense Federal University (UFF) and a research group of the Graduate Program in Systems and Computer Engineering of the Military Institute of Engineering (IME) [15].

SapevoWeb code was developed in python language, using the Django framework, and HTML [18]. The program allows the inclusion of a sufficiently large integer of decision-makers, criteria and alternatives, limited only by the processing capacity of the server. The computational tool can be accessed at www.sapevoweb.com [15].

3.2 VIKOR Method

The *ViseKriterijumska Optimizacija i Kompromisno Resenje* (VIKOR) method was developed as a method of commitment programming, to determine a ranking with weights of a set of alternatives [19]. To observe this weight that measures the ranking, a few steps must be followed [20, 21] such as:

Step 1: Determine the highest values f_i^* and the lowest values f_i^- of all the alternatives in each criterion, $i = 1, 2, ..., n$.

$$f_i^* = max_j f_{ij}$$

$$f_i^- = min_j f_{ij}$$

Where f_i^* is the highest value presented by the alternatives in each criterion;
f_i^- is the lowest value presented by the alternatives in each criterion; and.
f_{ij} is the value of the alternative in a given criterion.

Step 2: Calculate the values S (maximum utility group) (1) and R (minimum individual weight) (2), with j = 1, 2,..., j, in the relationships; where W_i is the weight of the criteria.

$$S_j = \frac{\sum_{i=1}^{n} W_i(f_i^* - f_i^-)}{(f_i^* - f_i^-)} \tag{1}$$

$$R_j = max_j\left(\frac{\sum_{i=1}^{n} W_i(f_i^* - f_i^-)}{(f_i^* - f_i^-)}\right) \tag{2}$$

S_j: maximum utility group of the alternative j;
W_i: weights of the criteria obtained by calculating entropy;
R_j: the minimum individual weight of alternative j.

Step 3: Calculate the values of Q_j (3), with j = 1, 2,..., j, by the ratio in which $S^* = min_j S_j$; S^- $max_j S_j$ and $R^* = min_j R_j$; $R^- = max_j R_j$. Parameter v is entered as a strategy weight, commonly used as v = 0.5.

$$Q_j = \frac{v(S_j - S^*)}{(S^- - S^*)} + \frac{(1 - V)(R_j - R^*)}{(R^- - R^*)} \tag{3}$$

Where v = 0.5 and Q_j is the final score of an alternative j;

Step 4: Classify the alternatives in a decreasing ordination, using the values obtained by S, R and Q. The results are three classification lists. However, only the values obtained by Q can be considered.

Step 5: Once the Ranking on the decreasing values of Q_i is established, the analysis on the verifications of stability conditions C_1 and C_2, which are obtained through (4) and (5):

C_1 - Acceptable difference/advantage:

$$Q(A_2) - Q(A_1) \geq DQ \tag{4}$$

Where $DQ = \frac{1}{j-1}$ and j is the number of alternatives.

C_2 - Acceptable stability in decision-making: A_1 should be the best rated in S and/or R.

If one of the conditions (C_1 and C_2) is not met, then a cluster of compromise solutions is proposed, which consists of:

Alternatives A_1 and A_2, if only C_2 is not satisfied; or.

Alternatives A_1, A_2,... A_M, if only condition C_1 is not satisfied; The M value is determined for the maximum value of M that satisfies the Eq. (5):

$$Q(A_M) - Q(A_1) < DQ \tag{5}$$

4   Application of SAPEVO-M and VIKOR Methods

First, the SAPEVO-M method will be applied to obtain the weights of the criteria, through the Software SapevoWeb [18]. After this, using the VIKOR method, through the Vikor software [22], the classification of cities will be obtained and stability conditions checked.

4.1 Evaluation of Criteria

Three Navy officers with more than 20 years of experience have already been invited to evaluate the criteria that compose the problem. After registering the decision-makers (DM), criteria and alternatives of the cities, the DM will evaluate, one by one, the importance of the criteria and alternatives [15].

After evaluating the criteria chosen by the Navy Officers, the weights of the criteria are obtained, according to Table 2.

Table 2. Weights of the criteria, obtained by applying the SAPEVO-M method.

Criterion	Weight
DS	1,4185
EW	2,9473
DRJ	1,4135
IM	2,6040
SS	0,0131

Analyzing the weights of the criteria, it is observed that the exit width was the criterion that obtained the greatest weight in the evaluation, followed by the infrastructure for maintenance. The state of the sea was considered the least important by the DM.

The SAPEVO-M method allows to know how much an alternative and/or criterion was better ordered, in relation to another, passing additional information to the DM [15]. In other words, this means that the exit width criterion (Weight 2.9473) was considered about 2 times better or more important than the distance from RJ by sea and distance to the sea.

4.2 Application of the VIKOR Method

From the decision matrix demonstrated in Table 3 and including the weight vector (Table 3), acquired by SAPEVO-M, it is necessary to determine the best value f_i* and the worst value f_i^- of each criterion:

It is worth note the differentiation between the cost criteria (the lower, the better) - Distance to the sea and state of the sea - and benefit (the bigger, the better) - Exit Width, Infrastructure for Maintenance and distance from RJ by sea.

4.3 Obtaining S_i and R_i

After applying (1) and (2), Table 4 is obtained.

It is observed that the city of Ladário obtained the highest value of S_i, while Belém presented the lowest value in this parameter. In relation to the parameter R_i, it is observed that Belém and Manaus obtained the lowest values, while Rio Grande and Natal presented the highest scores.

Table 3. Decision matrix of the criteria with their respective weights and the values of f_i^* and f_i^-.

Weights	1,4185	2,9473	1,4135	2,6040	0,0131
	DS	EW	DRJ	IM	S
Salvador/BA	27	450	745	4	2
Natal/RN	5,5	200	1233	3	2
Rio Grande/RS	18	500	755	1	5
Ladário/MS	10000	200	0	2	1
Belém/PA	100	1200	2218	2	1
Manaus/AM	10000	1500	0	2	1
fi*	5,5	1500	2218	4	1
fi-	10000	200	745	1	5

Table 4. Standard decision matrix and obtaining S_i and R_i.

	DS	EW	DRJ	IM	SS	S_i	R_i
Salvador/BA	0,0031	2,3805	0,9387	0,0000	0,0033	3,3256	2,3805
Natal/RN	0,0000	2,9473	0,6373	0,8680	0,0033	4,4559	2,9473
Rio Grande/RS	0,0018	2,2672	0,9323	2,6040	0,0131	5,8184	2,6040
Ladário/MS	1,4185	2,9473	1,4135	1,7360	0,0000	7,5153	2,9473
Belém/PA	0,0134	0,6801	0,0000	1,7360	0,0000	2,4295	1,7360
Manaus/AM	1,4185	0,0000	1,4135	1,7360	0,0000	4,568	1,7360

Source: Authors (2020).

4.4 Calculating Q_i and Ordering Alternatives

Applying (3), Table 5 is obtained.

From the data obtained, the decreasing ordering is made, by S, R and Q values. For this, the results compose three classification lists, but only the Q values can be considered (Table 6).

Analyzing the classification of the cities, it is observed that the best evaluated by the methods was the city of Belém/PA, which obtained the lowest Q value.

4.5 Analysis of Results

After establishing the classification of alternatives, the stability conditions should be checked. First analyzing condition C_1 - Acceptable difference/advantage - applying (4), DQ equal to 0.2 is obtained. The difference between the second-placed (Manaus) and the first (Belém) must be greater than or equal to DQ for this condition to be met:

$$Q(A_2) - Q(A_1) = 0.2102 - 0.0000 = 0.2102.$$

Table 5. Obtaining Q_i.

	S_i	R_i	Q_i
Salvador/BA	3,3256	2,3805	0,3541
Natal/RN	4,4559	2,9473	0,6992
Rio Grande/RS	5,8184	2,6040	0,6915
Ladário/MS	7,5153	2,9473	1,0000
Belém/PA	2,4295	1,7360	0,0000
Manaus/AM	4,568	1,7360	0,2102
S^*, R^*	2,4295	1,7360	
S^-, R^-	7,5153	2,9473	

Table 6. Ordering of alternatives.

	S_i	R_i	Q_i	Ranking
Salvador/BA	3,3256	2,3805	0,3541	3
Natal/RN	4,4559	2,9473	0,6992	5
Rio Grande/RS	5,8184	2,6040	0,6915	4
Ladário/MS	7,5153	2,9473	1,0000	6
Belém/PA	2,4295	1,7360	0,0000	1
Manaus/AM	4,568	1,7360	0,2102	2

Therefore, Condition C_1 is satisfied, since $0.2102 \geq 0.2$.

For condition C_2, it should be verified if the best-classified alternative also presents the best indices for R and/or S. It is observed that Belém, besides being the best alternative analyzing Q, presents the lowest values in R and S, also satisfying the stability condition C_2. As the two conditions were met, Belém can be considered as the most suitable city to be the headquarters of the Second Brazilian Squadron.

Analyzing Table which presents the normalized values of the alternatives in each criterion, it is observed that the first two placed in the proposed analysis (Belém and Manaus) present the best performance (note 0) in two criteria, which justifies their best classifications rather than the others.

The city of Manaus presents the best score in the criterion with greater weight (exit width) but obtained the worst performance in the distance to the sea and distance from RJ by sea. On the other hand, the city of Belém obtained the second score in the analysis of the exit width and the best performance in the criteria state of the sea and distance from RJ by sea, besides presenting regularity in the other criteria, which corroborates with its choice.

5   Conclusion

The application of the SAPEVO-M method considered the evaluation of three different specialists and made it possible to obtain the values of the weights of the criteria taking into account the opinion of each of them, making the analysis more robust and reliable. Considering that the objective of the article is to select a headquarters for the Second Squadron, the method proved to be quite efficient, since the criteria that obtained the highest weights are of high importance for fulfilling the mission of the city to be chosen as the headquarter.

Regarding the application of the VIKOR method, it was observed that the results obtained presented a consistent basis, with easily verifiable logical chaining, since the cities with the best classifications presented the best performances in the most important criteria. Besides, the verification of acceptable advantage and stability conditions have analyzed results more reliable and safer, providing the decision-making person with extremely relevant information for the decision.

Given the above, it was clear that the hybrid modeling using SAPEVO-M and VIKOR methods can be used to solve complex problems of the most varied types, thus being a combination of methods of great utility for the contribution of decision making, considering that it takes into account the evaluation of several criteria made by multiple decision-makers, besides allowing a rich and robust analysis of the results obtained, which makes the decision-making process more transparent.

It is also emphasized that, because there are web softwares available, including free of charge, the application of the methods were greatly facilitated, since their axiomatics involve calculations often complex and long, which implements a software, which facilitates the use of methods, a determining factor for the diffusion of this modeling.

Finally, the future papers could address comparative analyses between cities where there are no pre-structured bases, including other criteria such as ease of access to coast and area available for the construction of a naval complex.

References

1. Brasil. National Defense Strategy (END) [Internet] (2012). Accessed 10 Aug 2020, https://www.defesanet.com.br/defesa/noticia/32308/END---Estrategia-Nacional-de-Defesa/
2. Marinha do Brasil. Naval Policy [Internet] (2019). Accessed 10 Aug 2020, https://www.marinha.mil.br/politicanaval
3. Leal Ferreira, E.B.: The second fleet, the Amazon and the South Atlantic [Internet]. Bonifácio (2019). Accessed 14 Aug 2020, https://bonifacio.net.br/a-segunda-esquadra-a-amazonia-e-o-atlantico-sul/
4. dos Santos, M., da Costa, M., dos Reis, M.F.: Using the branch and bound algorithm to optimize the production of a plastic products industry. Rev. Trab. Acadêmicos Lusófona. **2**(2), 217–37 (2019)
5. de Teixeira, L.F.H.S.B., Ribeiro, P.C.C., Gomes, C.F.S., dos Santos, M.: Use of the SAPEVO-M method with SCOR 12.0 model parameters for ranking suppliers in a supply chain for hospital material in the Brazilian Navy. Revista Pesquisa Naval, no. 31, pp. 1–13 (2019)
6. Cinelli, M, Kadziński, M, Gonzalez, M, Słowiński, R.: How to Support the Application of Multiple Criteria Decision Analysis? Let Us Start with a Comprehensive Taxonomy. Omega. 2020;102261.

7. Bandeira, M.C.G.S.P., Mattos, R.I., Belderrain, M.C.N., Correia, A.R., Kleba, J.B.: Business model in an agricultural community: Application of Soft Systems Methodology and Strategic Choice Approach (2018)
8. Rosenhead, J., Mingers, J.: Rational Analysis for a Problematic World Revisited: Problem Structuring Methods for Complexity, Uncertainty and Conflict. Wiley, Chichester (2001)
9. dos Santos, M.: Proposta de Modelagem Atuarial aplicada ao setor militar considerando influências econômicas e biométricas. [Niterói, RJ]: Tese de Doutorado apresentada no Programa de Pós-Graduação em Engenharia de Produção da Universidade Federal Fluminense. Niterói, RJ (2018)
10. Checkland, P.B.: Systems theory. Systems Practice Wiley, Chichester (1981)
11. Rose, J.: Soft systems methodology as a social science research tool. Syst. Res. Behav. Sci. Off. J. Int. Fed. Syst. Res. **14**(4), 249–58 (1997)
12. Marinha. Escala Douglas do Estado do Mar [Internet]. Centro de Hidrografia da Marinha (2020). Accessed 14 Aug 2020, https://www.marinha.mil.br/chm/sites/www.marinha.mil.br. chm/files/u2035/estado_do_mar.pdf
13. Gomes, L.: Gomes CFS. Princípios e métodos para a tomada de decisão: Enfoque multicritério. São Paulo: Atlas (2019)
14. Gomes, L., Mury, A.-R., Gomes, C.F.S.: Multicriteria ranking with ordinal data. Syst. Anal. **27**(2), 139–46 (1997)
15. de Teixeira, L.F.H.S.B., dos Santos, M., Gomes, C.F.S.: Python proposal and implementation of the Simple Aggregation of Preferences Expressed by Ordinal Vectors-Multi Decision Makers method: a simple and intuitive web tool for Decision Support. Simpósio Pesqui Operacional e Llogística da Mar **19**, 1–12 (2019)
16. dos Santos, M., de Oliveira, N.C., de Oliveira, P.F.C., Gomes, C.F.S.: Application of the SAPEVO-M multicriteria method in the selection of equipment: a case study in a bakery in RJ. An do XIX Simpósio Pesqui Operacional e Logística da Mar Rio Janeiro, RJ (2019)
17. Grego, T., Santos, M., Gomes, C.F.S., Lima, A.R.: Choosing a Troop Landing Ship for the Argentine Navy using the SAPEVO Method with Multiple Decision Makers (SAPEVO M). An do XXI Simpósio Apl Operacionais em Áreas Defesa-SIGE (2019)
18. Teixeira, L.F.H.S.B., dos Santos, M., Gomes, C.F.S.: SapevoWeb Software (v.1), sob registro INPI: BR512020000667–1 [Internet] (2018). Accessed 14 Aug 2020, https://www.sapevo web.com/
19. Duckstein, L., Opricovic, S.: Multiobjective optimization in river basin development. Water Resour. Res. **16**(1), 14–20 (1980)
20. Opricovic, S., Tzeng, G.: Multicriteria planning of post-earthquake sustainable reconstruction. Comput. Civil Infrastruct. Eng. **17**(3), 211–220 (2002)
21. Opricovic, S., Tzeng, G.-H.: Compromise solution by MCDM methods: a comparative analysis of VIKOR and TOPSIS. Eur. J. Oper. Res. **156**(2), 445–455 (2004)
22. Vikor Software. Vikor Software [Internet] (2020). Accessed 14 Aug 2020, https://soft.onl ineoutput.com/vikor?entry=%2FK7rpuRJ5TI14mDYRROxySKDAxLHA7dlp67mE18o2 pg%3D

Adaptation of the Balanced Scorecard to Latin American Higher Education Institutions in the Context of Strategic Management: A Systematic Review with Meta-analysis

Mauricio F. Hinojosa V.(✉) 📵

Universidad de Santiago de Chile, 9160000 Santiago, RM, Chile
mauricio.hinojosa@usach.cl

Abstract. The objective of this research is to examine from a scientific perspective how the use of the Balanced Scorecard has been adapted in the various types of higher education institutions in Latin America within the context of their strategic management, through a systematic review with meta-analysis, in order to contribute to the development of strategic processes in these organizations, given the relevance they have acquired, based on the new state regulations and the speed of social and cultural changes in the region. To achieve this, a statistical analysis was carried out, first to identify the most frequent implementation techniques and the type of institution in which the studies were conducted; and, secondly, to examine the underlying interrelationships between the variables involved in the strategic context through the chi-squared test. The results showed a scarcity, but incipient evidence from the year 2000, of works mostly concentrated in universities of three countries, both in public and private organizations, with greater use of the original structure of Kaplan and Norton, and of the perspective "Customer". With the meta-analysis, no relationship was found between the ownership of the organization and, on the one hand, the structure of strategic dimensions used in the comprehensive scorecard and, secondly, with the perspectives used.

Keywords: Balanced Scorecard · Strategic management · Latin American higher education institutions · Systematic review · Meta-analysis

1 Introduction

In the same way that it has occurred in the most developed countries of the world, the industrial sector of higher education (HE) in Latin America is now going through a time of rapid and great changes, due in large part to the transition from an industrial society to the knowledge society [1]. Now higher education institutions (HEIs) must face greater competitiveness in their markets, more empowered students, universalization of education and a new legal and regulatory framework. There is a transition, with different nuances between countries, towards expansion and then towards quality[1], phenomena that developed countries began to experience at the beginning of the 20th century [2].

---

[1] In another aspect of diagnosis there is also evidence of a transition from the elite to inclusion.

© Springer Nature Switzerland AG 2021
D. A. Rossit et al. (Eds.): ICPR-Americas 2020, CCIS 1408, pp. 125–140, 2021.
https://doi.org/10.1007/978-3-030-76310-7_10

With regard to expansion, in most Latin American countries there is strong growth in access to HE and enrollment, especially in the lower socioeconomic strata [3], presenting a large variance, which depends on factors such as educational spending as a percentage of gross domestic product and the nation's budget, and on the Net Enrolment Rate in secondary education[2] [1].

Because the States are not able to satisfy all the need for SE, now this service is shared by public and private institutions [2], whose difference, according to international standards[3] [4], lies in whether it is public or private the entity that has the ultimate power to make decisions about the institution's affairs[4]. Worldwide, 33% of students are enrolled in private HEIs [5], and an accelerated growth of private HE [6] is reflected in a growing Gross Enrollment Rate[5] in Latin America [7], an area that has the highest participation worldwide, with 48% of its enrollment in private HEIs[6] [2]. At one end of the spectrum are countries such as Brazil, Chile, El Salvador, Peru and the Dominican Republic, concentrating their enrollment in the private sector, and at the other end Cuba, without private registration [8]. A reflection of the above is the decrease in public investment per student in Latin America [3].

On the other hand, the transition to quality has implied the establishment of new state policies, which tend to order and try to regulate an increasingly heterogeneous and complex market, which have resulted in changes at the legislative level, in the contribution of state funds, and in the creation of evaluating and supervising bodies [9]. Opposite examples of this reality are represented by Mexico, as a case of less regulation, and Chile, as an example of greater regulation.

Within this context are the accreditation and/or quality assurance systems[7], which oblige HEIs to carry out strategic processes, execute the minimum functions required by each State, carry out performance and quality evaluations based on standardized indicators, and accountability (delivery of information, compliance with standards, transparency) [10]. For example, in Chile, the HE law [11] establishes strategic management as one of the dimensions to be accredited, and also the obligation to carry out a process of institutional strategic planning, which generally it is documented in a strategic development plan.

One of the tools used by HEIs in their strategic processes is the Balanced Scorecard (BSC), which was presented by Kaplan and Norton 27 years ago [12], based on

---

[2] It is defined as the total number of students in the theoretical age group for the secondary level, enrolled at that level, expressed as a percentage of the total population in that age group [29].

[3] Each country provides information on private and public HEIs to international organizations such as UNESCO [5] according to its own taxonomy. In addition, HEIs differ in terms of owner, mission, main source of income, student aid system, quality control system, institutional governance mode, and academic staff management system, with a different ideological-cultural meaning depending on the country [30].

[4] Private institutions in turn are classified as "dependent" and "independent" of the government, which only refers to the degree of dependence of an institution on government financing [4].

[5] In higher education, the Gross Enrolment Rate is defined as the total number of students enrolled in tertiary education, expressed as a percentage of the population in the five-year group immediately after secondary education, typically 18 to 22 years old [27].

[6] In Europe in 2010 it was only 15% [31].

[7] Almost every nation in the region now has an accrediting agency.

the integration and complementation of the literatures on quality, efficient management, financial economics, and stakeholder theory. It started as a tool to measure the performance of a company through financial and non-financial measures that promote the creation of long-term value for its shareholders, based on four perspectives: financial, customers, internal business, and innovation and learning ago [12]. Soon its authors realized that the priority in companies was to implement their strategy and that, therefore, the first thing was not to generate the metrics, but rather the logic is to describe the strategy through strategic objectives, based on the structure of the four perspectives, which finally leads to the selection of the ideal indicators [13]. Then, with the incorporation of the strategy map, the focus is taken away from the indicators, and put on the objectives and the relationship between them [14].

Starting in the mid-90s, the BSC began to be applied in both public institutions and non-profit organizations, so Kaplan and Norton generated an adaptation of the BSC to this type of organizations [14], which do not measure their performance with financial indicators, but through effectiveness to generate benefits to the recipients. As seen in Fig. 1, the financial perspective disappears from the first level, as it is not its main or final objective, being replaced by objectives related to its impact and mission; and the client perspective is broadened to donors, who provide financial resources (pay for the service), and to clients or beneficiaries, who are the recipients of the service[8].

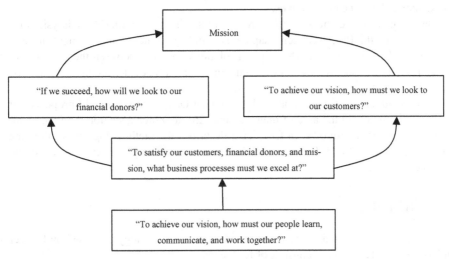

Fig. 1. Adaptation of the BSC to non-profit organizations (Source: Kaplan & Norton [14]).

Then came the creation of the concept of Organization focused on strategy [14], based on 5 principles [15], and an enrichment of the methodology in 2004 [16], but it is the adaptation to other types of organizations in 2001, which should lead to a greater use of the BSC in ES.

---

[8] Faced with the question of who is the client, a cross-cutting problem for non-profit and public organizations, instead of opting for one or the other, in practice the perspectives of the donor and the recipient have been placed at the same level [14].

A review of articles in ISI Web and Scopus [17] shows that in 2006 25% of the total studies found used the BSC in the context of strategic management, that only 3.8% of the total correspond to applications in educational institutions and no evidence of work was found in Latin America; there is little evidence on the application of the BSC in HEIs, compared to companies [18].

From the need to know how the BSC is being applied in HE, the idea of generating a systematic review[9] was born, whose advantage over a narrative review of the literature is that it reduces bias and error, allowing the generalization of scientific findings and supporting the decision making [19].

An Internet exploration through 8 databases (Scopus, Web of Sicence, Springer, Jstor, Science Direct, Wiley online Library, Emerald, Taylor & Francis online) and a search engine (Google Scholar), did not provide evidence of systematic or narrative reviews focused on the application of the BSC in Latin American HEIs. There are close studies, such as, for example, a narrative review on the practical application of the BSC in different industries [20], another on the application of the BSC considering environmental, social or ethical problems, called sustainability BSC [21], a systematic review of the strategies in non-profit organizations [22], and without being a review, an article from this year [23], in which the influence of the BSC on the research and innovation performance of six Latin American universities was studied. Of all the reviews found, the most related is a narrative review that focuses on the context of the BSC in IES [24], which identifies and contextualizes the perspectives.

Based on the above, this systematic review of the literature, with meta-analysis, aims to analyze how the BSC has been adapted as a tool for strategic management in Latin American HEIs, discovering what type of relationship exists between the variables of the tool and the characteristics of the organizations who have evidenced their application through publications.

This document is structured according to what Denyer and Tranfield proposed for a systematic review in the area of management [25], to which a section with the meta-analysis is added, as follows: in Chapter 2 describes the methodology used, then the results are presented and discussed, in Sect. 4 the meta-analysis is evidenced, to end with the conclusions.

2 Methodology

The systematic review was carried out based on the methodology proposed by Denyer and Tranfield[10] [25], which consists of four steps.

2.1 Question Formulation

The basic question is: in what way has the BSC been applied in strategic processes of different types of Latin American HEIs?

---

[9] Although there is no single definition, it can be established that a systematic review is an academic research article that synthesizes studies through planned, explicit, transparent, and repeatable procedures to answer predefined questions.

[10] This methodology is applied since it focuses on the field of management.

Reformulating the question according to the CIMO logic [25], acronym of Context, Intervention, Mechanism and Outcome, it results: Are the characteristics of an HEI (context) a determining factor (intervention) in the definition of the perspectives and objectives of the BSC (outcome)? How (mechanism) affect the design of the BSC (result)?

2.2 Location of Studies

The exploration was carried out using the 9 search engines mentioned in the previous section, to which the academic platform Academia.edu was added, using keywords, Boolean operators, truncation, and special characters, to the extent that the search engine allowed it Table 1.

Table 1. Techniques used to search for studies.

Search identifier	Search settings
A	balanced scorecard higher education strategic management
B	cuadro de mando integral educación superior gestión estratégica
C	(TITLE ("balanced scorecard") AND ABS (higher AND education AND strateg* AND management) OR TITLE-ABS-KEY (strateg* AND management AND universit*))
D	"strategic management" argentina perspectives financial learning internal process customer indicators "balanced scorecard" "higher education institution"
E	strategic management argentina perspectives indicators "balanced scorecard" "higher education institution"
F	strategic management perspectives indicators argentina "higher education institution" "balanced scorecard"
G	strategic management perspectives indicators chile "higher education institution" "balanced scorecard"
H	gestión estratégica perspectiva indicador "institución de educación superior" "cuadro de mando integral"
I	gestión estratégica perspectiva indicador educación OR superior "cuadro de mando integral" -empresa -salud

An exploration was carried out with the keywords in English and Spanish, and then a more detailed search was carried out, testing different combinations, discarding the first since they were very inefficient (produced more than 500 results), using one at a time the name of different countries, until the use of the Spanish language without country restriction generated positive search results.

2.3 Study Selection and Evaluation

The following three selection criteria were used.

The first is that the BSC was applied in a Latin American HEI. Latin American countries are defined as: Argentina, Bolivia, Brazil, Chile, Colombia, Costa Rica, Cuba, Ecuador, El Salvador, Guatemala, Honduras, Mexico, Nicaragua, Panama, Paraguay, Peru, Dominican Republic, Uruguay and Venezuela [10].

It is understood "application of the BSC" according to the definition made by its creators [12], in which perspectives, objectives and indicators are identified. Therefore, if any of the first two is not present, the document is not considered.

Finally, the application must be about an organization that provides HE services, regardless of the type of degree or title that it delivers, its owner/administrator, or its teaching modality; but parts of it are not considered, that is, academic units, departments, centers, institutes, faculties, among others, nor are the institutions that train members of the armed forces.

The second criterion is that the BSC was used within the context of a strategic process. Applications that include strategic objectives are considered.

And the last criterion was that the HEI was identifiable in terms of the owner, destination of its benefits and type of institution. Are considered both public and private institutions, for-profit and non-profit institutions, and those that teach study programs levels 5 and/or 6 according to international standards [26].

2.4 Analysis and Synthesis

A form was used for data extraction, with the following aspects:

- Publication: Search engine - Year - Research name - Author (s) - Type of publication (paper, thesis, conference proceeding, degree or professional seminar, monograph, working paper, or dissertation) – Country.
- Research design: Research objective - Name of the Institution - Type of organization according to its owner (public or private) - Type of HEI according to its mission (university, technical institute or professional institute) - Number of students – Institution's age - Participation of representatives of the organization in the study.
- Application of the BSC: - Map (yes/no) - Perspectives - Strategic objectives - Number of indicators.

2.5 Report and Use of the Results

Finally, the report is generated, in which the research is contextualized and justified, the results obtained are presented, the analysis is carried out and the research questions are answered.

To this is added a meta-analysis, in which the studies are statistically combined, which allows a more technical analysis with a greater scope.

3 Findings and Discussion

Once the search was carried out through the procedure described, it yielded different results with the different search engines.

From the review of the documents applying the selection and evaluation criteria described above, positive results were obtained only through the Google Scholar search engine and the Academia.edu platform. The exploration revealed a total of 17 documents that met the established selection criteria, using H and I techniques (Table 1).

In publication aspect, the distribution of researches over time (Fig. 2) shows that it was not until 2000 that the first evidence of a study was found in which the BSC was applied to a strategic planning process in a Latin American HEI, with a slight increasing trend towards 2018.

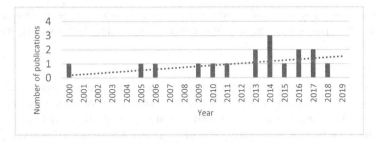

Fig. 2. Number of documents found (January 2000–October 2019)

Regarding the country of origin of these documents, the majority came from Colombia, Ecuador and Venezuela, representing almost 80% of the publications found (Fig. 3).

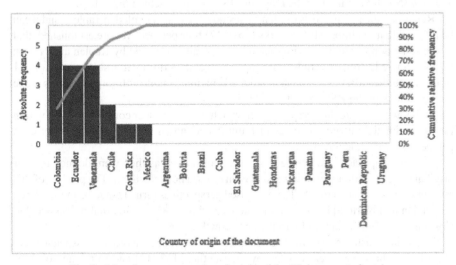

Fig. 3. Most frequent countries of origin of the 17 documents found.

The documents found are of the following classes: Degree Seminar, Paper, Thesis, Monograph, Professional Seminar, and Conference Publication. In Fig. 4 it can be observed that most of the works (64%) correspond to degree seminars and papers.

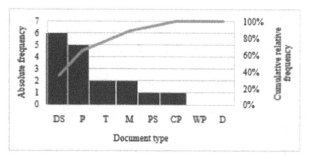

Fig. 4. Most frequent types of document (DS = degree seminar, P = paper, T = thesis, M = monograph, PS = professional seminar, CP = conference proceeding, WP = working paper, D = dissertation).

Categorizing according to the level of education of the organizations in which the research was carried out, by a large majority the researches were carried out in universities, with 70.6%, and the rest correspond to non-university HEIs. And according to the classification by type of property, 59% of the researches were carried out in public institutions and the remaining 41% in private organizations.

Within the aspects related to the research design, such as the number of students, the age of the institution and the participation of representatives of the organization in the study, they could not be considered because only partial information was found. Something similar with aspects of the application such as the strategy map or the indicators; the analysis could not be made considering that very few valid documents were found, and that 35% did not include the map, and 18% did not include indicators.

Regarding the perspectives used in each study, these are detailed in Table 2, and it can be inferred that in almost all the works (16 of 17) four perspectives are established, that is, the same number of dimensions as those originally proposed by Kaplan and Norton [12, 13]; there is only one study that proposes five. In addition, it is striking that only 1 of the 16 studies after 2001 considers the adaptation to non-profit organizations made by the authors of the BSC in that year [14].

To know which are the perspectives in which there is greater consensus to use within the institutional strategic processes of Latin American HEIs, the frequencies for each perspective were calculated.

In a first approach, the perspectives were considered as they were defined in their original studies, without applying association criteria between them. The product of this raw analysis is shown in Fig. 5. The most used perspectives turned out to be the 4 defined by Kaplan and Norton [12, 13], and in the following order: 1st Internal processes, 2nd Costumer, 3rd Financial, 4th Learning and growth.

Then, an adjustment was made to the perspectives found, in order to associate them according to how they were defined by their authors [12, 13] and to the approach or explanation given for each of them by the authors of the respective studies, obtaining the result of Table 3. From these it can be inferred that all the perspectives used in the researches can be associated with some perspective of the original Kaplan and Norton BSC. This link is summarized in Figs. 6 and 7.

Table 2. Documents found of the application of the BSC in strategic processes of Latin American HEIs.

Year	Author(s)	Name of research	Perspectives
2018	Barrera, Juan.	Propuesta de Diseño de una herramienta de gestión para fortalecer la administración y facilitar la toma de decisiones Universidad San Buenaventura Cali.	1) Financial 2) Customer 3) Internal processes 4) Learning and innovation
2017	Boglio, Alejandro.	Propuesta de un sistema de Control de Gestión para el Centro de Formación Técnica Juan Bohón - La Serena	1) Financial 2) Customer 3) Internal processes 4) Learning and knowledge
2017	Guevara, Olimar.	Cuadro de Mando Integral como herramienta estraté-gica para la efectividad de la gestión universitaria	1) Financial legal 2) Customers External/internal 3) Internal processes 4) Learning and growth
2016	Mendivil, Guillermo	Diseño de un modelo de gestión con las herramientas del Cuadro de Mando Integral (CMI) o Balanced Scorecard para el Instituto Nacional de Formación Técnica Profesional de San Andrés, INFOTEP	1) Community 2) Customer 3) Internal processes 4) Learning and growth
2016	Panesso, Carlos; Jaramillo, Ana.	Modelo Balanced Scorecard para el Instituto de Educación Técnica Profesional INTEP de Roldanillo Valle del Cauca	1) Society 2) Customer 3) Internal processes 4) Learning and growth
2015	Gamarra, Mauricio; Poveda, Eidis.	Actualización del Balanced Scorecard de la Funda-ción Universitaria Tecnológico COMFENALCO	1) Customer 2) Processes and projects 3) Learning and growth 4) Financial
2014	Villacís, Allison.	Diseño del Prototipo para un Sistema Automatizado de Administración de Indicadores de Gestión de Calidad aplicado en una Institución de Educación Superior (IES)	1) Student training 2) Teacher development 3) Research 4) Relationship with society
2014	Samaniego, Víctor.	Medir la capacidad del desempeño administrativo y académico basado en el plan estratégico del Instituto Bolivariano.	1) Financial 2) Customer 3) Internal processes 4) Learning and growth
2014	Arias, David; Vélez, Yadira.	Plan estratégico y Balanced Scorecard para la PUCE SD, periodo 2014-2018	1) Financial + Customer 2) Internal processes 3) Learning and growth
2013	Armas, Gladys.	Cuadro de Mando Integral como herramienta para el seguimiento del Plan Estratégico de la Universidad José Antonio Páez	1) University management 2) Students and community 3) Internal processes 4) Infrastructure and technology 5) Human talent
2013	Alvarado, Luis; Aguilar, Alfredo; Cabral, Agustín; Alvarado, Tomás.	Diseño de un sistema de planificación estratégica basado en el Balanced Scorecard: El caso de la Antonio Narro	1) Finance 2) Customer 3) Internal processes 4) Training and growth
2011	Muñoz, Jorge.	Balanced scorecard-BSC o cuadro de mando integral-CMI para la Universidad Nacional de Loja	1) Students 2) Financial 3) Processes 4) Learning and growth
2010	Vega, Mayela.	El modelo de planificación estratégica en la Universi-dad Nacional 2004-2009 y una propuesta de un nuevo modelo basado en el uso del Cuadro de Mando Integral	1) Priority population 2) Financial management 3) Substantive rationality 4) Human talent
2009	Alana, Bolivar.	Planificación Estratégica Para La Universidad Adventista De Chile, 2005-2007	1) Institution 2) Customer 3) Internal processes 4) Learning and development
2006	Bastidas, Eunice; Moreno, Zahira.	El Cuadro de Mando Integral en la gestión de las organizaciones del sector público: El caso Universi-dad Centroccidental Lisandro Alvarado.	1) State 2) Society 3) Internal processes 4) Organizational learning and devel-opment
2005	Arias, Leonel; Castaño, Juan; Lanzas, Angela.	Balanced Scorecard en instituciones de educación superior	1) Society 2) Customer 3) Processes 4) Training and growth
2000	López, Yoleiza.	Hacia un Cuadro de Mando Integral para el Instituto Universitario Experimental de Tecnología "Andrés Eloy Blanco"	1) Financial 2) Customer 3) Internal processes 4) Training and growth

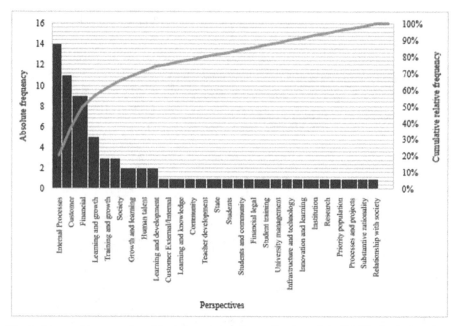

Fig. 5. Perspectives most found in documents on the application of the BSC in Latin American HEIs in the context of their strategic planning.

It can be deduced, from these graphs, that the "Customers" dimension is the most considered of the four perspectives (33%) to generate strategic objectives, and at the same time the "Financial" perspective is the least considered (16%).

In addition, if the structure of perspectives is analyzed, that is, at what levels are located each of the strategic areas (Fig. 7), the most frequently used is the one originally suggested by Kaplan and Norton for the generation of the BSC in companies, this is: First level: Financial, Second level: Costumer, Third level: Internal processes, Forth level: Learning, development, and innovation.

The only case (mentioned above) that structures its perspectives according to the suggestion of Kaplan and Norton for non-profit organizations stands out within all the studies, that is: First level: Financial + Costumer, Second level: Internal processes, Third level: Learning, development, and innovation.

Table 3. Linking the perspectives established in the BSC applications found with the 4 original perspectives of Kaplan and Norton.

Kaplan and Norton perspectives	Associated perspectives of documents
Learning, development, and innovation	Learning and knowledge Learning and growth Learning and development Growth and learning Training and growth Innovation and learning Teacher development Human talent Infrastructure and technology
Customer	Customers Community Society Relationship with society State Students Students and community Institution Research Priority population
Financial	Financial Financial management University management
Internal processes	Internal processes Processes and projects Student training Substantive rationality

Fig. 6. Use of the original 4 perspectives of Kaplan and Norton in the documents found.

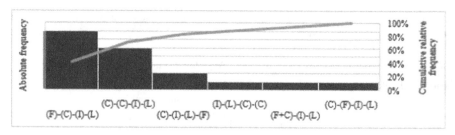

Fig. 7. Structure of perspectives most used, where: F = Financial, C = Customers, I = Internal processes, and L = Learning, development, and innovation.

4 Meta-analysis

Given the need to relate nominal qualitative variables, contingency tables were developed to generate two analyzes based on the chi-square test.

4.1 Relationship Between the Property of the HEI and the Strategic Level of the Financial Perspective

The first goal was to find out if there is a relationship between two variables: the "property of the HEI" and the "level at which the Financial perspective is situated within the structure of the BSC".

For this, the following null hypothesis was raised:

H0 = The ownership of the HEI and the strategic level at which the financial perspective is situated are independent variables.

First the corresponding contingency table was designed Table 4, composed of the observed frequencies fij, for the financial perspective in public and private HEIs.

Table 4. Contingency table for the location of the financial perspective in public and private HEIs.

Owner of the HEI	Location of the Financial perspective			
	Level 1	Level 2	Level 3	Level 4
Public	3	2	0	0
Private	5	0	0	1

From this, the expected frequencies eij were obtained, to finally calculate the observed chi-square statistic using the expression (1):

$$\chi^2 = \sum_i \sum_j \frac{\left(f_{ij} - e_{ij}\right)^2}{e_{ij}} \tag{1}$$

The theoretical value of chi-square, 5,99, was obtained for a probability of 5%, and only 2 degrees of freedom of the original 3, since the third column was not considered, in which the expected frequencies were zero, product that the financial field in none of the applications was considered for level 3 of the BSC.

The observed result was, then, $\chi^2 = 3{,}44$, less than 5,99, so the null hypothesis is not rejected and, therefore, it can be stated that the variables "Property of the HEI" and the "Level at which the Financial perspective is situated within the structure of the BSC", are independent.

4.2 Relationship Between the Property of the HEI and the Perspectives Used

In a second analysis, the objective was to find out if there is a relationship between the "Property of the IES" and the "Perspectives used".

For this, the following null hypothesis was raised:

H0 = The ownership of the HEI and the perspectives used are independent variables.

Table 5 shows how the different original perspectives of Kaplan and Norton were used in private and public HEIs. From this, the corresponding contingency table was constructed (Table 6), composed of the observed frequencies fij for the different perspectives in public and private HEIs, which served as the basis to obtain the expected frequencies eij.

Table 5. Absolute frequencies of use of the perspectives in the different levels of the BSC according to the property of the HEI.

	Financial				Costumers				Internal processes				Learning, development, and innovation			
	1	*2*	*3*	*4*	*1*	*2*	*3*	*4*	*1*	*2*	*3*	*4*	*1*	*2*	*3*	*4*
Owner of the HEI/Level																
Public	3	2	0	0	6	7	1	1	1	0	9	0	0	1	0	9
Private	5	0	0	1	3	5	0	0	0	2	5	0	0	0	2	5

Table 6. Contingency table for the use that is given to the different perspectives in public and private HEIs.

Owner of the HEI/Perspective	*Financial*	*Customer*	*Internal processes*	*Learning, development, and innovation*
Public	5	15	10	10
Private	6	8	7	7

The chi-square statistic, calculated through (1), gave a value of 1,20.

Then, the theoretical value of chi-square 7.81 is obtained, for a probability of 5% and 3 degrees of freedom.

The result, then, is that $\chi^2 = 1,20$ is less than 7,81, so the null hypothesis is not rejected and, therefore, the variables "Property of the IES" and "Perspectives used", are independent.

5 Conclusion

There is a growing use of the BSC in support of the strategic processes of the institutions, which are key to the consolidation and sustainability of them, considering the demands posed by the new regulations in the HE systems (accountability, quality, strategic development plans, among others).

The association of the perspectives used with the original ones by Kaplan and Norton shows that the perspective that was most considered in the different strategic levels of the BSC was that of "Costumers" (in its broadest sense), evidencing the concern that HEIs have for their stakeholders[11].

HEIs tend strangely to reproduce the original structure of the Kaplan and Norton BSC model, originally aimed at companies, and not at educational services organizations, where the "Financial" perspective is established at the first strategic level. At the opposite hand, there were cases that did not consider the financial dimension, which, according to the meta-analysis, is not related to the ownership of the HEI; based on the application of the Chi-square test, there was no evidence of a relationship between ownership of Latin American HEIs (public or private) and the level at which it is situated, or whether or not it represents the cornerstone of the strategy the "Financial" dimension; nor is there a specific structure of perspectives considered that reflects the differences between public and private institutions of HE.

Based on the above, it is striking that the adaptation to non-profit organizations, carried out by their authors, which recognizes the importance of mission in HEIs, has not had a great impact on the strategic planning processes of HEIs, since its appearance 19 years ago. This opens up a field of development for these processes and shows that more experience is needed in Latin America and, above all, knowledge of the results obtained from their use, in order to generate scientific conclusions about the impact that the BSC is having in this strategic and fundamental sector for the development of people and societies.

On the other hand, the systematic review and meta-analysis allowed to summarize part of the strategic planning of Latin American HEIs and, adapted to their context, they are projected as very useful and reliable tools to observe how management in higher education will continue to develop.

It is concluded, then, that the incipient use observed of the BSC in Latin America is independent of the type of HEI or the system in which it is immersed, becoming an instrument of transversal management, in a strategic process that is fundamental for an adequate management of these organizations in the current complex and challenging context.

---

[11] Internal stakeholders: students, academics, administrators, and managers. External stakeholders: employers, former students, parents and guardians, society in general, the government and its agencies, education providers, and non-governmental organizations [28].

References

1. Bellei, C.: Situación Educativa de América Latina y el Caribe: Hacia la educación de calidad para todos al 2015. OREALC/UNESCO, Santiago (2013)
2. Orellana, V.: Current situation and challenges of higher education in Latin America and the Caribbean. Oficina de Santiago **6**, 1–14 (2014)
3. Ferreyra, M.M., Avitabile, C., Botero Álvarez, J., Paz, F.H., Urzúa, S.: At a Crossroads: Higher Education in Latin America and the Caribbean. Directions in Development. World Bank, License: Creative Commons Attribution CC BY 3.0 IGO, Washington, D.C. (2017). https://doi.org/10.1596/978-1-4648-0971-2
4. UIS/OECD/EUROSTAT: UOE Data Collection on Education Systems - Manual: Concepts, definitions and classifications. UOE, Paris, vol. 1 (2005)
5. Levy, D.: Private Higher Education. In: Shin J., Teixeira P. (eds.) Encyclopedia of International Higher Education Systems and Institutions. Springer, Dordrecht (2016). https://doi.org/10.1007/978-94-017-9553-1_28-1.
6. Barberá de la Torre, R., González, S.J.: Diagnóstico de la Educación Superior en Iberoamérica. Organización de Estados Iberoamericanos para la Educación, la Ciencia y la Cultura (OEI), Madrid (2019)
7. UNESCO: Global education monitoring report. Accountability in education: meeting our commitments, París: 1st ed. UNESCO (2017)
8. García de Fanelli, A.: Panorama de la educación superior en Iberoamérica - Edición 2019. Observatorio Iberoamericano de la Ciencia, la Tecnología y la Sociedad de la Organización de Estados Iberoamericanos (OCTS-OEI), Buenos Aires (2019)
9. Villanueva, E.: Perspectivas de la educación superior en América Latina: Construyendo futuros. Perfiles Educativos, XXXII **129**, 86–101 (2010)
10. UNESCO: Latin America and the Caribbean: Education for All 2015 regional review. Santiago (2014)
11. Ley 21091: Sobre Educación Superior (2018)
12. Kaplan, R.S., Norton, D.P.: The balanced scorecard - Measures that drive performance. Harvard Bus. Rev. 70–80 (1992)
13. Kaplan, R.S., Norton, D.P.: The Balanced Scorecard: Translating the strategy into action, 1st edn. Harvard Business School Press, Boston (1996)
14. Kaplan, R.S., Norton, D.P.: The Strategy-Focused Organization: How Balanced Scorecard Companies Thrive in the New Business Environment, 1st edn. Harvard Business School Publishing Corporation, Boston (2001)
15. Kaplan, R.S.: Conceptual foundations of the Balanced Scorecard. In: Handbook of Management Accounting Research. 1st ed. Elsevier, Amsterdam, vol. 3, 1253–1269 (2009)
16. Kaplan, R.S., Norton, D.P.: Strategy Maps: Converting Intangible Assets into Tangible Outcomes, 1st edn. Harvard Business Review Press, Boston (2004)
17. Massón, J.L., Truñó, J.: La Cuarta generación del Balanced Scorecard: Revisión Crítica de la Literatura Conceptual y Empírica. Universidad Autónoma de Barcelona, May 2006
18. Beard, D.F.: Successful applications of the balanced scorecard in higher education. J. Educ. Bus. **84**(5), 275–282 (2009)
19. Cook, D.J., Mulrow, C.D.: Systematic reviews: synthesis of best evidence for clinical decisions. Ann. Intern. Med. **126**(5), 376–380 (1997)
20. Hasan, R.U., Chyi, T.M.: Practical application of balanced scorecard - a literature review. J. Strategy Perform. Manage. **5**(3), 87–103 (2017)
21. Hansen, E., Schaltegger, S.: The sustainability balanced scorecard: a systematic review of architectures. J. Bus. Ethics **133**, 193–221 (2016)

22. Laurett, R., Ferreira, J.J.: Strategy in nonprofit organizations: a systematic literature review and agenda for future research. Voluntas: Int. J. Voluntary Nonprofit Organ. **29**(5), 881–897 (2017)
23. Peris-Ortiz, M., García-Hurtado, D., Devece, C.: Influence of the balanced scorecard on the science and innovation performance of Latin American universities. Knowl. Manage. Res. Pract. **17**(4), 373–383 (2019)
24. Al-Hosaini, F.F., Sofian, S.: A review of balanced scorecard framework in higher education institutions (HEIs). Int. Rev. Manage. Mark. **5**(1), 26–35 (2015)
25. Denyer, D., Tranfield, D.: Producing a systematic review. In: The SAGE handbook of organizational research methods. SAGE Publications Ltd., London, pp. 671–689 (2009)
26. UIS: International Standard Classification of Education: ISCED 2011. UNESCO Institute for Statistics, Paris (2012)
27. OECD: Education at a Glance 2019: OECD Indicators. OECD Publishing, Paris (2019)
28. Marshall, S.J.: Shaping the University of the Future: Using Technology to Catalyse Change in University Learning and Teaching. 1st ed. Springer, Wellington (2018). https://doi.org/10.1007/978-981-10-7620-6.pdf
29. UIS: UNESCO Institute of Statistics (2020). https://uis.unesco.org/en/glossary. Accessed 14 Aug 2020
30. Brunner, J.J., Miranda, D. (eds.): Educación Superior en Iberoamérica: Informe 2016. 1st ed. Ril editores, Santiago (2016)
31. PROPHE: The Program for Research on Private Higher Education (2010). https://www.prophe.org/en/global-data/regional-tables/regional-summary-overview-2010/. Accessed 14 Aug 2020

Project Management Process Resilience: Assessing and Improving the Project Review Process Using FRAM

Vinícius Bigogno-Costa$^{(\boxtimes)}$ ⓘ, Moacyr Machado Cardoso Jr. ⓘ,
and Ligia Maria Soto Urbina ⓘ

Instituto Tecnológico de Aeronáutica, Pça. Mal. Eduardo Gomes, 50, São José dos Campos
12228-900, Brazil
{viniciusvbc,moacyr,ligia}@ita.br

Abstract. Technology development projects are constantly subject to risks and uncertainties and their performance are key to deliver the organizational value. Failing to meet project goals and objectives is critical, therefore, projects must be able to cope with an environment of complexity and pressure. This paper aims to assess and improve project resilience by using the Functional Resonance Analysis Model (FRAM) to look at project review process, a stage-gate milestone typical to Systems Engineering process. FRAM is supported by the Resilience Engineering perspective that things go wrong by the same reasons they go right, and everyday work is subject to variability and small adjustments, which is key to address such project environments. After assessing sources of variability and aggregating them, means to manage variability are proposed order to increase process resilience.

Keywords: Project Management · Resilience Engineering · FRAM

1 Introduction

Technology-intensive projects are subject to complex environments and high levels of risk and uncertainty and require multidisciplinary approaches to meet their strategic objectives [1]. Resilience engineering focuses on helping people to achieve success while coping with complexity under pressure [2]. If risk management is entirely associated with Project Management and Systems Engineering [3], project resilience is one of the emerging topics in the area [4] and, along with Resilience, Resilience Engineering is described by Aven [5] as recent advances in the foundations of risk assessment and management.

This work aims to assess and improve project resilience in the context of the Institute of Aeronautics and Space (Instituto de Aeronáutica e Espaço, IAE), an organization from the Brazilian Air Force that defines its mission as "to develop technological and scientific solutions to strengthen Brazilian Aerospace power through research, development, innovation […] in aeronautical, space and defense systems" [6]. IAE, as a project-oriented organization, uses Systems Engineering, a process-oriented product life

D. A. Rossit et al. (Eds.): ICPR-Americas 2020, CCIS 1408, pp. 141–155, 2021.
https://doi.org/10.1007/978-3-030-76310-7_11

cycle management approach that enable developers to evolve from customers' needs to product delivery [7]. Systems Engineering may be seen either as a function within the Project Management process [8], or a whole process that intersect with Project Management functions [9]. Project reviews, or design reviews, are gates that set decision-making milestones and determine whether the project is heading towards its original goals and the design has met criteria for the corresponding Systems Engineering process phase, aiding project control process and reducing risks and uncertainties during the development [10]. This work aims exclusively on project reviews.

All these processes are subject to variability and small adjustments, given that every project is unique. Functional Resonance Analysis Method (FRAM) aims to describe everyday performance model performance variability in complex dynamic socio-technical systems [11] and may be used in order to assess and find ways to manage variability and improve resilience within an interest system. Although originally imagined by Eric Hollnagel for modelling socio-technical systems [12], in this work FRAM will be applied in an essentially organizational process: the project review process.

This paper is divided into 6 sections. In Sect. 1, a brief introduction to Resilience, Variability and FRAM will set the scene for the case study, the project review process in IAE, which will be discussed in Sect. 2. Initial steps of FRAM will be applied for modelling work-as-done and assessing variability in Sect. 3, then different scenarios for instantiations will be considered for later steps to establish ways to manage and improve resilience management of the process in Sect. 4. Finally, an overview of the work as done in the conclusion and some insights for future works are provided.

2   Resilience, Variability and FRAM

2.1   Resilience and Resilience Engineering

Resilience and Resilience Engineering address issues with respect to the safety of systems, not by looking to things that go wrong, as usually do traditional approaches to safety, but by also looking to things that go right. It emerges as a consequence of understanding that both situations derive from the same circumstances: performance in sociotechnical contexts vary with time [13] and, as a conclusion, Resilience Engineering defines safety as "the ability to succeed under varying conditions" [14]. The perspective given by resilience is key to address modern context, in which traditional risk frames alone may not be enough and, therefore, both approaches should be considered together [15].

Resilience is supported upon four cornerstones, as seen on Fig. 1, that can be improved within the context of a system [14]: responding (actual), which is dealing with what is already expected or considered normal behavior within minimal adjustments; monitoring (critical), that is observing the environment and the system in order to find threats; anticipating (potential): expecting potential developments, opportunities, and threats; and learning (factual): acquiring the right experience from both successes and failures.

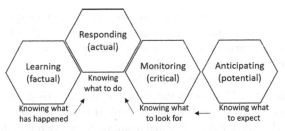

Fig. 1. The four cornerstones of resilience engineering [14, p. xxxvii]

2.2 Functional Resonance Analysis Method– FRAM

Functional Resonance Analysis Method (FRAM) was proposed by Erik Hollnagel in 2004 and developed in congruence with the principles of Resilience Engineering [11]. It introduces two key concepts: work-as-imagined (WAI) and work-as-done (WAD). Work-as-imagined can be understood as the task list or the process flowchart, in which things must be done the way it was meant to, in a linear thinking. Work-as-done perspective, on the other hand, provides the understanding that operators adjust their everyday operation with variations and small adjustments conditioned to the environment and other circumstances and, therefore, perform their work not necessarily according to the WAI, yet do not incur in a failure, from a traditional safety perspective [12].

FRAM defines system as "a set of coupled or mutually dependent functions" [12, p. 17] and model the relationship between these functions from a systemic perspective, which means properties may emerge from variability and resonate within the system. Functions in FRAM are coupled to each other through six aspects: input (the which is processed through the function), time (which determines a temporal constraint), preconditions (which must be met for the function to be performed), resources (which are consumed during the performance), control (which determines control or monitoring conditions during the performance), and the output (which may be an entity or a state change). Figure 2 shows how a function is represented when modelled with FRAM.

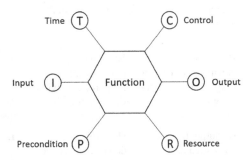

Fig. 2. A hexagon representing a function [12, p. 48]

FRAM is built upon four principles [12]: "equivalence of failures and successes": things go wrong by the same reason they go right; "approximate adjustments": everyday performance adjusted to match its conditions; "emergence": in the systemic sense, that the outcomes, predicted or not, must be described as emergent rather than resulted; and "resonance": relations and dependencies among functions are not cause-effect links, but they develop from a specific situation and context.

Hollnagel proposes four steps for using FRAM and one, which is necessary prior to the analysis, called "Step 0", in which the purpose of the analysis is determined. FRAM may be used either for event investigation (to look at what have happened) or for risk assessment (to predict what may happen in the future). Step 1 identifies the functions and describes how something is done through connecting the functions' aspects, then Step 2 will characterize potential and actual variability in FRAM. Then, Step 3 will aggregate variability by looking at specific instantiations of the model and, finally, Step 4 allows the user to propose ways to manage performance variability.

2.3 Evaluating the Process Performance Variability

Hollnagel proposes in step 3 a method for the aggregation of variability as a result of the upstream-downstream couplings for the aspects of every function [12]. His method, however, was essentially qualitative. In 2010, Luigi Macchi [16] proposed an increment for Hollnagel's approach for agregating variability through an application of the FRAM giving integer scores for aspect's qualities. For every function, its upstream aspects are assessed with respect to its quality, then this is related to how performance variability may be dumped or amplified. Table 1 links temporal and precision to quality of aspects, and Fig. 3 relates quality of aspect to potential for increasing or decreasing variability.

Table 1. Characterization of functions' output [16, p. 73]

		Temporal characteristics		
		Too early	On time	Too late
Precision	Precise	**A:** output to downstream functions is precise but too early	**B:** output to downstream functions is precise with the right timing	**C:** output to downstream functions is precise but delayed, reducing available time
	Appropriate	**D:** output to downstream functions is appropriate but too early	**E:** output to downstream functions is appropriate with the right timing	**F:** output to downstream functions is appropriate but delayed, reducing available time
	Imprecise	**G:** output to downstream functions is imprecise and too late	**H:** output to downstream functions is imprecise but correctly timed	**I:** output to downstream functions is imprecise as well as delayed, reducing available time

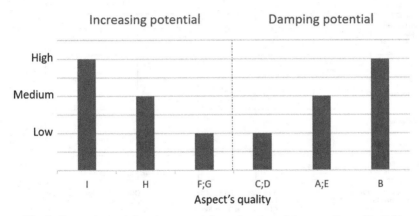

Fig. 3. Increasing and damping potential as a function of the aspect's quality [16]

Scores corresponding to the potential for increasing are positive (+) and, for damping, are negative (−), and the values for high, medium, and low are respectively +/− 3, +/− 2 and +/− 1. The quality of the output of a function would, then, be determined by the median of the quality of the n aspects provided by functions upstream as in (1).

$$q_o = \text{median}(q_{a_1}, q_{a_2}, \ldots, q_{a_n}) \qquad (1)$$

For this work, another factor considered was complexity of a function, rather than precision and timely aspects. Complexity will be a source of variability if operator is not well trained or used to the function they are supposed to perform. Scores are also given considering low, medium, or high potential for damping or increasing variability.

3   Case Study: Systems Engineering Project Reviews in IAE

3.1   Systems Engineering Process in IAE

The Instituto de Aeronáutica e Espaço, as a project-oriented organization, has developed its own processes for Project Management and Systems Engineering, aligned with standards and good practices and the procedures for product lifecycle [17] and project lifecycle [18] at the Brazilian Air Force. The results are two internal standards for project and product lifecycle phases, and one for project review procedures. For aeronautical and defense systems, the phases of product development and their main project reviews are described in Fig. 4.

3.2   Project Review Process

Project reviews are milestones in which an independent group of experts evaluate whether the project is heading towards its original goals and meeting the phase criteria or not [20, 21]. If the project meets the requirements as stated in the standards, or a subset of requirements, justified by a tailoring process, the project is considered approved to proceed to the next phase. If the project fails to meet all the requirements, it may either

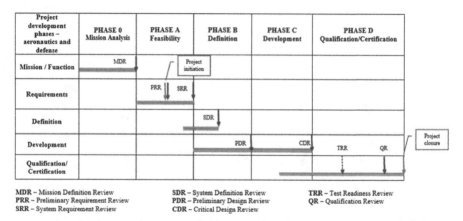

Fig. 4. Phases of product development and typical project reviews in IAE [19].

be approved, conditioned to meeting the requirements in time for the next milestone, or it is considered reproved and review must be retaken.

The "Project Review Procedures" internal standard [22] was developed by the Systems Engineering Office (SEO), derived from some international standards and procedures [23–25], and it represents a description for the work-as-imagined (WAI). Its last appendix shows a process flow using Business Process Model Notation. As already discussed in Sect. 2.2, this is a result of a straight linear thinking and likely not to describe the real work-as-done (WAD), but it is helpful when modelling the WAD.

There are 3 main groups or actors involved with the process: the Project Review Team (PRT), who will be responsible for representing the whole project team and giving all the information necessary; the Review Committee (RC), which consists of invited and independent experts, who will be responsible for assessing the project and the evidence provided by the PRT; and the Review Authority, who acts as a moderator and is responsible for process quality assurance and ensuring that Review Committee (RC) works independently and free of political pressure.

The Project Review Process consists of 4 subprocesses:

1. *Prepare project review*: consists of assigning the Project Review Team (PRT), gathering the evidence that the criteria for the selected milestone were met, inviting experts for the Review Committee (RC), communicating to the SEO the intention of conducting a review and planning the next steps (communications, meetings, etc.)
2. *Assess evidence*: PRT discloses the initial data pack with documents and evidence, which shall be assessed by the RC. Comments and questions are issued in "Comments and Questions Register" (CQR), which will be relied by an appointed PRT member. This goes on until register is scheduled to be consolidated by the PRT.
3. *Assess project*: RC presents its initial conclusions at a Coordination Meeting. From RC's considerations, they issue Review Item Discrepancies (RIDs), to be addressed by the PRT. Answers to the RIDs are assessed by the RC, who will consider whether answers are adequate or not to fix the discrepancies.

4. *Close review process*: RC shall present its final conclusions after evaluating RID's answers and approve if all milestone criteria were met, recommend conditional approval, or deny approval. Conditional approval needs Portfolio manager's approval. After Review Authority consent, review process shall be evaluated by its participants and a final report is issued.

4   Modelling the Work-as-Done

Before initiating Step 1, it is necessary to determine the purpose of the analysis. As stated in the introduction, this work aims to assess and find ways to improve the resilience by looking at the project review process. Therefore, the purpose is to predict what may happen in the future, but with a particularity: project reviews process are supposed to happen several times during a project life cycle, so monitoring work actually "as–done" will give further information and provide feedback to validate the model.

4.1   Step 1: Identifying and Describing Functions

FRAM modelling of the process was aided by the process flow given by the project review standard [22] and by monitoring the Mission Definition Review (MDR) for the Project IFF Mode 4 NSM, which is currently under development by IAE [26]. The model was done using the FRAM Model Visualizer (FMV) [27].

All four subprocesses described in Sect. 3.2 were modelled from work-as-imagined (WAI) into work-as-done (WAD) and their functions were collapsed into a high-level perspective, as seen in Fig. 5. Functions in green are the 4 subprocesses, and functions in purple and red are background functions, which are not a direct part of the main process, yet they are sources of variability and must be considered for the next steps.

4.2   Step 2: Characterizing Potential and Actual Variability

From the observed MDR process, subprocess 2, "Assess evidence" raised a flag and, therefore, was taken into consideration for further analysis. The interaction between PRT and RC did not work as imagined, which considered a loop of comments and questions

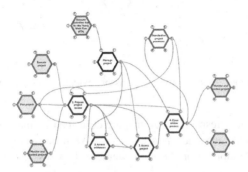

Fig. 5. FRAM model of the high-level project review process.

continuously being replied. Although the interaction happened, there was little feedback to the process, and few comments and questions were reassessed by RC. Figure 6 shows WAI for subprocess 2, and Fig. 7 show its WAD, modelled with FRAM, in which green functions are organizational considered in WAI model, yellow functions are human functions considered in WAI, red functions are human functions not considered in WAI and grey functions are background functions.

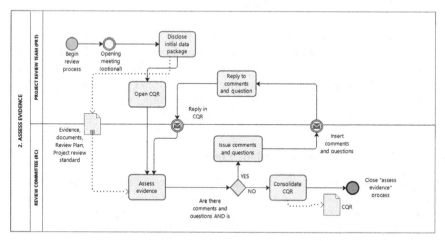

Fig. 6. Work-as-imagined (WAI) for subprocess "2. Assess evidence".

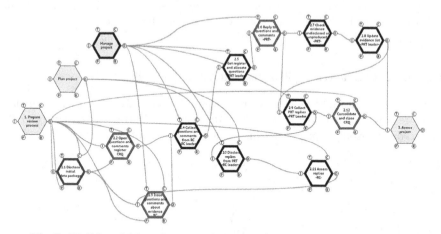

Fig. 7. FRAM model for subprocess "2. Assess evidence" work-as-done (WAD).

Functions, with their respective outputs, are described in Table 2. The output flow goes downstream up to a feedback, as seen in the WAI, in which replies are assessed by the RC and give them more information to interactively assess the evidence.

Table 2. Functions in subprocess 2 and their outputs

N°	Function	Output	
2.1	Disclose initial data package	Data package disclosed	
2.3	Issue questions and comments about evidence	Comments and questions	
2.4	Collect questions and comments from RC	CQR with questions and comments	
2.5	Get CQR and assign questions to PRT	Questions and comments allocated	FEEDBACK
2.6	Reply questions and comments	Questions and comments replied	
2.9	Collect PRT replies	CQR with replies	
2.10	Disclose replies to RC	CQR with replies disclosed	
2.11	Assess replies	Replies assessed	
2.12	Consolidate and close CQR	CQR consolidated	

Here is a brief description of each function:

- 2.1– Disclose initial data package: as the review process begin, the RC shall receive the data package to assess the evidence. Review plan helps to set a deadline for disclosing, but this may vary. RC members should also sign a confidentiality agreement if the project data access is restricted, which is the case for most projects in IAE.
- 2.3 – Issue comments and questions about evidence: RC members, experts invited due to their experience with other development projects, shall assess the evidence. Usually, there is a need for clarification, therefore comments and questions are issued. The quality of this process depends on the RC members expertise, which is defined in subprocess 1. Prepare Review.
- 2.4 – Collect questions and comments from RC – Review Committee leader shall gather every RC members' comments and questions and send them to the Project Review Team leader.
- 2.5 – Get CQR and assign questions to PRT: PRT leader shall assign questions and comments to be replied by the project team member that deals with the matter.
- 2.6 – Reply questions and comments: PRT member assigned by the PRT leader shall reply to their comments and questions and send them back to the PRT leader.
- 2.9 – Collect PRT replies: as PRT members reply to their assigned questions and comments, PRT leader shall collect them to send a new version of the CQR to RC leader.
- 2.10 – Disclose replies to RC: RC leader discloses the latest CQR version to the other RC members.
- 2.11 – Assess replies: RC members assess replies given by PRT members, considering if further information is required and, so, issuing new comments or questions.

Table 3. Background functions and their outputs

Function	Output	Description
0.1 Manage project	Pressure upon review schedule (C)	Determined by Project Manager's expertise and approach to the team. A healthy pressure may lightly damp variability (−1), whereas lack of expertise or excessive pressure may amplify variability (+2 or +3)
1. Prepare review	A. Project review plan (C)	Project Review Plan has little effect over the process, although it establishes a schedule and enlightens stakeholders with respect to their roles and responsibilities, which may be regarded as a low damp factor (−1)
	B. RC members expertise (C)	Depending on the expertise of the invited members of RC, good knowledge of the industry or with past similar projects damp variability (−1 or −2), whereas lack of expertise or knowledge of the subject of the review, or even the review process itself, may be a factor that amplifies variability (+1 or +2)
	C. Initial list of evidence (P/T)	Usually delivered on time (D, −1). If project manager is unexperienced or do not pay attention to producing the evidence prior to the milestone, then output quality may be imprecise (H, +2)
	D. Signed confidentiality agreements (P/T)	Usually, delivered too late. There is no record of inappropriate output, which would mean RC member would not take part in the process. (F, +1)
	E. Data package ready for distribution (P/T)	Usually, delivered on time, but not usually precise, because typically there is a need to add new information to the initial data package (E, −1). Level of access may couple and delay the distribution (F, +1)
0.2 Plan project	Documentation level of access (C)	R&D projects at IAE usually are classified and this adds complexity to the process, especially regarding data transmission. Medium or high potential to amplify variability (+2 or +3)

- 2.12 – Consolidate and close CQR: Review Plan defines, in its schedule, a deadline for the subprocess 2, whether assessment had been successfully done or not. After the deadline, review process proceeds to the next step, at the Coordination Meeting.

Background functions must also be considered for the resilience assessment. In this organizational process, they play a major role as a source of variability with respect to the quality of their outputs, either from a precision/timely (P/T) perspective or from a complexity (C) perspective. Table 3 gives further detail on background functions and quality of their outputs and Table 4 relates background functions and functions of subprocess 2 with respect to their aspects (I – input, P – precondition, R – resource, T – time, and C – control).

Table 4. Relationship between background functions outputs and functions' aspects at subprocess "2. Assess evidence"

	2.1	2.3	2.4	2.5	2.6	2.9	2.10	2.11	2.12
0.1		T	T	T	T	T	T		
0.2	C	C	C				C		
1.A	C	C							T
1.B		C						C	
1.C		I							
1.D	P								
1.E	I								

5 Improving the Process Resilience

5.1 Step 3: Aggregating Variability

After steps 1 and 2, step 3 will aggregate variability and step 4 will propose ways to manage variability. Once it was possible to monitor one round during the Project IFF Mode 4 NSM Mission Definition Review, one of the scenarios for aggregating variability will be the work-as-done. Other four scenario will be considered, using two variables: Project Manager expertise, which impacts output of background function 0.1, and RC members expertise, which impacts output B for background function 1. "Prepare Review". Both variables will have a "best-case" and a "worst-case" scenario. For instantiation where both are "best-case" scenario, a "worst-case" scenario will be added for level of access, adding more complexity to the process through output of background function 0.2. Table 5 gives all five instantiations scenarios and Table 6 gives relative scores with respect to potential for damping or amplifying variability for each scenario. Using the variability aggregation method as in Sect. 2.3, scores were propagated through the functions and the results can be seen in Fig. 8.

Table 5. Instantiation scenarios

SCENARIOS	PM expertise & attitude		RC expertise		Level of access
	BEST	WORST	BEST	WORST	
WAD	WORK AS DONE				
Scenario 1		X		X	REGULAR
Scenario 2	X		X		HIGH
Scenario 3	X			X	REGULAR
Scenario 4		X	X		REGULAR

Table 6. Scores for the instantiation scenarios

Function/output		Scenario				
		WAD	1	2	3	4
0.1	Schedule pressure	2	3	– 1	– 1	3
1.A	Review plan	– 1	– 1	– 1	– 1	– 1
1.B	RC expertise	– 1	2	– 1	2	– 2
1.C	Initial data list of evidence	1	2	– 1	– 1	2
1.D	Confidentiality agreements	1	1	1	1	1
1.E	Data pack ready	1	1	1	1	1
0.2	Level of access	2	2	3	2	2

5.2 Step 4: Managing Variability

From the results in Fig. 8, three main findings may be drawn. Firstly, the results from the WAD scenario reflect how the process was performed. As stated in Sect. 4.2, there was little interaction after comments and questions were replied by the PRT. This was partially due to the comments being issued with precision, but too late (F, +1), which is exactly the model quality output for function 2.3. Functions downstream were performed out of time or not even performed, which is in accordance with the results given by the model.

Secondly, Comments and Questions Register (CQR) consolidation is an event with a clear deadline, given by the Review plan, and is a pre-requisite to execute the Coordination Meeting, a mandatory event within the process. Output quality score of function 2.12 drops in almost every scenario (except in scenario 4), which was expected. Output

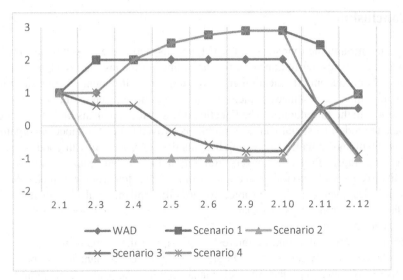

Fig. 8. Aggregation of variability and propagation through the process

quality +1 may indicate issues with precision, which can be translated into information quality. This is the case for both scenarios in which RC expertise is at worst-case scenario, and, hence, justified.

Finally, the Project Manager attitude and expertise play a major role in the variability aggregation. A skilled, well-trained project manager, with a good attitude, will cope with adjustments necessary after planning is done and help their team towards meeting the process objectives with more ease. This impacts other functions, mostly related to the Project Management processes, such as risk management and leadership, which will keep team motivation and readiness levels high. If, on the other hand, the project manager is unexperienced, lack of attitude or puts too much pressure on their team, variability will be amplified, as seen in scenarios 1 and 4. This last conclusion could have been found in other sources that discuss the role of the project manager and good project management practices [4, 28, 29], but it shows that the model gives results that are coherent with the main bodies of knowledge.

In order to close step 4 and the whole FRAM analysis, there are two actions which are suggested to manage variability. Firstly, to provide better training and environment for Project Managers. Well trained PM can manage their own team and to oversee the whole process from both project management and systems engineering, which will help them to increase the whole Systems Engineering process resilience. Also, good PMs will allow room for small adjustments to the original plans, should things work differently than expected (which they do). The second action, which is coupled to the first, is to give clearer instructions to participants within the project review process. This will help participants to clearly understand their roles and responsibilities, especially those external experts invited to the Review Committee, who may not be familiar with the whole review process. Resilience principles, however, must be considered: there must be room for small adjustments whenever they arise.

6 Conclusion

After going through the four steps of FRAM, plus its step 0, the project review process may now have its resilience improved by the findings of the last step, such as investing in Project Management qualification and providing an environment where project may adapt within the circumstances of their execution, therefore reaching the work's objective as stated in the Introduction. The Functional Resonance Analysis Method, although imagined to modelling social-technical problems, proved to be fit to model an organizational problem and provided interesting insights that are coherent with good practices, such as investing in Project Managers' training and career, in order to improve their capability to cope with variability, and establishing clearly to participants their roles and responsibilities. Everyday work is subject to variability and minimal adjustments, so it is important to determine ways to manage this variability and avoid resonances that would crush system stability.

For future works, it would be interesting to expand the scope of the analysis and assess relationship between Systems Engineering and Project Management processes, respectively product and project life cycle management tools. In an organization such as IAE, technology development projects are its day-to-day operations, which are subject to variability, and may have its resilience improved with the help of FRAM.

Acknowledgements. The authors would like to thank the Instituto de Aeronáutica e Espaço and Project IFF Mode 4 NSM manager and team for the support, collaboration, and providing information to produce this paper. The authors would also like to thank the Risk Analysis Research Group – Grupo de Estudos em Análise de Riscos (www.gear.ita.br) at ITA.

References

1. Thamhain, H.J.: Challenges of managing projects in a technology world. In: Thamain, H.J.: Managing Technology-Based Projects, pp. 1–18. Wiley, Hoboken (2014)
2. Woods, D.D., Hollnagel, E.: Prologue: Resilience Engineering Concepts. In: Hollnagel, E., Woods, D.D., Leveson, N. (eds.) Resilience Engineering: Concepts and Precepts, pp. 1–6. Ashgate, Surrey (2006)
3. Department of Defense, DoD: Risk, Issue, and Opportunity Management Guide for Defense Acquisition Programs. Defense Acquisition University, Washington (2017). https://www.dau.edu/tools/Lists/DAUTools/Attachments/140/RIO-Guide-January2017.pdf Accessed 11 Jun 2020
4. Project Management Institute, PMI: A Guide to the Project Management Body of Knowledge: PMBoK Guide. 6th edn. PMI: Newton Square (2017)
5. Aven, T.: Risk assessment and risk management: review of recent advances on their foundation. Eur. J. Oper. Res. **253**, 1–13 (2016). https://doi.org/10.1016/j.ejor.2015.12.023
6. Instituto de Aeronáutica e Espaço, IAE: Mission, Vision and Values, http://www.iae.cta.br/index.php/mission. Accessed 11 Jun 2020
7. Department of Defense, DoD: Systems Engineering Fundamentals. Defense Acquisition University Press, Fort Belvoir (2001)
8. Kossiakoff, A., et al.: Systems engineering management. In: Kossiakoff, A., et al.: Systems Engineering Principles and Practice, pp. 111–135. Wiley, Hoboken (2011)

9. National Aeronautics and Space Administration, NASA: Fundamentals of Systems Engineering. In: NASA, Systems Engineering Handbook (NASA/SP-2007–6105Rev1), pp. 3–17. NASA: Washington (2007)

10. Johansson, C.: Managing Uncertainty and Ambiguity in Gates: Decision Making in Aerospace Product Development. Int. J. Innov. Technol. Manage. **11**(2), 1–21 (2014). https://doi.org/10.1142/S0219877014500126

11. Patriarca, R., et al.: Framing the Fram: a literature review on the functional resonance analysis method. Saf. Sci. 129 (2020). https://doi.org/10.1016/j.ssci.2020.104827

12. Hollnagel, E.: FRAM: Functional Resonance Analysis Method – Modelling Complex Socio-Technical Systems. Ashgate, Surrey (2012)

13. Patriarca, R., et al.: Resilience engineering: Current status of the research and future challenges. Saf. Sci. **108**, 79–100 (2018). https://doi.org/10.1016/j.ssci.2017.10.005

14. Hollnagel, E.: Prologue: The scope of resilience engineering. In: Hollnagel, E., et al. (eds.): Resilience Engineering in Practice: A Guidebook, pp. xxix-xxxix. Ashgate, Surrey (2011)

15. Aven, T.: The Call for a Shift from Risk to Resilience: What Does it Mean? Risk Anal. **39**(6), 1196–1203 (2019). https://doi.org/10.1111/risa.13247

16. Macchi, L.: "A Resilience Engineering approach for the evaluation of performance variability: development and application of the Functional Resonance Analysis Method for air traffic management safety assessment", Business administration, École Nationale Supérieure des Mines de Paris, 2010. English. https://pastel.archives-ouvertes.fr/pastel-00589633. Accessed 11 Jun 2020

17. BRASIL, Estado-Maior da Aeronáutica.: Ciclo de Vida de Sistemas e Materiais (DCA 400–6). EMAER: Brasília (2007)

18. BRASIL, Departamento de Ciência e Tecnologia Aeroespacial: Gestão de Projetos de Ciência, Tecnologia e Inovação no DCTA (ICA 80–12). DCTA: São José dos Campos (2019)

19. BRASIL, Instituto de Aeronáutica e Espaço: Fases de Desenvolvimento de Sistemas Aeronáuticos e de Defesa (NPA-IAE 134/2019A). IAE: São José dos Campos (2019)

20. Blanchard, B.S., Blyler, J.E.: Design review and evaluation. In: Blanchard, B.S., Blyler, J.E.: Systems Engineering Management, pp. 251–274. 5th edn. Wiley, Hoboken (2016)

21. International Council on Systems Engineering, INCOSE: Generic Life Cycle Stages. In: INCOSE, Systems Engineering Handbook: A guide for system life cycle processes and activities (INCOSE-TP-2003-002-04), pp. 25-46. Wiley: Hoboken (2015)

22. BRASIL, Instituto de Aeronáutica e Espaço: Procedimento de Revisão de Projetos (NPA–IAE 144/2019). IAE: São José dos Campos (2019)

23. Associação Brasileira de Normas Técnicas, ABNT: NBR/ISO 21349:2010: Sistemas Espaciais – Revisões de Projeto (in Portuguese). ABNT: (2010)

24. European Comission for Space Standardization, ECSS: ECSS-M-ST-10-01C: Space Management – Organization and conduct of reviews. ECSS: Noordwijk (2008)

25. National Aeronautics and Space Administration, NASA: Appendix N: Guidance on technical peer reviews/inspections. In: NASA, Systems Engineering Handbook (NASA/SP-2007-6105Rev1), pp. 3-17. NASA: Washington (2007)

26. Instituto de Aeronáutica e Espaço, IAE: Sistema IFFM4BR – Fase 2 (in Portuguese). http://www.iae.cta.br/index.php/todos-os-projetos/projetos-aeronautica/iffm4br-fase-2 Accessed 11 Jun 2020

27. Hollnagel, E.: FRAM Model Visualiser (FMV). https://www.functionalresonance.com/the%20fram%20model%20visualiser/index.html. Accessed 11 Jun 2020

28. Thamhain, H.J.: Contemporary Project Management: Concepts and Principles. In: Thamain, H.J.: Managing Technology-Based Projects, pp. 19–38. Wiley, Hoboken (2014)

29. Kerzner, H.: Overview. In: Kerzner, H.: Project Management: A systems approach to planning, scheduling, and controlling, pp. 1–38. 12th edn. Wiley, Hoboken (2017)

Proposal for a Biorefinery in a Cuban Sugar Industry Taking Advantage the Biomass

Ana Celia de Armas Martínez[1] (ID), Yailet Albernas Carvajal[1(✉)] (ID),
Gabriela Corsano[2] (ID), and Erenio González Suárez[1] (ID)

[1] Department of Chemical Engineering, Faculty of Chemistry and Pharmacy,
Central University "Marta Abreu" of Las Villas, Santa Clara, Cuba
yailetac@uclv.edu.cu

[2] Instituto de Desarrollo y Diseño (CONICET-UTN), Avellaneda 3657, S3002GJC
Santa Fe, Argentina

Abstract. Cuban sugar industry opens possibilities for a biorefinery development due to its flows characteristics and its facilities for obtaining different products, co-products and energy. This work proposes the development of a biorefinery considering the production of sugar, energy, ethanol, xylose, torula cream and biogas as main products, using sugar cane and leftover bagasse as biomass. When obtaining second-generation ethanol, bagasse hydrolyzate, filter juice extracted from sugar production and molasses are used as sugar sources, which allows a reduction of 24% in molasses consumption in fermentation stage. The profit of proposed scheme was optimized considering 120 and 150 harvest days, and the results indicates that superfine alcohol production can increase to 800 and 900 hL/d respectively, keeping constant the rectified ethyl alcohol production. An economic analysis shows that profitability is only achieved when only ethanol is produced, although the integrated scheme is profitable for both analyzed harvest periods.

Keywords: Biorefinery · Sugar cane · Integration · Economic feasibility

1 Introduction

The term biorefinery is one of the most important concepts that encompasses processes integration and technologies for an efficient use of biomass as the main feedstock source. It follows the same principle as a petroleum refinery, but energy and chemicals are produced from biomass through different processes [1, 2].

One biorefinery definition is: "a facility where, through various biomass transformation processes, bioenergy (heat, electricity, biofuels) and a wide range of bioproducts (materials, chemicals, food and feed) are generated, requiring the integration of different processes and technologies in a single facility" [3].

The heterogeneity of biomass and its numerous conversion possibilities multiply the possible operation schemes that can be developed in a biorefinery. This variety means that their classification is based on fundamental aspects such as: raw material used, obtained products, used process and degree of integration [4, 5]. The degree of integration is

© Springer Nature Switzerland AG 2021
D. A. Rossit et al. (Eds.): ICPR-Americas 2020, CCIS 1408, pp. 156–163, 2021.
https://doi.org/10.1007/978-3-030-76310-7_12

one of the most commonly used and identifies them as first, second or third generation biorefineries.

Countries such as Belgium, Holland, France, Austria and Germany produce bioethanol, starch or animal feed, in first generation biorefineries on a commercial scale, using wheat and corn as raw materials. The second generation models include the use of lignocellulosic biomass such as forest residues, with the cellulosic ethanol biorefinery being the greatest exponent. Finally, there are third generation biorefineries, which the main developed schemes are for obtaining fertilizers, omega three and six, and biofuels such as biodiesel and biogas. These are mostly based on microalgae processing and have been developed at pilot plant scale in the United States, Austria, Australia and Argentina, to cite some examples [6].

Several authors have studied the possible biorefinery schemes that can be adapted to a sugar industry [7]. These schemes have a flexibility degree in processing capabilities, allowing a base of biological production and coproduction, economically sustainable, with sugar, ethanol and energy as its main products [8]. This can represent a sustainable solution for energy reduction from fossil sources, besides contributing to make sugar and alcohol industry less vulnerable to market fluctuations [7]. [9] suggest that the sugarcane biorefinery requires an integration of distillery with sugar process and sugarcane as the main raw material, making possible the use of final molasses, intermediate juices and molasses.

In the present work, different integration alternatives are proposed with the objective of taking advance of the use of sugar process streams in the sugar industry as a first and second generation biorefinery.

2 Development

The primary process objective is to obtaining raw sugar from sugar cane, where molasses and energy are also obtained. Other products are derived from this process, in this case bagasse and Xylose hydrolyzate, ethanol, torula yeast and biogas.

2.1 Identification of the Different Processes

Raw Sugar Production
Raw sugar production begins with cane that is fed to the mills preparation, which facilitates more efficient juice extraction. During milling, juice is obtained, as well as bagasse that represents approximately 25% of milled cane weight, which is used in boilers to generate steam. Juice from mills, with values between 15 and 16°Bx, goes through of alkalization, clarification, evaporation, boiling, crystallization and centrifugation stages. On this latter stage, commercial sugar is obtained [10].

Production of Bagasse Hydrolyzate, Xylose and Ethanol
A second generation scheme implies using bagasse as lignocellulosic biomass, not only to produce energy, but also to take advantage of its glucose and xylose content. Therefore, no water in fermentation stage is employed for ethanol production, but a mix of sugaring

substrates formed by bagasse hydrolyzate, filters juices (separated from sugar production) and molasses is proposed. In addition, this process obtains the xylose contained in bagasse fibers.

Ethanol production from molasses, filter juices and bagasse hydrolyzate has four fundamental stages: pretreatment, prefermentation, fermentation and distillation. The raw material used to obtain bagasse hydrolyzate is the bagasse from the sugar mill not used as fuel for steam generation.

The incorporation of enzymatic hydrolysis stage in a sugar mill, with a traditional ethanol plant, allows a reduction in purchased molasses in non-harvest time, increasing unused production capacity, as well as sugar industry adaptation as an integrated biorefinery [6].

In first pretreatment stage, to obtain bagasse hydrolyzate, a liquid containing mostly xylose, glucose and arabinose is obtained. This can be subjected to different treatment stages to obtain a stream with a higher xylose content that can be used as raw material for xylitol production [6].

Torula Yeast and Biogas Production

From ethanol production, specifically in distillation stage, stillage or distillery must is separated, which is the waste of the process. This stream can be used to produce protein concentrates using *Candida utilis,* exploiting its organic matter content. Once this process is completed, the residual stillage can be used in biogas production [11]. This is an alternative that makes possible to use stillage as a source of sugaring substrates, representing a more economical raw material than final molasses. In this way, the use of molasses in this process is avoided, using it for ethanol production, and a protein cream is obtained that can be used as animal feed.

This process also offers a COD removal of approximately 40%, by reducing it to 30000 mg/L, according to the results of [12]. With these organic load values, biogas production can be carried out by anaerobic digestion, reducing the retention time in biodigester. This allows reducing pollutant organic load, obtaining fertilizer sludge and wastewater to be used as irrigation [13]. The produced biogas can be used for steam generation in the distillery [14], saving fuel amount. Meanwhile, effluent liquid from biodigesters can be used for sugarcane system irrigation.

Figure 1 shows a summary of the integration scheme.

2.2 Material and Energy Balances

The considered productions in the biorefinery scheme were characterized starting from mass and energy balances, in order to know the streams availability to be used in the different productions. The main used equations are shown below, and the obtained the results are displayed in Table 1.

Sugar Process

$$Water + Cane = Bagasse + Mill\ Juice \tag{1}$$

$$Filters\ Juice = 15\% * Mill\ Juice \tag{2}$$

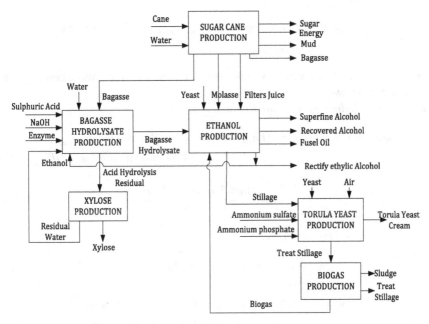

Fig. 1. Proposed biorefinery scheme.

$$Sugar = \frac{\left(Purity_{syrup} - Purity_{final\ molasses}\right) * Solid\ Mass_{syrup} * 100}{\left(Purity_{sugar} - Purity_{final\ molasses}\right) * (100 - Humidity_{sugar})} \tag{3}$$

$$Molasse = \frac{\left(Sugar - Purity_{syrup}\right) * SolidMass_{syrup}}{\left(Purity_{sugar} - Purity_{final\ molasses}\right) * 4.88} \tag{4}$$

where *syrup* is the molasses, which has not been subjected to the crystallization stage in the sugar process.

Ethanol Process

$$Molasses = Ethanol\ production * Molasses\ consumption\ index \tag{5}$$

$$Dilute\ molasses = Water + Molasses \tag{6}$$

$$Brix_{Molasses} * Molasses = Brix_{Dilute\ Molasses} * Dilute\ Molasses \tag{7}$$

$$Alcohol_{total} = Alcohol_{prefermentation} + Alcohol_{fermentation} \tag{8}$$

$$Wine_{distillation\ column} + Steam = Alcoholic\ Steam + Stillage \tag{9}$$

$$Wine_{distillation\,column} * Composition_{wine\,alcohol} = Alcoholic\,Steam$$
$$* Composition_{alcoholic\,steam} \qquad (10)$$

where *Wine* represents the way in which the raw material of distillation, coming from fermentation, is known.

Bagasse and Xylose hydrolyzate Process

$$Bagasse\,hidrolyzate = (Enzymatic\,Liquid - Bagasse\,water_{out\,HE})$$
$$+ Separated\,Glucose\,in\,HE \qquad (11)$$

$$Xylose = \frac{WasteLiquid * \%Xylose}{Purity\,Xylose} \qquad (12)$$

Torula cream and biogas Process

$$Total\,Flow = Cream + Effluent \qquad (13)$$

$$Total\,Flow * concentration_{total\,flow} = Cream * concentration_{cream}$$
$$+ effluent * concentration_{effluent} \qquad (14)$$

$$Methane = Stillage * COD * (1 - bacterial\,growth) * efficiency * 0.35$$
$$* \left(\frac{Temperature + 273}{273}\right) \qquad (15)$$

$$Biogas = \frac{Methane}{\%\,de\,Methane\,in\,biogas} \qquad (16)$$

2.3 Economic Analysis

The economic analysis is carried out applying the methodology proposed in [15]. Considering the necessary equipment needed for each process, the total investment cost is determined (Table 2), except for sugar production, since it is considered as an already installed facility. The rest of the plants are added in this work.

The total production costs (TPC) are obtained calculating the raw materials consumption, labor and requirements, among other costs. Profit was determined based on production value. Total production cost was calculated for 300 operation days per year (Fig. 2).

In order to assess the investment feasibility of the proposed biorefinery, dynamic profitability indicators are determined: NPV, IRR and payback period, taking an interest rate of 15%. Initially, they are carried out for each process separately. The results are summarized in Table 3.

Table 1. Main data and mass balance results

Process	Raw material	Value	Product	Value
Sugar	Cane (t/d)	2700	Sugar (t/harvest)	35 640
	Water (t/d)	791	Final molasses (t/harvest)	29 400
			Electricity (kW-h/harvest)	6 340 320
Bagasse and Xylose hydrolyzate	Bagasse (t/d)	24.8	Xylose (kg/d)	4 248
			Bagasse hydrolyzate (t/d)	131
Ethanol	Final molasses (t/d)	128	Superfine alcohol (hL/d)	800
	Filters juice (t/d)	379	Alcohol phlegm (kg/h)	352.5
	Bagasse hydrolyzate (t/d)	131	Fusel oil (kg/h)	376
Torula cream	Stillage (m³/h)	32.12	Torula cream (t/d)	7
Biogas	Treat stillage (m³/d)	1028	Biogas (m³/d)	12 625

Table 2. Total investment costs

Cost ($)	Bagasse and Xylose hydrolyzate	Ethanol	Torula cream	Biorefinery
Equipment cost	3 268 080	1 238 201	1 051 998	5 558 279
Directs cost	8 987 221	3 405 053	2 892 994	14 656 570
Indirect cost	2 911 860	1 103 237	937 330	4 826 687
Total investment cost ($)	12 258 570	4 644 492	3 946 044	20 094 668

As can be seen, a positive return (profitability) is only achieved for ethanol production as an independent process. However, when the complex is analyzed in integral way, i.e. considering it as a biorefinery, profitability is reached, with a recovery period of 5.5 years for 120 days of harvest. These results demonstrate that the proposed biorefinery scheme is economically viable, considering sugar, energy, ethanol, Xylose, torula in cream and biogas as final products production. In this way, the biorefinery concept is fulfilled by integrating different technological processes, using sugar cane and bagasse as biomass sources, and obtaining different products and energy.

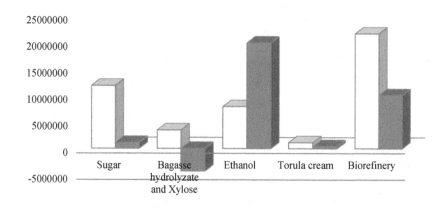

☐ Total Production Cost ($/year)   ■ Profitability ($/year)

Fig. 2. Total production cost and profit for each analyzed technology.

Table 3. Dynamic profitability indicators

Indicator	Bagasse and Xylose hydrolyzate	Ethanol	Torula cream	Biorefinery
NPV ($)	−81 781 666	90 585 492	−14 297 833	48 389 965
IRR (%)	-	74	-	34.3
Payback period (year)	-	3	-	5.5

3 Conclusions

- The use of bagasse as a lignocellulosic raw material to obtain hydrolyzate in a second-generation biorefinery integrated scheme reduces 24% the molasses consumption and no water is necessary to make the dilution.
- From residual liquid of bagasse pretreatment by acid hydrolysis, Xylose can be obtained, reducing the consumption of this stage in 51%.
- Torula yeast production from distillery stillage allows increasing the added value of this residual before to be used in biogas production.
- When analyzing the processes separately, only ethanol production is economically feasible, but the general biorefinery scheme is profitable for the analyzed harvest period.

References

1. Temmes, A., Peck, P.: Do forest biorefineries fit with working principles of a circular bioeconomy? a case of Finnish and Swedish initiatives. For. Policy Econ. (2019). https://doi.org/10.1016/j.forpol.2019.03.013

2. Hassan, S.S., Williams, G.A., Jaiswal, A.K.: Emerging technologies for the pretreatment of lignocellulosic biomass. Bioresour. Technol. J. **262**, 310–318 (2019). https://scholar.goo gle.com.cu/scholar?q=Emerging+technologies+for+the+preatment+of+lignocellulosic+bio mass+bioresour+technol&hl=es&as_sdt=0&as_vis=1&oi=scholart#d= gs_qabs&u=%23p%3DP64O2Afme6YJ
3. Manual sobre biorrefinerías en España. BioPlat, SUSCHEM. (2017). https://www.bioplat. org/setup/upload/modules_docs/content_cont_URI_4020.pdf
4. Martín, P., Martín, J.: Biorrefinerías basadas en explotaciones agropecuarias y forestales. Materiales elaborados como parte del Proyecto de Innovación Docente de la Universidad de Zaragoza PIIDUZ_16_276. España (2017). https://ocw.unizar.es/ocw/pluginfile.php/915/ mod_resource/content/1/Manual%20del%20curso.pdf
5. Castilla-Archilla, J., O'Flaherty, V., Lens, P.: Biorefineries: Industrial innovation and tendencies. In: Bastidas-Oyanedel, Juan-Rodrigo., Schmidt, Jens Ejbye (eds.) Biorefinery, pp. 3–35. Springer, Cham (2019). https://doi.org/10.1007/978-3-030-10961-5_1
6. de Armas, A.C.: Evaluación de esquemas de biorrefinería de segunda y tercera generación en una industria azucarera cubana. PhD. Thesis, Universidad Central "Marta Abreu" de las Villas, Cuba (2019)
7. Furtado, J., et al.: Biorefineries productive alternatives optimization in the Brazilian sugar and alcohol industry. Appl. Energy **259**(C), 2–19 (2019). https://www.sciencedirect.com/sci ence/article/abs/pii/S0306261919307329
8. Aguilar, N.: A framework for the analysis of socioeconomic and geographic sugarcane agro industry sustainability. Socio-Econ. Plann. Sci. **66**, 149–160 (2019). https://www.sciencedi rect.com/science/article/abs/pii/S0038012117302379
9. Fito, J., Tefera, N., Van Hulle, S.: Sugarcane biorefineries wastewater: bioremediation technologies for environmental sustainability. Chem. Bio. Technol. Agric. **6**(1), 1–13 (2019). https://doi.org/10.1186/s40538-019-0144-5
10. de Armas, A.C., Morales, M., Albernas, Y., González, E.: Alternativas para convertir una industria azucarera cubana en biorrefinería. Centro Azúcar **45**(3), 65–77 (2018). http://centro azucar.uclv.edu.cu/index.php/centro_azucar/article/view/73
11. Torres, A., Díaz, M., Saura, G.: Factibilidad económica de alternativas de inversión para reducir el costo de producción de la levadura torula. Centro Azúcar **43**(1), 10–17 (2016). http://centroazucar.uclv.edu.cu/index.php/centro_azucar/article/view/125
12. Santos, R.: Estrategia de análisis de alternativas para la reactivación de instalaciones actuales de levadura torula. PhD. Thesis, Universidad Central "Marta Abreu" de las Villas, Cuba (1999)
13. Chanfón, J., Lorenzo, Y.: Alternativas de tratamiento de las vinazas de Destilería. Experiencias nacionales e internacionales. Centro Azúcar **41**(2), 56–67 (2014). http://centroazucar.uclv. edu.cu/index.php/centro_azucar/article/view/255
14. Lorenzo, Y.: Nueva tecnología de producción en etanol y biogás de menor costo e impacto ambiental negativo para la UEB "Derivados Heriberto Duquesne". PhD. Thesis, Instituto Superior Politécnico "José Antonio Echeverría" (2016)
15. Peters, M.S., Timmerhaus, K.: Plant Design and Economics for Chemical Engineers, 4th Editions, pp. 140–141. McGrall-Hill International Editions, New York (2006)

Evaluation of the Economic and Volumetric Performance of Strategic Planning in a Forest Plantation in the North of Misiones Province, Argentina

Diego Broz[1]([✉]) [ID], Mathías López[2], Enzo Sanzovo[2], Julio Arce[3] [ID], and Hugo Reis[4]

[1] CONICET, FCF, Universidad Nacional de Misiones, N3382GDD
Eldorado, Misiones, Argentina
[2] FCF, Universidad Nacional de Misiones, N3382GDD Eldorado, Misiones, Argentina
[3] DECIF, Universidad Federal de Paraná, Curitiba, Paraná 80210170, Brazil
jarce@ufpr.br
[4] Pindó SA, Jefe de Área Forestal, 3378 Puerto Esperanza, Misiones, Argentina
hugoreis@pindosa.com.ar

Abstract. Forest planning, based on mathematical optimization, makes it possible to model the complexity of a forest system. In this work two approaches to forest plantation planning were evaluated. The aim was, firstly, to regulate production and, secondly, to maximise economic benefit. The models proposed give satisfactory results and allow different forest management scenarios to be explored efficiently.

Keywords: Strategic planning · Linear programming · Forest management

1 Introduction

Strategic planning is particularly important as it seeks to regulate forest production in the long-term and increase forest income in order to ensure sectorial competitiveness [1]. The multidimensionality of the system makes this process complex and the aid of mathematical models is essential [2]. In this sense, linear programming is a robust tool to solve this type of problem. The first models applied to forest sciences were developed in the 60s by [3, 4] and [5]. However, the approach of [6] defines the bases of strategic planning based on an alternative formulation of the linear programming harvest scheduling problem which they called Model type I and II. The first models were relatively small due to the computational limitation to solve them, but, at present, it is possible to tackle large problems with standard computers.

In this work a forest planning model was developed based on the Type I Model of [6] in order to evaluate two production scenarios, minimize the fluctuation of interannual production and maximize the net present value (NPV).

© Springer Nature Switzerland AG 2021
D. A. Rossit et al. (Eds.): ICPR-Americas 2020, CCIS 1408, pp. 164–170, 2021.
https://doi.org/10.1007/978-3-030-76310-7_13

2 Materials and Methods

The study area consists of stands of *Pinus taeda* (Pt), *Pinus elliottii var. elliottii x Pinus caribaea var. hondurensis* or hybrid pine (Hp) and *Araucaria angustifolia* (Aa) belonging to a forest heritage located in the north of the province of Misiones. This case corresponds to a total of 162 stands representing an area of 3,056 ha, with an age range from 1 to 36 years old.

The forest has a traditional forest management scheme, which is 3–4 thinnings and rotation age older than 18 years whose objective is the production of large logs. In Table 1 the silvicultural action for each species is summarized. In the management activities, the following costs were considered: planting and maintenance for 3 years: 951.51 USD/ha; pruning 1: 114.59 USD/ha; pruning 2: 117.1 USD/ha; pruning 3: 147.82 USD/ha; administrative costs: 50 USD/ha/year [7].

Table 1. Traditional management regimes for the three species planted.

Species	Intervention	Intensity of intervention (%)	Age (years)
Araucaria angustifolia - Initial density 2,222 trees/ha - 3 pruning	1° thinning	50	8
	2° thinning	40	11
	3° thinning	40	15
	4° thinning	40	20
	Clearcutting	100	25
Pinus taeda - Initial density 1,667 trees/ha - 2 pruning	1° thinning	55	6
	2° thinning	40	9
	3° thinning	40	13
	Clearcutting	100	18
Hybrid pine - Initial density 1,333 trees/ha - 2 pruning	1° thinning	50	7
	2° thinning	40	10
	3° thinning	40	14
	Clearcutting	100	18

The current forest management scheme is very rigid; this does not allow an adequate balance of interannual log production. Furthermore, the production of large logs does not correspond to the demand of the modern installed sawmill which demands logs with maxim small end diameter (SED) of 24 cm. Based on this, in this work we proposed, firstly, a more flexible clearcutting age; secondly, eliminate a thinning operation and, in addition, establish a new management policy, replace Aa and Pt stands by Hp. This species allows obtaining skills equal or superior structural factors with their products and remanufactures and, further, logs with less taper and, consequently, with higher industrial performance.

Table 2 shows log characteristics obtained from the forest according to the species, SED, length and market prices put at the sawmill subtracting average costs of thinning, harvest and transport [7]. In other words, these prices are denominated standing timber prices.

Table 2. Forest products prices, in foot.

Species	SED (cm)	Log length (ft)	Price (USD/t)
Pt	35–99	14	13,88
Pt	29–35	14	8,23
Pt	25–29	14	3,7
Pt	18–25	10	2,37
Pt	14–18	10	1,99
Pt	<14	8	1,21
Hp	35–99	14	13,88
Hp	29–35	14	8,23
Hp	25–29	14	3,7
Hp	18–25	10	2,37
Hp	14–18	10	1,99
Hp	<14	8	1,21
Aa	35–99	14	36,3
Aa	29–35	14	20,68
Aa	25–29	14	20,68
Aa	18–25	10	9,33
Aa	14–18	10	5,66
Aa	<14	8	1,21

Currently, the company carries out the planning process empirically, with a spreadsheet. From the situation presented, a mathematical optimization model is proposed, and two objective functions (F1 and F2) are evaluated in the same way as [8]. Using the objective function F1 (1) we seek to maximize NPV, while with the objective function F2 (2) we intend to stabilize the global production of the forest. These models are evaluated in a planning horizon (PH) of 50 years and 5 scenarios in which the clearcutting age (CA) is relaxed. The basis scenario uses the CA proposed by the company, that is, without flexibility. Then, the CA is relaxed by ± 1 (that is, 3 clearcutting age options) and next CA is relaxed by ± 2 years (5 clearcutting age options). In each one, it is replaced with the implantation of Ph to the Aa and Hp stands at the time of clearcutting. In addition, for the calculation of profitability, a discount rate of 10% is used.

$$MaxNPV = \sum_{i=1}^{M} \sum_{j=1}^{N} NPV_{ij} \times X_{ij}; \forall k \qquad (1)$$

$$MinDIF = Min(MinMax - MaxMin) \qquad (2)$$

To prevent the mathematical model from being infeasible, a set of constraints are defined. Constraint (3) is used to indicate the maximum area of each stand. This ensures that the sum of the area of each stand is equal to its total area, under different management alternatives, in F1 and F2. In addition, F2 requires restrictions (4) and (5), in order to minimize the difference between the maximum and minimum annual log production.

$$\sum_{j=1}^{N} X_{ij} \leq A_i; \forall i \tag{3}$$

$$\sum_{i=1}^{M} \sum_{j=1}^{N} (VOL_{ijt} X_{ij}) \leq MinMax; \forall t \tag{4}$$

$$\sum_{i=1}^{M} \sum_{j=1}^{N} (VOL_{ijt} X_{ij}) \geq MaxMin; \forall t \tag{5}$$

In these equations, NPV_{ij} is the net present value (USD) of the i-th stand after the j-th forest management alternative; VOL_{ijt} is the volume (m³) produced in the i-th stand following the j-th forest management alternative in t period; $MinMax$ is the lowest volume from maximum production volume in the planning horizon; $MaxMin$ is the highest volume of the minimum possible production in the planning horizon; X_{ij} is the area fraction (ha) of the i-th stand following the j-th forest management alternative; A_i is the total area (ha) of the i-th stand at the beginning of the planning horizon; M is the total number of stands; N is the total number of forest management alternatives for the i-th stand.

These models were implemented in Optimber-LP, a specific software to strategic forest planning developed by Brazilian enterprise Optimber Otimização e Informática Ltda. For the simulation of forest management regimes, Optimber-LP uses the SIS Pinus simulation package, developed by the Empresa Brasileira de Pesquisa Agropecuária (EMBRAPA), for the projection of the growth of the forest. On the other hand, to solve mathematical optimizations models, Optimber-LP uses LINGO, a software developed by Lindo System Inc.

The data required to model are: stand, species, age, site index, area, year of planting, initial density, basal area, pruning height and number of thinnings carried out. All data are provided by the company through forest inventories for commercial purposes. The implementation was carried out on a computer with Intel Core i5-2310M, CPU @ 2.10 GHz, 8 GB of RAM and 64-bit OS.

3   Results

In Fig. 1 you can see the volumes obtained for each product (see Table 2) and without CA relaxing in a 50-year forestry planning horizon. In this case, a significant oscillation is observed in the log production, this is between 60,000 and 200,000 m³/year. The main problem here lies in the instability of log supply. This scenario is unrealistic, especially in integrated companies, and some strategies must be established to avoid the year-on-year oscillation.

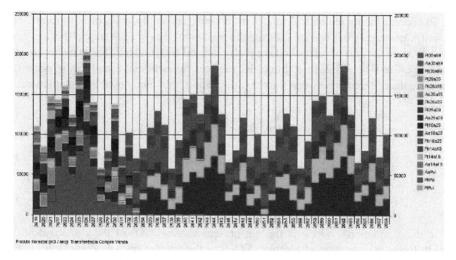

Fig. 1. Global volumes of the control scenario.

Figure 2 and Fig. 3 show the result of relaxing the CA using only the objective function F1. In Fig. 2 the result is shown when the CA relaxes in ± 1 year, obtaining from this a volumetric variation between 30,000 and 235,000 m³/year. On the other hand, Fig. 3 shows the result when the CA relaxes in ± 2 years. In this case, a greater volumetric variation is observed, which is between 30,000 and 300,000 m³/year. This large oscillation of log production in model based on objective function F1 is because it seeks to obtain the highest economic return, regardless of the fluctuation of its production.

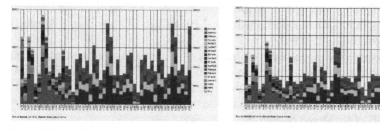

Fig. 2. Volume behavior applying model with objective function F1 and CA ± 1.

Fig. 3. Volume behavior applying model with objective function F1 and CA ± 2.

On the other hand, when the objective function F2 is used, it seeks to stabilize the production of logs, in other words, to minimize its interannual variation. This result can be seen in Fig. 4 whit CA ± 1, where the production for the first 10 years is 135,000 m³/year, and 105,000 m³/year in the following years. There are also sudden increases in production that do not exceed 135,000 m³/year. In Fig. 5 with CA ± 2, the volumetric production is 125,000 m³/year in the first 12 years and 110,000 m³/year in the following 38 years.

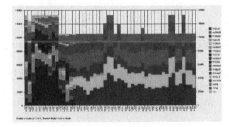

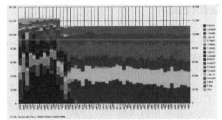

Fig. 4. Volume behavior applying model with objetive function F2 and CA ± 1.

Fig. 5. Volume behavior applying model with objetive function F2 and CA ± 2.

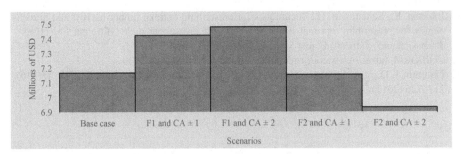

Fig. 6. NPV calculated for each scenario.

The average economic performance of each evaluated scenario is presented in Fig. 6. The base scenario generated NPV of USD 7.17 million of USD. On the other hand, the F1 and CA ± 1 scenarios presented a 3.5% higher NPV. In turn, the F1 and CA ± 2 scenarios were 4.2% higher than the base scenario. However, the F2 scenarios were 0.2% and 3.4% lower than the base scenario for CA ± 1 and CA ± 2 respectively.

4 Conclusions

The flexibility of the CA leads to a greater stability of production, mainly when using objective function F2. However, the highest economic performance is obtained with objective function F1. The use of one or the other will depend on the overall objective of the organization.

Regarding profitability, the results of F2 were exposed to determine the cost of stabilizing the interannual production and, consequently, ordering the forest. Possibly, this situation triggers a significant reduction mainly in the fixed operating costs if a more detailed cost scheme was used.

On the other hand, it is coherent to find higher economic returns in F1, where the model aims at longer felling shifts, thereby reducing the area affected by silvicultural activities, consequently resulting in lower costs and higher income with higher diameter products, with a higher price-volume ratio.

References

1. Andersson, D.: Approaches to Integrated Strategic/Tactical Forest Planning. Licentiate thesis. Swedish University of Agricultural Sciences. Umeå (2005)
2. Broz, D., Durand, G., Rossit, D., Tohmé, F., Frutos, M.: Strategic planning in a forest supply chain: a multigoal and multiproduct approach. Can. J. For. Res. **47**, 297–307 (2017)
3. Gilmore, P.C., Gomory, R.E.: A linear programming approach to the cutting stock problem. Oper. Res. **9**(6), 848–859 (1961)
4. Curtis, F.: Linear programming the management of a forest property. J. For. **60**(9), 611–616 (1962)
5. Pnevmaticos, S., Mann, S.: Dynamic programming in tree bucking. For. Prod. J. **22**(2), 26–30 (1972)
6. Johnson, K., Scheurman, H.: Techniques for prescribing optimal timber harvest and in techniques for prescribing optimal timber harvest and investment under different objectives - discussion and synthesis. Forest Science. Monograph 18 (1977)
7. COIFORM, https://www.coiform.com.ar. ultimo acceso 17 Aug 2020
8. Fiorentin, L.D., et al.: Strategies for regulating timber volume in forest stands. Sci. For. **45**(116), 717–726 (2017)

Process Mining and Value Stream Mapping: An Incremental Approach

Mário Luis Nawcki, Gabriel Nogueira Zanon(✉), Licia Cristina de Paula Santos, Eduardo Alves Portela Santos, Anderson Luis Szejka, and Edson Pinheiro de Lima

Pontifical Catholic University of Paraná - PUCPR/PPGEPS, Imaculada Conceição Street, 1155, Curitiba, PR 80.215-901, Brazil
mario.nawcki@pucpr.edu.br, {eduardo.portela, anderson.szejka,e.pinheiro}@pucpr.br

Abstract. Organizations of all sizes and complexities are increasingly facing a challenging environment. Competition is increasingly fierce and market dynamism requires agility in emerging issues to adapt and survive. Understanding and promoting process improvements is a necessity that has troubled managers and, these use methods and tools to support their work teams in the search to eliminate losses and waste. This article aims, from a theoretical and conceptual approach, to understand VSM, its application, main positive aspects, and problems, as well as to present concepts and premises of process mining. To study two process mapping tools that are Value Stream Mapping (VSM) and Process Mining PM showing a way to improve these approaches. Seek its foundations in a quick literature review, propose a method to apply the two tools empirically and in a case study provide the authors with evidence that indicates an incremental improvement in the mapping process.

Keywords: Process mining · Value stream mapping · Process improvement

1 Introduction

Organizations of all sizes and complexities need continuous adaptation to changes in the global, economic, and market environments. This adaptation requires solutions that consider the organization in a systemic way, that is, its role, people, knowledge, information, processes, strategy and technology [1].

A widely used tool for seeking improvement in the process is the Value Stream Mapping (VSM), which consists of an analysis performed with a pen and paper to understand the process flows and that, at the end of the analysis, a future map can be drawn, considered ideal for the process. VSM, which has an easy language, makes it possible to evaluate both the production and information flow, allowing a real view of the process [2]. Despite the many benefits of the tool, in complex and dynamic environments the level of difficulty in the application of VSM increases. For [3] and [4], for example, the problems of applying the tool in these environments appear mainly in the collection and updating of data, as complexity can lead to a high time for the construction of the

© Springer Nature Switzerland AG 2021
D. A. Rossit et al. (Eds.): ICPR-Americas 2020, CCIS 1408, pp. 171–183, 2021.
https://doi.org/10.1007/978-3-030-76310-7_14

mapping and even so some information can be neglected. Also, continuous monitoring of flows through VSM in highly dynamic environments is often not feasible.

With the rapid evolution of technology and the constant search for innovation in organizations, new perspectives and combinations of actions are developed in search of increasing business actions and executions, among them we can mention the computerization and digitalization of equipment, activities, or processes. One of the aspects of evolution, which can be considered an increase in process improvement tools, such as VSM, is Process Mining. The idea of process mining is to discover, monitor and improve real processes (that is, unassumed processes) by extracting knowledge from the event logs available in current systems [5].

This article aims, from a theoretical and conceptual approach, to understand VSM, its application, main positive aspects, and problems, as well as to present concepts and premises of process mining. Then, process mining is proposed as an incremental approach to VSM. Based on the theoretical analysis, a practical case study was then carried out to evaluate the applicability of the obtained perceptions and the proposed method.

The structure of the article has a brief review of the literature regarding the value stream mapping and process mining in Sect. 2. In Sect. 3, the proposed method and the framework developed for the application of process mining are demonstrated. Section 4 shows the application of the method in a case study in the automotive industry. Finally, in Sect. 5, discussions about the work developed are presented, followed by the conclusions in Sect. 6 of the article.

2   Literature Review

The literature review aims to provide theoretical support about value flow mapping (VSM) and process mining (PM), raising its concepts, perspectives, application, assumptions, and vulnerabilities.

2.1   Value Stream Mapping

Value stream mapping is a paper and pen tool that helps you understand the flow of material and information as a product progresses through the value stream, helping to identify waste. For mapping, one must follow the production flow of a product, from the supplier to the customer, and carefully draw a visual representation of each process in the material and information flow. Then, a future map can be drawn of how the product should flow optimally, which will serve as the basis for an action plan. VSM concerning other techniques stands out for being able to demonstrate, in an easy language, the relationship between production and information flow both within the organization and within the supply chain. Also, the tool can provide important information such as stock levels and production time, helping to make decisions regarding production flows [2].

From the studies by [6], it can be inferred that although it has numerous advantages, VSM has some limitations. First, it can be a big challenge when it comes to a complex product or process. The level of difficulty in preparing the VSM can be high due to the high effort involved in data collection and the time spent in building the mapping of the current state. Also, the existing complexity and possible invisibilities of the process can prevent the collection of sufficient data by direct observation. Another factor is that VSM is a static method, based on paper and pen, so its precision level and dynamic capture capacity are limited. Thus, many companies fail to apply the value flow mapping continuously, as there are always changes in products and processes. However, to obtain improvement effects, continuous monitoring is necessary for several months.

To better understand the difficulties experienced with the use of VSM, the present work brings some important information taken from a set of articles where the VSM was applied, and from which it was possible to remove some problems identified when applying the tool. These works were found through a brief survey on the problems related to VSM and brought several characteristics that were grouped into 7 categories.

1.   Non-integrated processes: lack of integration between processes makes it difficult to analyze the value flow.
2.   Lack of standardization and stability: processes are not clear. Existence of more than one material flow in the same line.
3.   Poor understanding of the Lean principles and VSM: there is no deep understanding of the basic concepts of the tools.
4.   Difficulty in measuring in-process data: problems with layout, complexity of product and process, which end up making it difficult to measure data in-process.
5.   Current flow map out of date: Process that changed over time without updating documents.
6.   Very flexible production: constant change to meet market demand: production mix with small batches, and many products assembled on the same line.
7.   Process performed intuitively: process flow depends on the tacit knowledge of the operator, deciding in real-time how the product enters the flow.

In addition to the problems reported in the categories presented, some authors still cite possible causes for the problems encountered when applying VSM as; the increase in the product portfolio offered by the companies, the difficulty in choosing one of the products to be evaluated by the VSM, and the lack of a holistic view of the managers, who are unable to see the importance of making their processes clear to the partners. Other factors pointed out are the lack of process stability, errors in data measurements, product/process complexity and their obsolescence [3, 4, 7].

As can be seen in the problems pointed out by the authors, for an effective VSM that can provide support for decision making, it is necessary that the company has a well-structured process and can update the VSM frequently, as today companies have an increasingly short product life cycle, which makes VSM obsolete quickly. Besides, Table 1 shows that one of the most pointed problems in the literature is the difficulty in measuring data in-process. Because of this, this article proposes the use of process mining as a tool capable of assisting in data collection and process mapping, allowing a faster and more assertive update of the value flow map.

2.2 Process Mining

Clearly, process models play an important organizational role. The modeling of processes and/or workflows, as in the case of VSM, seeks to reflect an alignment between what is

Table 1. List of problems cited by authors in the application of VSM.

Problems	Authors
Non-integrated processes	[3, 4, 7–11]
Lack of standardization and stability	[3, 10–12]
Poor understanding of the Lean principles and VSM	[2, 7–10, 13–18]
Difficulty in measuring in-process data	[2, 3, 7, 8, 11–15, 17, 19–21]
Current flow map out of date	[2, 3, 7, 13]
Very flexible production	[2–4, 7–11, 14, 16, 17, 20, 22]
Process performed intuitively	[3, 8, 10, 15]

modeled and what actually occurs in real age. Unfortunately, many of the "hand-made" methods do not have this desired connection, showing only an idealized view of the facts [23]. The re-design and process innovation approaches can present a failure indicator of 60–70%, since most of these approaches are subjective, incomplete and require high availability of time and resources, since they are based on interviews and observations of actors involved in the process [24].

Organizations face constant challenges to maintain competitiveness and process improvement initiatives, such as VSM, are often conducted in search of improvement in the efficiency and effectiveness of processes. Process mining presents itself as an important facilitator of continuous process improvement, as the use of the event log can be used as information evidence of processes [25].

Process mining offers an interesting approach that seeks to solve or mitigate these problems. As organizations depend on information systems that support activities, they can record a large amount of data, for example: which activities were performed, who performed them, when they were performed. This event data can be organized as a path that contains the history of what, in fact, occurs during the execution of a process. This history can be analyzed using process mining techniques [26]. It is a relatively young research discipline, between machine learning and data mining, on the one hand, and process modeling and analysis, on the other. The idea of process mining is to discover, monitor, and improve real processes (that is, unassumed processes) by extracting knowledge from the event logs available in current systems [5].

The collection of information about the processes, from the information systems, when performed automatically reduces the time of analysis of the processes. Even so, the main benefit of process mining techniques is the possibility of analyzing a model based on the actual execution of the steps. Therefore, the insights obtained during the analysis, as well as the knowledge acquired, can be used to effectively improve the processes submitted to mining techniques. The event data can be analyzed according to different perspectives: (1) the flow-control perspective; (2) the organizational perspective; (3) the data perspective; and (4) performance perspective. The flow-control perspective is concerned with the behavior of the process, namely, the process and its order of execution. The organizational perspective focuses on the relationships between users who performed the activities, as if they belong to the same group or to different groups or organizational units. The data perspective is related to the data objects that serve as input and output for the activities in a case. The performance perspective aims to detect bottlenecks or calculate performance indicators, such as production times and, therefore, recording times [26].

Another possible use for process mining is conformity assessment, which consists of comparing a hand-drawn model (eg, VSM, BPM, VBM) with its event log. The compliance check can be used to determine efficiency and raise proposals for improvement in the face of an established process [27]. The quality of the process, generally, takes into account factors of case behavior, precision, generalization, and simplicity [23].

This mining application supports decision makers on issues such as time spent passing cases through the process, transitions and intermediary decision making, time consumed at each stage, critical activities and resources and time between activities [28]. Decision-makers in organizations show great interest in mining processes by realizing the high added value for the business and the possibility of insights brought about during process improvement studies [29].

3 Methodology

The methodology seeks to establish empirically, through a case study, improvements in the mapping of the value stream, which, despite being widespread and with satisfactory results, can still evolve with the increment of process mining. The study is carried out in two phases, being: first the execution of the VSM according to the steps in Fig. 1, which are: specify the customer data; place the functions and departments involved; identify units that trigger the process; record the main steps; evaluate the stages, interfaces and information flows of the process; register flows of external material and services; insert the timeline.

The second phase consists of the application of process mining in the process previously analyzed by VSM. The steps of process mining are performed according to Fig. 1, which are: extraction of the event log; analysis of the event log; process analysis; and results. The methodology can also be used in similar processes where it is possible to remove the event log from the database.

4 Case Study

The company used for the case study is part of the Brazilian automotive sector and has a good level of maturity in lean manufacturing practices. The area to which the methodology is applied is Quality, which has experience in using value stream mapping. In order to increment the VSM with the benefits of process mining, both approaches were conducted. VSM by technicians, Quality engineers, and an external consultant, and the process mining relied on Quality engineering. In both processes, there was also the accompaniment and support of the authors for understanding, applying the proposed method, and analyzing the results.

The execution had two distinct phases. In the first phase, a mapping of the value flow was carried out to handle of non-conformity, using paper and pen. The process was analyzed in-depth following the steps in Fig. 1, requiring five meetings of 3 h to map

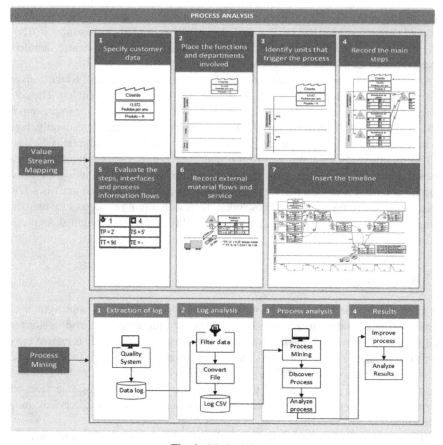

Fig. 1. Methodology

the process. With this activity, it was possible to perceive the flow of the Handling of Non-Conformity (HNC) as well as to see weaknesses in the process (Fig. 2).

In the second phase, with the process modeled by VSM, the focus is on the existing Quality system, and the analysis is done through a log of events that was extracted via the Quality Information System. This step aims to bring the analysis under the lens of process mining in the same process analyzed by VSM, providing an increase in analysis and other possibilities for improvement.

The next subsections of the article detail the application of process mining from log extraction to processing analysis in the software. The results step, shown in Fig. 1, does not apply to the scope of this article, since the present study intends to verify what the process mining can add to the VSM. However, with the analysis of the process found, the team responsible for quality management in the company will continue to improve the process and further evaluate it.

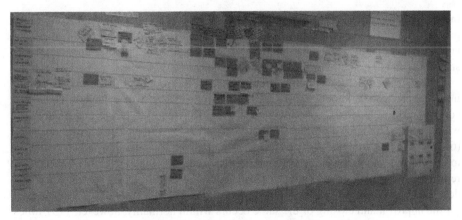

Fig. 2. Value stream mapping (VSM)

4.1 Extraction and Analysis of Log

The structure of the case study follows the methodology presented in Fig. 1. It starts with the step of extracting the Quality System data log, and after the data collection, a first sorting was carried out on the file, removing unnecessary information for the study. The file was then converted and mined in the DISCO software, where the process could be analyzed by checking its adherence to the VSM model.

Although process mining is a very important tool for analyzing and understanding a process, the starting point for mining is not always easy. More than extracting data from a system, it is necessary to prepare it for the processing in question. Thus, the first stage needs the support of the IT area to filter the activities and the important attributes for analysis. For the study, the case ID and records such as date, activities, and employees were selected. After identifying the attributes of the log, it was converted to the standard format of the DISCO software.

4.2 Process Analysis

The analysis of the process included 27,181 events, which in turn were contained in 7714 cases, 4 activities, over 13 years and 11 months. The activities of the process are: Opening, Plan in execution, Analysis of effectiveness, and Completed. It is expected to observe that activities follow this order within the process model found by data mining.

Process Discovery

Through the event log, the Software automatically builds a process model, where it is already possible to find out how the process actually behaves, that is, in this case, it could analyze itself from the process flow found if what was mapped in the VSM corresponds to the reality. Process mining, therefore, allows companies to confirm or correct their ideas about the existing process. Besides, the tool allows an in-depth analysis of the process performance.

The Fuzzy Miner algorithm was used to analyze the process model, this algorithm is capable of analyzing models with a large number of events. Figure 3, shows the model of the uncovered process, indicating the real behavior of the handle of the non-conformity process. As can be seen, the process model is a spaghetti model, which contains frequent and infrequent paths. Analyzing the process, we can see that in most cases the events start with plan in execution, skipping the Open stage.

The extracted model showed differences from the proposed ideal model. Taking into account that the model phases are similar to a PDCA, where: Plan is Open, Do corresponds to Plan in execution, Check is the Effectiveness Analysis and finally, Act would be the Completed profile. We have the following situations: (1) most analyzes (6495/7714) start from the plan in execution, not making clear the study and preparation to understand and propose the appropriate action plans for the HNC; (2) 47 analyzes return from the plan in execution to Open, implying that the plans were created and deleted; (3) 56 times was looped in the Open status; (4) The stage of the Plan created for the approval and closing of the same occurs within normality, however, 1737 times the plan is not effective and returns to replan the actions, it can also be understood as the action is taken and the problem to recur; (5) The effectiveness analysis when the analysis is approved receives the status of completed. Some non-compliant flows are also observed, showing great potential for improvement in the system regarding compliance.

In summary, the model did not show the expected behavior, where the handle of non-conformity would start with the Open activity, move to the Plan in execution, after the Analysis of effectiveness, and finally, the handle would be concluded.

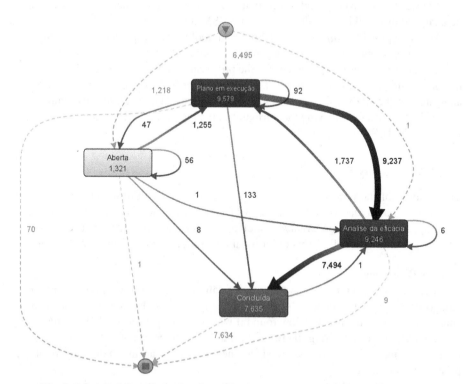

Fig. 3. Model of the non-conformity treatment process using the DISCO software.

Process Performance Analysis

This section presents the analysis of the performance of the Quality System, which intends to discover the bottleneck of the process and whether the events presented are within the time limit estimated by VSM. It was possible to evaluate the events since when the system was created, for comparison, the average time between events was chosen. Figure 3, shows the performance of the process presented in the log, with the thickness of the arc and the color representing the average times between the activities. The thicker and stronger color indicates the bottleneck of the process. As can be seen in Fig. 3, the bottleneck of the process is between the activity of Plan in execution and Analysis of effectiveness with an average of 18 weeks and with 9237 plans created. The 61.8-day effectiveness analysis shows that technicians wait this time to understand what action was effective. The accuracy of the information extracted, the time used, as well as the ease in analyzing the data, were characteristics evaluated by practitioners as very positive in the use of process mining.

Another analysis made was in relation to the conformity of the process, which showed that of the 7714 events that started, 7635 appear finished. 1737 action plans were not effective, and in the effective analysis, they were not approved, returning to the previous phase to redo the same. One must question the importance of the Open activity, which in most cases starts with the plan being executed. Taking into account that the time taken to handle of non-conformity is aimed at 20 days, it can be seen that in most cases this time is exceeded. For a better analysis of this fact, a filter was applied based on the 20-day estimate for the resolution of the handle of non-conformity, where only the events that are within the time limit were left. As a result, 18% of the events in the log are within the stipulated time limit. Within these cases, the average time between the execution plan activities and the effective analysis was 4.8 days.

5 Discussion

During the realization of the case study, it was observed that the VSM methodology analyzed the current state of the process flow of the handle of non-conformity. The approach aimed to provide the Quality manager with a detailed mapping of activities that involve non-conformities, the human resources used, and the time consumed. In the Quality department, the main reactive activity is the problem-solving process, which includes registering the problem in the Quality system, analyzing the root cause in the Kaizen model (A3/project), and defining and monitoring the countermeasure. The technicians are responsible for this activity and it was detected that they were overloaded, therefore, the quality of the handling was not as good as expected. VSM was able to identify and eliminate double work and waste from the process, improve performance and the workload can be revised.

However, the work mobilized the entire department with many hours of work to arrive at a photograph of the process. The energy to map the flow was so great that repeating the process with the required frequency proved to be unfeasible. The discussions for the construction of the map were based on people's intuition, which in addition to taking a long time, often generated inaccurate and unreliable data.

To conduct the activities, it was necessary to hire an external consultant, because although the methodology is quite widespread, many doubts arose on how to map the activities and mainly the fact that people often present different, but legitimate points of view. VSM achieved some of its objectives, but for the manager, it was not possible to promote more profound changes because it would be necessary to run the cycle several times, which proved to be impracticable due to the energy and resources used. The comparison of the two VSM and process mining approaches is summarized in Table 2.

In relation to VSM, it is possible to verify that process mining allows a visualization of the process as it actually occurs, thus requiring less human effort and mitigating the subjectivity existing in the discovery phase. In addition, process mining allows for quick and constant updating of process models, as it occurs from readings, from systems, extracted in real-time and stored in an event log.

Still, to be able to execute process mining projects, there is a need for data storage, requiring computational structure and acquisition systems. These data can be extracted with dirt and found scattered in several different systems, which impacts the need for processing these data. In addition to that, structuring to collect data from systems can imposing the organization with implementation costs.

Table 2. The comparison of approaches: VSM and process mining.

	VSM	Process mining
Numbers of people involved	22 – the entire Quality team	2–1 engineer and 1 manager
Hours used	5 meetings of 3 h × 10 people = 150 h	To remove the event log from the system: 8 h, to pass to the DISCO software: 10 h, analysis time: 10 h = 28 h
Level of detail	Relation to people's capacity for abstraction	Logs written to the system
Ability to repeat the process	Difficult - difficulty in bringing together all involved	Easy - whenever needed
Reliability of information	Estimated times - People have different ways of seeing the process and many disagreements regarding the duration of activities	Times collected from the system - The data are in accordance with the process performed in the Quality system
Period analyzed	It is a photograph of the process	It is possible to analyze different moments of the process
Ability to assess system compliance	It's not possible	It is possible to see all occurrences, even those that have not been completed or that have not followed the required process

6   Conclusion

This study presents a method for using process mining to increment VSM. In general, processes, such as the non-conformity approach, are dynamic and complex. Thus, the data provided by systems is highly valuable so that, when applied to process mining, and analysis of the real situation of the process is carried out, enabling a deep understanding of the behavior of the activities.

From the evaluation of the contents present in the literature, in addition to the applied case study, it is understood that currently, process mining can be an incremental tool for process improvement initiatives such as VSM. Process Mining allows the discovery and updating of models that are true to reality, supporting the analysis and decision-making phases present in the VSM. Decision making and improvement planning is not carried out by the machines and is not foreseen in mining models. Thus, the models are complementary. In the future, with greater technological maturity and mining models, they can become prognostic, making the model itself propose and set up the improvements, making the process mining complete and enabling the replacement of the VSM.

One of the objectives achieved with the use of the PM after the VSM, from the case study, was to understand the real lead time of the dealings, making it clear that the times before suggested by the people in most cases were very different from what actually happened. It was also possible to prove that many analyzes did not follow the recommended path, and worse than that, many were not even concluded, being lost in the system. It is possible to infer that this incremental approach can extend to other types of processes, as long as the system requirements and event log extraction are met.

According to Slack and Lewis (2009) the excessive "quality bureaucracy" associated with TQM is expensive and time-consuming. This affirmation with the PM was proven with a very high lead time of the negotiations and with the percentage of analyzes that were started and were not finished being lost in the system, characterizing a total waste.

Future work can be developed through event logs containing a greater wealth of details so that the additional information provides an analysis of compliance regarding the stages and times of handle of non-conformity processes, as well as the inclusion of incremental improvements in the process. Also, these works can address other types of processes to validate the proposal developed against new applications.

References

1. Bellamy, M.A., Basole, R.C.: Network analysis of supply chain systems: a systematic review and future research. Syst. Eng. 16(2), 235–249 (2013). https://doi.org/10.1002/sys.21238
2. Braglia, M., Frosolini, M., Zammori, F.: Uncertainty in value stream mapping analysis. Int. J. Logist. Res. Appl. 12(6), 435–453 (2009). https://doi.org/10.1080/13675560802601559
3. Forno, A., Pereira, F., Forcellini, F., Kipper, L.: Value stream mapping: a study about the problems and challenges found in the literature from the past 15 years about application of Lean tools. Int. J. Adv. Manuf. Technol. 72(5–8), 779–790 (2014). https://doi.org/10.1007/s00170-014-5712-z
4. Eisler, M., Horbal, R., Koch, T.: Cooperation of lean enterprises — techniques used for lean supply chain. In: Olhager, Jan, Persson, Fredrik (eds.) APMS 2007. ITIFIP, vol. 246, pp. 363–370. Springer, Boston, MA (2007). https://doi.org/10.1007/978-0-387-74157-4_43

5. van der Aalst, W.: Process mining: discovering and improving Spaghetti and Lasagna processes, no. c, pp. 1–7 (2012). https://doi.org/10.1109/cidm.2011.6129461
6. Knoll, D., Reinhart, G., Prüglmeier, M.: Enabling value stream mapping for internal logistics using multidimensional process mining. Expert Syst. Appl. **124**, 130–142 (2019). https://doi.org/10.1016/j.eswa.2019.01.026
7. Salzman, R.A.: Manufacturing system design : Flexible manufacturing systems and value stream mapping, p. 126 (2002)
8. Bauch, C.: Lean Product Development: Making waste transparent Diploma thesis. Technical University, Munich, p. 140 (2004)
9. Frenkel, Y.: B.S. Chemical Engineering, Rensselaer Polytechnic Institute and Submitted, 1999 (2004). Accessed 1 July 2004
10. Grove, A.L., Meredith, J.O., Macintyre, M., Ange, J., Neailey, K.: UK health visiting: challenges faced during lean implementation. Leadersh. Heal. Serv. **23**(3), 204–218 (2010). https://doi.org/10.1108/17511871011061037
11. Kurilova-palisaitiene, J., Sundin, E.: Remanufacturing : Challenges and Opportunities to be Lean. In: EcoDesign 2013 International Symposium, pp. 1–6 (2013)
12. de O. Oliveira, F.S., Antonioli, A.F., de Q. e Silva, A.: Mapeamento do fluxo de valor: Uma revisão e classificação da literatura em publicações nacionais. In: XXXV Encontro Nac. Eng. ProduçãoEncontro Nac. Eng. Produção, vol. II, no. 3, pp. 369–389 (2008). https://www.abepro.org.br/biblioteca/TN_STO_206_219_27613.pdf
13. Kato, J.: Development of a process for continuous creation of lean value in product development organizations, no. 1996, p. 206 (2005)
14. Alves, T.D.C., Tommelein, I.D., Ballard, G.: Value stream mapping for make-to-order products in a job shop environment. In: Construction Research Congress, 2005 Broadening Perspectives - Proceedings Congress, vol. 40754, no. May 2014, pp. 13–22 (2005). https://doi.org/10.1061/40754(183)2
15. Hines, P., Rich, N., Esain, A.: Value stream mapping: a distribution industry application. Benchmarking Int. J. **6**(1), 60–77 (1999). https://doi.org/10.1108/14635779910258157
16. McManus, H.L., Millard, R.L.: Value stream analysis and mapping for product development. Technology **20**(3), 8–13 (2002). https://doi.org/10.1016/S0142-694X(98)00018-0
17. Mcmanus, H.L.: Product development value stream mapping (PDVSM) manual, no. September (2005_
18. de Queiroz, J.A., Rentes, A.F., de Araujo, C.A.C.: Transformação Enxuta: Aplicação do Mapeamento do Fluxo de Valor em uma Situação Real. In: XXIV Encontro Nac. Eng. Produção (2004)
19. Childerhouse, P., Towill, D.R.: Reducing uncertainty in European supply chains. J. Manuf. Technol. Manag. **15**(7), 585–598 (2004). https://doi.org/10.1108/17410380410555835
20. Ross, J.D.: Plant Efficiency. Trans. Am. Inst. Electric. Eng. **XXXI**(1), 471–484 (1912). https://doi.org/10.1109/T-AIEE.1912.4768425
21. Klotz, L., Horman, M.: Transparency, process mapping and environmentally sustainable building projects. In: Lean Construction: A New Paradigm Management Capacity Project - 15th IGLC Conference, no. January 2007, pp. 322–331 (2007)
22. Ikovenko, S., Bradley, J.: TRIZ as a lean thinking tool. J. Appl. Phys. **91**(10), 7526–7528 (2004)
23. van der Aalst, W.: Process Mining: Data Science in Action. Springer, Heidelberg (2016). https://doi.org/10.1007/978-3-662-49851-4
24. Park, S., Kang, Y.: A study of process mining-based business process innovation. Procedia Comput. Sci. **91**, 734–743 (2016). https://doi.org/10.1016/j.procs.2016.07.066
25. Low, W.Z., van der Aalst, W.M.P., ter Hofstede, A.H.M., Wynn, M.T., De Weerdt, J.: Change visualisation: analysing the resource and timing differences between two event logs. Inf. Syst. **65**, 106–123 (2017). https://doi.org/10.1016/j.is.2016.10.005

26. Rebuge, Á., Ferreira, D.R.: Business process analysis in healthcare environments: a methodology based on process mining. Inf. Syst. **37**(2), 99–116 (2012). https://doi.org/10.1016/j.is. 2011.01.003

27. Van Der Aalst, W.: Process mining: overview and opportunities. ACM Trans. Manag. Inf. Syst. **3**(2), 1–17 (2012). https://doi.org/10.1145/2229156.2229157

28. Alves de Medeiros, A. K., van der Aalst, W. M. P.: Process mining towards semantics. In: Dillon, Tharam S., Chang, Elizabeth, Meersman, Robert, Sycara, Katia (eds.) Advances in Web Semantics I. LNCS, vol. 4891, pp. 35–80. Springer, Heidelberg (2008). https://doi.org/ 10.1007/978-3-540-89784-2_3

29. dos Santos, C., et al.: Process mining techniques and applications – A systematic mapping study. Expert Systems with Applications **133**, 260–295 (2019). https://doi.org/10.1016/j.eswa. 2019.05.003

Minimizing Total Implementation Cost of a 2-D Indoor Localization System with a Constant Accuracy for Underground Mine Tunnels

Andrea Teresa Espinoza Pérez[1], Daniel A. Rossit[2,3], and Óscar C. Vásquez[1(✉)]

[1] Department of Industrial Engineering, Universidad de Santiago de Chile, Santiago, Chile
{andrea.espinozap,oscar.vasquez}@usach.cl
[2] Engineering Department, Universidad Nacional del Sur, Bahía Blanca, Argentina
[3] INMABB UNS CONICET, Bahía Blanca, Argentina
daniel.rossit@uns.edu.ar

Abstract. In this paper, we introduce the problem of minimizing total implementation cost of a 2-D indoor localization system for underground mine tunnels, guaranteeing a constant accuracy. To address this problem, we propose a system based on Cell-ID technique and visible light communication (VLC) technology with square panels of light-emitting diode (LED) lights, which fixes the maximum position error in each cell by considering the distribution of overlapping and non-overlapping cells. Formally, our system is mainly defined by an easy-to-implement algorithm based on a simple order rule. In order to illustrate the usefulness of the proposal, an example is provided. Finally, potential applications in the industrial environments are discussed and future research is proposed.

Keywords: Visible light communication · Indoor localization system · Accuracy · Cost minimization

1   Introduction

In the last decade, visible light communication (VLC) systems for positioning has gained the attention of researchers due to its merits in terms of accuracy, cost, safety, and reliability for indoor environments [13]. Several works are related to channel design models [8,10], while others researches assessed signal propagation and interference mitigation [11] . Most of researches related to the physical design of VLC [5,12,15] seeks to improve the system accuracy, considering accuracy as a result of the optimized location of the LED lights. Nevertheless, the research carried out by [6] seeks to determine the distribution of circular light-emitting diode (LED) lamps, mounted on the ceiling, in order to provide a 1-D linear positioning system with a constant accuracy along the tunnel. This perspective has large advantages in order to improve safety in mining operations.

© Springer Nature Switzerland AG 2021
D. A. Rossit et al. (Eds.): ICPR-Americas 2020, CCIS 1408, pp. 184–189, 2021.
https://doi.org/10.1007/978-3-030-76310-7_15

Complementary to the previous works in this field, our research integrates the cost and 2-D positioning systems with overlapping and non-overlapping cells, which allows to explore Cell-ID technique and VLC applications with square panels of LED light in larger mining tunnels.

2 Description of the Indoor Positioning System

2.1 General Overview

We introduce the problem of minimizing total implementation cost of a 2-D indoor localization system with a constant accuracy. Consider underground mine tunnels, where the location is determined into the two coordinate axis (along the tunnel). In this scenario, we assume a 2-D indoor localization system based on Cell-ID technique and VLC technology. A practical implementation of Cell-ID and VLC technology is detailed in [6]. An unique code is assigned to each cell, in order to identify the lamp location. Each LED device is identical and produces a square illumination pattern subject to the used beam angle, which constraints on the minimum and maximum lighting area at the floor level and the position defined by it. Considering that the calculation of the accuracy depends on the amount of error present in the localization [16], several areas could be not necessary covered with a specific and constant error at minimal cost by using the same size of squares LED device.

To face this situation, a solution alternative emerges from the distribution of overlapping and non-overlapping cells. Some similarities with other problems in the family of *packing* are found. Considering the overlapping condition, our problem is close to the *three-dimensional bin packing problem* [9] and the *container loading problem* [3]. However, the most of these problems on 2D do not allow the overlapping elements to be positioned [7]. In addition, these problems consider the number of elements (or boxes) and their dimensions as parameters, no decision variables, with the goal of minimizing the free space [4]. Therefore, the results do not necessarily have to completely cover the surface.

To address our problem, without loss of generality we consider a rectangular area of sides a and b, $a \leq b$.

In the case of non-overlapping cells, a LED device produces a square illumination pattern of side ℓ with a maximum distance error δ between the real and estimated positions. This above value is expressed by $\delta = \frac{\ell\sqrt{2}}{2}$. Then, the projection of nine LED lamps has a coverage area of $(3\ell)^2$. Without loss of generality, we fix the maximum distance error $\delta = \frac{\sqrt{2}}{2}$, with $\ell = 1$. A VLC positioning system projection of nine square LED lamps in a rectangular area with $b = a$ with non-overlapping cells is illustrated in Fig. 1.

To consider the case of overlapping cells, we introduce the concept of *complete cell* as follows:

Definition 1. *A complete cell is a set of equal overlapping cells that defines a maximum distance error $\delta = \frac{\sqrt{2}}{2}$ independent of the number of cell identification codes received.*

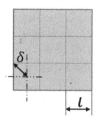

Fig. 1. VLC positioning system projection of nine square LED lamps with non-overlapping cells.

Figure 2 illustrates a scenario with a complete cell with four overlapping cells of side $\ell = 2$, assuming equal distance between LEDs and identical beam angles. In this case, the mobile node estimated position will depend on the number of identification binary codes received from each cell and the maximum distance error δ between the real and estimated positions still remain the same, as it is depicted in Fig. 1.

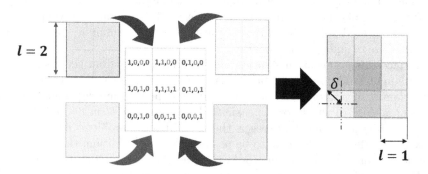

Fig. 2. VLC positioning system with a complete cell defined by four overlapping cells. The binary number with colors represents the possible cell identification codes received. (Color figure online)

Theorem 1. *Consider a positive integer number ℓ. A complete cell of side $2 \cdot \ell - 1$, requires a set of ℓ^2 equal overlapping cells of side ℓ.*

The proof is based on geometric Euclidean analysis. The detail is omitted by space constraint.

From these above results, we propose an easy-to-implement algorithm in the next section.

2.2 Algorithm

Let α_ℓ be the cost of a unit cell of side ℓ, $\ell \geq 1$. First, we obtain a feasible solution for the rectangular area $a \cdot b$ by a number $a \cdot b$ of non-overlapping cells of side $\ell = 1$

with a total implementation cost equal to $a \cdot b \cdot c_1$. The previous value represents an upper bound for the objective value of the problem. Second, we compute the complete cell of maximum side ℓ_{max} and the maximum number of each feasible complete cell β_ℓ defined by cells of side $\ell \leq \ell_{max}$, comparing the implementation cost of a complete cell defined by cells of side ℓ and the non-overlapping cells of side $\ell = 1$ subject to the same coverage area. Finally, the algorithm considers a simple rule defined by the decreasing cost order $(2 \cdot \ell - 1)^2 / (\ell^2 \cdot \alpha_\ell)$, locating the cells configuration according to it when possible, and completing the remaining area with non-overlapping cells of side $\ell = 1$.

3   An Illustrative Example

Table 1 shows the parameters considered for an illustrative example, considering a rectangular area.

Table 1. Illustrative example parameters.

Rectangular area	Sides	$a = 10$		$b = 20$	
	Area	200			
Unit cell side (ℓ)	1	2	3	4	5
Unit cell cost (α_ℓ)	3.00	5.63	8.85	6.13	12.96
Complete cell side ($2 \cdot \ell - 1$)	1	3	5	7	9
Complete cell cost ($\ell^2 \cdot \alpha_\ell$)	3.0	22.5	79.7	98.1	324.0

The initial feasible solution is given by 200 unit cells of side $\ell = 1$ with a total implementation cost equal to 600. Finally, Fig. 3 represents the solution obtained

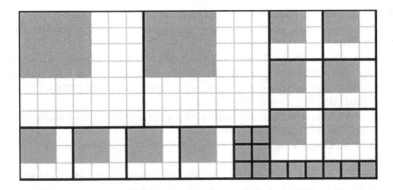

Fig. 3. Optimal VLC positioning system with overlapping and non-overlapping cells for the illustration example. The total implementation cost is equal to 457.36. The number of the considered complete cells defined by the overlapped unit cells of side ℓ equal to 4 and 2; are 2 and 10 respectively. The remaining space is completed with 12 unit cells of side $\ell = 1$.

by the easy-to-implement algorithm, where the gray squares corresponds to unit cells that define the complete cells. This solution is optimal.

4   Conclusions

This article introduced the problem to minimize the total implementation cost of a 2-D indoor localization system based on Cell-ID technique and VLC technology for underground mine tunnels, guaranteeing a constant accuracy. The system is based on easy-to-implement algorithm based on simple order rule. The usefulness of the proposal is illustrated by an example.

From the obtain results, the benefits of our proposal for transmitting positional information about elements within indoor scenario could be extended to the industrial environment, in particular to conceive and design warehouse management systems (WMS), one of the central links in any supply chain [14]. In these systems, the part traceability control is based on detection by RFID systems or manual picking performed by operators with barcode reading devices [1]. However, it can lead to somewhat limited information on the real-time positioning of parts or inventory items. That is, once an item has been removed from its shelf (or rack) and is moved to another location, we only have information about the item being removed, and not about the evolution or path of that item. If there is movement control, it is on the carrier of the element [2]. Thus, our proposal would greatly improve the availability of information, with an easier implementation since it is directly linked to the lighting system.

Finally, future research work could be focused on the formulation of mathematical models and algorithms in order to obtain optimal or approximate solutions for the problem.

References

1. Atieh, A.M., et al.: Performance improvement of inventory management system processes by an automated warehouse management system. Procedia Cirp **41**, 568–572 (2016)
2. Fragapane, G., Ivanov, D., Peron, M., Sgarbossa, F., Strandhagen, J.O.: Increasing flexibility and productivity in industry 4.0 production networks with autonomous mobile robots and smart intralogistics. Ann. Oper. Res. 1–19 (2020). https://doi.org/10.1007/s10479-020-03526-7
3. George, J.A., Robinson, D.F.: A heuristic for packing boxes into a container. Comput. Oper. Res. **7**(3), 147–156 (1980)
4. Hougardy, S.: On packing squares into a rectangle. Comput. Geom. **44**(8), 456–463 (2011)
5. Iturralde, D., Seguel, F., Soto, I., Azurdia, C., Khan, S.: A new VLC system for localization in underground mining tunnels. IEEE Lat. Am. Trans. **15**(4), 581–587 (2017)
6. Krommenacker, N., Vásquez, Ó.C., Alfaro, M.D., Soto, I.: A self-adaptive cell-ID positioning system based on visible light communications in underground mines. In: 2016 IEEE International Conference on Automatica, ICA-ACCA 2016 (2016)

7. Lodi, A., Martello, S., Vigo, D.: Approximation algorithms for the oriented two-dimensional bin packing problem. Eur. J. Oper. Res. **112**(1), 158–166 (1999)
8. Mansour, I.: Effective visible light communication system for underground mining industry. Indones. J. Electr. Eng. Inf. **8**(2), 331–339 (2020). https://doi.org/10.11591/ijeei.v8i2.1978
9. Martello, S., Pisinger, D., Vigo, D.: The three-dimensional bin packing problem. Oper. Res. **48**(2), 256–267 (2000)
10. Miramirkhani, F., Uysal, M.: Channel modelling for indoor visible light communications. Philos. Trans. Royal Soc. A: Math. Phys. Eng. Sci. **378**(2169), 20190187 (2020)
11. Palacios Játiva, P., Azurdia-Meza, C.A., Román Cañizares, M., Zabala-Blanco, D., Saavedra, C.: Propagation features of visible light communication in underground mining environments. In: Botto-Tobar, M., Zambrano Vizuete, M., Torres-Carrión, P., Montes León, S., Pizarro Vásquez, G., Durakovic, B. (eds.) ICAT 2019. CCIS, vol. 1195, pp. 82–93. Springer, Cham (2020). https://doi.org/10.1007/978-3-030-42531-9_7
12. Seguel, F.: Robust localization system using Visible Light Communication technology for underground mines. Ph.D. thesis, Université de Lorraine; Universidad de Santiago de Chile (2020)
13. Soto, I.: A hybrid VLC-RF portable phasor measurement unit for deep tunnels. Sensors (Switzerland) **20**(3), 1–17 (2020)
14. Van den Berg, J.P., Zijm, W.H.: Models for warehouse management: classification and examples. Int. J. Prod. Econ. **59**(1–3), 519–528 (1999)
15. Yang, S.-H., Kim, H.-S., Son, Y.-H., Han, S.-K.: Three-dimensional visible light indoor localization using AOA and RSS with multiple optical receivers. J. Lightwave Technol. **32**(14), 2480–2485 (2014)
16. Zia, M.T.: Visible light communication based indoor positioning system. TEM J. **9**(1), 30–36 (2020)

An Integrated Approach for Solving the Bucking and Routing Problems in the Forest Industry

Maximiliano R. Bordón(✉) ⓘ, Jorge M. Montagna ⓘ, and Gabriela Corsano ⓘ

Instituto de Desarrollo Y Diseño (INGAR), Avellaneda 3657, Santa Fe, Argentina
{mbordon,mmontagna,gcorsano}@santafe-conicet.gov.ar

Abstract. In the forest supply chain, decisions related to bucking (cross sectional cuts) and distribution (vehicle routing) activities are usually considered separately and in different planning horizons. This separation between closely related problems gives rise to suboptimal solutions that, if they are simultaneously considered, could bring significant benefits. A tight coordination between the type and amount of material to be harvested and the actually material required to satisfy the daily demand would allow improvements in the efficiency of the production system (both in the forest and in the plant) and significant cost savings. To address this problem, in this work, a mixed integer linear programming (MILP) model that simultaneously considers both decisions is presented. Through a case study, the performance of the developed model is analyzed and the respective conclusions are detailed.

Keywords: Forest industry · Bucking · Vehicle routing · MILP

1 Introduction

Forest supply chain management integrates different operations, such as harvesting, distribution, production, inventory management, among others, which are generally addressed in different planning horizons [1–3]. In a context where decisions are made considering one activity, a negative impact can be obtained on another; therefore, it is essential to improve the efficiency of all operations as a whole. A tight coordination between the type and amount of material to be harvested and the actually material required to fulfill the daily demand will allow improvements in the efficiency of the production system and significant cost savings.

In the existing literature, harvesting and transportation problems are often treated with a decoupled approach. In general, harvesting decisions are first addressed in a tactical planning horizon and later, assuming a certain supply from each harvest area to each plant, transportation is organized [4, 5]. There are situations where, during harvesting, transportation is considered as an additional cost (or a capacity constraint), without detailing how many vehicles will be required for transportation or how distribution will take place [6–8]. Although most of these models are developed for a tactical planning horizon, in daily operations it is generally unknown the effective availability of raw materials in the forests in that period.

© Springer Nature Switzerland AG 2021
D. A. Rossit et al. (Eds.): ICPR-Americas 2020, CCIS 1408, pp. 190–204, 2021.
https://doi.org/10.1007/978-3-030-76310-7_16

The effective availability of raw material is determined through the bucking activity. This activity consists of the transverse cutting of a stem to produce logs of a certain length and diameter, and it can be performed at the destination of the wood (plant or temporary storage sector) or directly in the forest when the stem is harvested. The associated transport activities will then depend on the raw material being transported, stems or smaller logs. The first case is the simplest to address, where it is enough to determine the destination of the harvested stem, while the second case is more complex since transport decisions must necessarily consider aspects related to the load of each truck (logs of the same length must be transported on the same trip, although not necessarily the same diameter). The second approach is addressed in this work.

In this work, a mixed integer linear programming (MILP) model that considers both harvesting and transportation/distribution decisions is proposed. More precisely, it is determined for a weekly planning horizon: the working periods for the harvesting crews, the bucking patterns to be used in each harvest area and the number of logs of a given length and diameter to be obtained, decisions related to log storage in both harvest areas and plants, the truck fleet required to perform the distribution, the routes to be followed by each truck, as well as the composition of the load of each truck (i.e., length of logs to be transported on each trip and the total transported weight). To the best of our knowledge, this is the first time that transportation/distribution decisions and bucking decisions are jointly addressed.

In the following section the problem statement is presented, while in Sect. 3 the proposed MILP model is detailed. In Sect. 4 a case study to analyze the model performance is developed, and finally, in Sect. 5 the final remarks are highlighted.

2   Problem Statement

The main assumptions that are taken into account to model the considered problem are described below.

It is considered a set of plants I whose raw material requirements must be covered. It is assumed that at a higher level of planning (that is, annual harvesting planning) the set of harvest areas F to be exploited during the planning period T considered has already been defined.

2.1   Problem Characteristics Related to Harvesting Activities

The initial number of standing stems (trees) in each open harvest area $f, f \in F$, $qini_f$, is assumed to be known. In turn, the stems that belong to the same harvest area have identical characteristics (planted forest). It should be noted that the harvesting activities considered in this work are those related to stumpage: thinning activities are not included.

At the end of the planning horizon, there is a desired goal of standing stems in the harvest areas, $qend_f$, which depends on the objectives of the company; in other words, the target level for the harvesting activity at the end of the planning horizon.

The harvesting capacity in a certain period is given by the average productivity of the harvesting crews assigned to that area (previously determined in the annual harvesting planning), that is, the minimum and maximum number of stems that can be harvested in a period ($qcsmin_f$ and $qcsmax_f$, respectively) depends on the number harvesting crews assigned to that harvest area. The cost associated with harvesting crew is not considered because it is included in the annual harvest planning and does not depend on the decisions made in this formulation.

For each harvest area f, there is a set of applicable bucking patterns b, $b \in B$, which was previously determined. Each bucking pattern b is characterized by parameters $nu_{l,d,b}$, that correspond to the number of obtained logs of diameter d and length l when bucking pattern b is applied. Also, the level of losses generated when bucking pattern b is applied, $loss_b$, is considered, and it includes both stem fine tip (i.e. wood not suitable for use) and losses due to the taper of the log (see Fig. 1). These losses have an associated cost per ton of unused wood, $closs_b$.

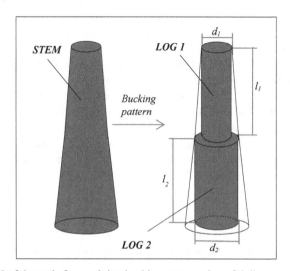

Fig. 1. Useful wood after applying bucking pattern. d: useful diameter, l: length.

Furthermore, it is assumed that if a bucking pattern is applied in a certain harvest area then it must be used at least a minimum number of times, $qminb_b$, in order to avoid setup times and costs. The types of logs that can be obtained by applying the bucking patterns are classified by length l, $l \in L$, and useful diameter d, $d \in D$.

Logs generated in a harvest area can remain on the roadside throughout the planning period without losing quality, in other words, there is no deterioration of the raw material during the planning horizon. However, logs generated in the harvest area f and not distributed at the end of the planning horizon are considered wasted material with an associated cost, $cqrl_{l,d,f}$.

2.2 Problem Characteristics Related to Production Activities

Each plant has associated a weekly demand for logs (by length and by diameter), $dtot_{l,d,i}$, which can be covered during any period of the planning horizon, as long as the maximum processing capacity in each plant is respected, $cap_{i,t}$. However, there is a minimum demand for logs of a certain length and diameter that must be satisfied in each period, $dmin_{l,d,i,t}$. This demand is related to the production commitments assumed by the plants. The minimum processing capacity of plants per period is closely related to the minimum demand for logs.

Each plant has a storage yard for logs. Each type of log has an assigned sector in the storage yard, with maximum capacity $maxstock_{l,d,i}$. Each log stored at the end of the planning horizon has an associated cost $cstock_{l,d,i}$ because they may not be required in the future and lose quality, or on the contrary, be used for the elaboration of products for which they were not intended, affecting the productivity levels of the plant. At the beginning of the planning horizon there is an initial inventory of logs available from previous planning periods, $stockini_{l,d,i}$.

2.3   Problem Characteristics Related to Distribution Activities

Regarding the distribution of the logs generated in the harvest areas, there is a heterogeneous truck fleet C to perform this activity. Each truck has a minimum and maximum load capacity ($capmin_c$ and $capmax_c$, respectively), therefore the quantity of logs to be transported will be limited by their corresponding weight, $weight_{l,d}$.

Each truck has a limited working day ($maxt_{c,t}$) and a maximum number of trips $v, v \in V$ to be completed in each period t. Each segment of the road network admits an average travel speed, which depends on whether the truck is loaded or unloaded. In turn, each truck has a stipulated loading and unloading time ($load_{c,f}$ and $unload_{c,i}$, respectively). Regarding the composition of the load, each truck can only transport logs of the same length per trip, allowing logs of different diameters to be loaded at the same time. Each truck is located at a regional base $p, p \in P$, from where its route begins and where it must end after completing all trips. The composition of routes presented in Bordón et al. (2018) [9] is considered in this work, where basically a route is made up of different types of movements.

The fixed cost for using a truck in a certain period is given by $ctruck_{c,p}$. The company cost policy considers a fixed cost (base) per use of trucks per period and a variable cost per kilometer traveled loaded and unloaded.

2.4   Optimization Variables

The variables considered in this work are the following:

- The number of harvested stems and the number of stems left standing in each harvest area in each period, $Q_{f,t}^{CS}$ and $Q_{f,t}^{SS}$, respectively.
- The bucking patterns selection and the number of times they are used in each harvest area and in each period, $Y_{b,f,t}^{BP}$ and $Q_{b,f,t}^{BP}$, respectively.
- The number of generated logs of each type (length-diameter combination), per period, in each harvest area, $Q_{l,d,f,t}^{CL}$.
- The number of transported logs of each type, per period, from each harvest area to each plant, $Q_{l,d,f,t}^{TL}$.
- The inventory levels of logs of each type, per period, on the roadside in harvest areas, $Q_{l,d,f,t}^{RL}$.

- The number of processed logs of each type (above the minimum demand) in each plant in each period, $Q^{PL}_{l,d,i,t}$.
- The inventory levels of logs of each type, per period, in the storage yard of each plant, $Q^{SL}_{l,d,i,t}$.
- The truck selection in each period, $Y^{T}_{c,p,t}$.
- The type of loaded logs in each trip of each truck and the number of transported logs (of the same length but not necessarily the same diameter) on each trip of each truck from each harvest area to each plant in each period, $Y^{TT}_{l,c,v,f,i,t}$ and $Q^{TT}_{l,d,c,v,f,i,t}$, respectively.
- The route to be performed by each truck in each period, that is, its initial, loaded, unloaded and return movements, ($X^{D}_{c,p,f,t}$, $X^{L}_{c,v,f,i,t}$, $X^{U}_{c,v,i,f,t}$ and $X^{R}_{c,v,i,p,t}$, respectively).

The decisions listed above must be made seeking the minimization of the total costs (Z) related to harvesting and transport activities, guaranteeing the fulfillment of the demand of the plants and satisfying all the operative restrictions.

3 Mathematical Model

The proposed MILP model is the following:

$$\min Z = CV^{Transp} + CF^{Transp} + CV^{Loss} + CV^{Stock} + CV^{Logs} \tag{1}$$

where,

$$CV^{Transp} = \sum_c \sum_{p \in PC_{p,c}} \sum_f \sum_t dpf_{p,f} cd_{p,f} X^{D}_{c,p,f,t} +$$

$$\sum_c \sum_v \sum_f \sum_i \sum_t dfi_{f,i} clf_{,i} X^{L}_{c,v,f,i,t} +$$

$$\sum_c \sum_v \sum_i \sum_f \sum_t dif_{i,f} cu_{i,f} X^{U}_{c,v,i,f,t} +$$

$$\sum_c \sum_{p \in PC_{p,c}} \sum_v \sum_i \sum_t dip_{i,p} cr_{i,p} X^{R}_{c,v,i,p,t} \tag{2}$$

$$CF^{Transp} = \sum_c \sum_{p \in PC_{p,c}} \sum_t ctruck_{c,p} Y^{T}_{c,p,t} \tag{3}$$

$$CV^{Loss} = \sum_b \sum_{f \in BF_{b,f}} \sum_t closs_b loss_b Q^{BP}_{b,f,t} \tag{4}$$

$$CV^{Stock} = \sum_l \sum_d \sum_i cstock_{l,d,i} Q^{SL}_{l,d,i,t=|T|} \tag{5}$$

$$CV^{Logs} = \sum_l \sum_d \sum_f cqrl_{l,d,f} Q^{RL}_{l,d,f,t=|T|} \tag{6}$$

The objective function is given by the minimization of (1), where: (2) establishes the variable cost per traveled kilometer, (3) represents the fixed costs per use of trucks, (4) defines the cost for raw material loss when applying a certain bucking pattern, (5) sets the cost of keeping logs in plant storage yards at the end of the planning horizon, and (6) establishes the cost of not transported logs which remain on the roadside in the harvest areas at the end of the planning horizon.

The restrictions associated with the problem are the following:

$$qini_f = Q^{CS}_{f,t} + Q^{SS}_{f,t} \qquad \forall f \in F, t = 1 \tag{7}$$

$$Q^{SS}_{f,t-1} = Q^{CS}_{f,t} + Q^{SS}_{f,t} \qquad \forall f \in F, t > 1 \tag{8}$$

$$Q^{SS}_{f,t} \leq qend_t \qquad \forall f \in F, t = |T| \tag{9}$$

$$Q^{CS}_{f,t} \geq qcsmin_f\, CUT_{f,t} \qquad \forall f \in F, \forall t \in T \tag{10}$$

$$Q^{CS}_{f,t} \leq qcsmax_f\, CUT_{f,t} \qquad \forall f \in F, \forall t \in T \tag{11}$$

$$Q^{CS}_{f,t} = \sum_{b \in BF_{b,f}} Q^{BP}_{b,f,t} \qquad \forall f \in F, \forall t \in T \tag{12}$$

$$\sum_{b \in BF_{b,f}} nu_{l,d,b} Q^{BP}_{b,f,t} = Q^{CL}_{l,d,f,t} \qquad \forall l \in L, \forall d \in D, \forall f \in F, \forall t \in T \tag{13}$$

$$Q^{BP}_{b,f,t} \geq qminb_b Y^{BP}_{b,f,t} \qquad \forall (b,f) \in BF_{b,f}, \forall t \in T \tag{14}$$

$$Q^{BP}_{b,f,t} \leq qcsmax_f Y^{BP}_{b,f,t} \qquad \forall (b,f) \in BF_{b,f}, \forall t \in T \tag{15}$$

$$Y^{BP}_{b,f,t} \leq CUT_{f,t} \qquad \forall (b,f) \in BF_{b,f}, \forall t \in T \tag{16}$$

$$CUT_{f,t} \leq \sum_{b \in BF_{b,f}} Y^{BP}_{b,f,t} \qquad \forall f \in F, \forall t \in T \tag{17}$$

$$CUT_{f,t} \leq CUT_{f,t+1} - CUT_{f,k} + 1 \qquad \forall f \in F, \forall t \in T, k \geq t+2 \tag{18}$$

$$Q^{TL}_{l,d,f,t} + Q^{RL}_{l,d,f,t} = Q^{CL}_{l,d,f,t} \qquad \forall l \in L, \forall d \in D, \forall f \in F, t = 1 \tag{19}$$

$$Q^{TL}_{l,d,f,t} + Q^{RL}_{l,d,f,t} = Q^{CL}_{l,d,f,t} + Q^{RL}_{l,d,f,t-1} \qquad \forall l \in L, \forall d \in D, \forall f \in F, t > 1 \tag{20}$$

$$Q^{TL}_{l,d,f,t} = \sum_c \sum_v \sum_i Q^{TT}_{l,d,c,v,f,i,t} \qquad \forall l \in L, \forall d \in D, \forall f \in F, \forall t \in T \tag{21}$$

$$Q^{TT}_{l,d,c,v,f,i,t} \leq qttup_{l,d,c,v,f,i,t} Y^{TT}_{l,c,v,f,i,t}$$

$$\forall l \in L, \forall d \in D, \forall c \in C, \forall v \in V, \forall f \in F, \forall i \in I, \forall t \in T \tag{22}$$

$$\sum_l Y^{TT}_{l,c,v,f,i,t} = X^L_{c,v,f,i,t} \qquad \forall c \in C, \forall v \in V, \forall f \in F, \forall i \in I, \forall t \in T \tag{23}$$

$$\sum_l \sum_d weight_{l,d} Q^{TT}_{l,d,c,v,f,i,t} \geq capmin_c X^L_{c,v,f,i,t}$$
$$\forall c \in C, \forall v \in V, \forall f \in F, \forall i \in I, \forall t \in T \tag{24}$$

$$\sum_l \sum_d weight_{l,d} Q^{TT}_{l,d,c,v,f,i,t} \leq capmax_c X^L_{c,v,f,i,t}$$
$$\forall c \in C, \forall v \in V, \forall f \in F, \forall i \in I, \forall t \in T \tag{25}$$

$$\sum_c \sum_v \sum_f Q^{TT}_{l,d,c,v,f,i,t} + stockini_{l,d,i} = dmin_{l,d,i,t} + Q^{SL}_{l,d,i,t} + Q^{PL}_{l,d,i,t}$$
$$\forall l \in L, \forall d \in D, \forall i \in I, t = 1 \tag{26}$$

$$\sum_c \sum_v \sum_f Q^{TT}_{l,d,c,v,f,i,t} + Q^{SL}_{l,d,i,t-1} = dmin_{l,d,i,t} + Q^{SL}_{l,d,i,t} + Q^{PL}_{l,d,i,t}$$
$$\forall l \in L, \forall d \in D, \forall i \in I, t > 1 \tag{27}$$

$$Q^{SL}_{l,d,i,t} \leq maxstock_{l,d,i} \qquad \forall l \in L, \forall d \in D, \forall i \in I, \forall t \in T \tag{28}$$

$$\sum_l \sum_d \left(dmin_{l,d,i,t} + Q^{PL}_{l,d,i,t} \right) \leq cap_{i,t} \qquad \forall i \in I, \forall t \in T \tag{29}$$

$$\sum_t \left(dmin_{l,d,i,t} + Q^{PL}_{l,d,i,t} \right) = dtot_{l,d,i} \qquad \forall l \in L, \forall d \in D, \forall i \in I \tag{30}$$

$$\sum_f X^D_{c,p,f,t} = Y^T_{c,p,t} \qquad \forall (c,p) \in PC_{p,c}, \forall t \in T \tag{31}$$

$$\sum_v \sum_i X^R_{c,v,i,p,t} = Y^T_{c,p,t} \qquad \forall (c,p) \in PC_{p,c}, \forall t \in T \tag{32}$$

$$X^D_{c,p,f,t} = \sum_i X^L_{c,v,f,i,t} \qquad \forall (c,p) \in PC_{p,c}, \forall f \in F, \forall t \in T, v = 1 \tag{33}$$

$$X^L_{c,v,f,i,t} \leq Y^T_{c,p,t} \qquad \forall (c,p) \in PC_{p,c}, \forall v \in V, \forall f \in F, \forall i \in I, \forall t \in T \tag{34}$$

$$X^U_{c,v,i,f,t} \leq Y^T_{c,p,t} \qquad \forall (c,p) \in PC_{p,c}, \forall v \in V, \forall f \in F, \forall i \in I, \forall t \in T \tag{35}$$

$$Y^T_{c,p,t} \leq \sum_v \sum_f \sum_i X^L_{c,v,f,i,t} \qquad \forall (c,p) \in PC_{p,c}, \forall t \in T \tag{36}$$

$$\sum_f X^U_{c,v,i,f,t} \leq \sum_f X^L_{c,v,f,i,t} \qquad \forall c \in C, \forall v \in V, \forall i \in I, \forall t \in T \tag{37}$$

$$\sum_i X^U_{c,v-1,i,f,t} \leq \sum_i X^L_{c,v,f,i,t} \qquad \forall c \in C, \forall f \in F, \forall t \in T, v > 1 \tag{38}$$

$$\sum_f X^L_{c,v,f,i,t} = \sum_f X^U_{c,v,i,f,t} + X^R_{c,v,i,p,t} \qquad \forall (c,p) \in PC_{p,c}, \forall v \in V, \forall i \in I, \forall t \in T \tag{39}$$

$$\sum_f \left(\frac{dpf_{p,f}}{vd_{p,f}}\right) X^D_{c,p,f,t} + \sum_v \sum_f \sum_i \left(\frac{dfi_{f,i}}{vl_{f,i}}\right) X^L_{c,v,f,i,t} +$$

$$\sum_v \sum_i \sum_f \left(\frac{dif_{i,f}}{vu_{i,f}}\right) X^U_{c,v,i,f,t} + \sum_v \sum_i \left(\frac{dip_{i,p}}{vr_{i,p}}\right) X^R_{c,v,i,p,t} +$$

$$\sum_v \sum_f \sum_i \left(load_{c,f} + unload_{c,i}\right) X^L_{c,v,f,i,t} \leq maxt_{c,t} \qquad \forall (c,p) \in PC_{p,c}, \forall t \in T \tag{40}$$

$$CUT_{f,t}, X^D_{c,p,f,t}, X^L_{c,v,f,i,t}, X^R_{c,v,i,p,t}, X^U_{c,v,i,f,t}, Y^{BP}_{b,f,t}, Y^T_{c,p,t}, Y^{TT}_{l,c,v,f,i,t} \in \{0,1\} \tag{41}$$

$$Q^{BP}_{b,f,t}, Q^{CL}_{l,d,f,t}, Q^{CS}_{f,t}, Q^{KT}_{c,t}, Q^{PL}_{l,d,i,t}, Q^{RL}_{l,d,f,t}, Q^{SL}_{l,d,i,t}, Q^{SS}_{f,t}, Q^{TL}_{l,d,f,t}, Q^{TT}_{l,d,c,v,f,i,t}, Z \geq 0 \tag{42}$$

Restrictions (7) and (8) establish the balance of stems in each harvest area in each period (the number of stems that are cut and left standing at the end of the period). Constraints (9) determine the maximum number of standing stems at the end of the planning horizon. Restrictions (10) and (11) establish the minimum and maximum quantities of stems to be cut, by period, where $CUT_{f,t}$ is the binary variable that indicates whether the harvesting crew is cutting stems in the harvest area or not. Constraints (12) state that the number of stems cut in a given harvest area is equal to the number of times all cutting patterns are applied. In Eqs. (13) the quantity of logs of each type that is generated by using the bucking patterns is determined. Constraints (14) and (15) determine the minimum and maximum number of times that the bucking patterns can be applied, respectively. Restrictions (16) and (17) establish that a bucking pattern must be used in a harvest area during a period if and only if a harvesting crew cuts stems in the harvest area during that period. Constraint (18) ensures that harvesting crew cuts in successive periods. Restrictions (19) and (20) establish the balance of logs in each harvest area in each period. Equation (21) states that the logs transported from a harvest area to the plants in a given period must match the number of logs loaded on trucks in the same period. Constraint (22) determines the maximum number of logs of a certain length that a truck can transport on a given trip, while constraints (23) state that only logs of the same length can be loaded in one trip at a time. Restrictions (24) and (25) establish that the load of a truck must satisfy the conditions of minimum and maximum load capacity, respectively. Constraints (26) and (27) set the inventory balances of logs of each type in each plant. Constraint (28) determines the maximum inventory capacity of logs of each type in each sector of the storage yard, while restrictions (29) establish the processing capacity limit of each plant in each period. Equation (30) forces the total demand for each plant to be covered. Through the restrictions (31) to (40) the routes that each truck must perform are built, these constraints are based on the work of Bordón et al. [9]. Finally, (41) and (42) establish the nature of the variables involved.

4　Case Study

In order to assess the capabilities of the developed model, a case study consisting of a sawmill that requires a certain number of logs of different types is considered. In Table 1 the minimum and total demands for a planning horizon of 5 days are shown. In addition, the total weight of the raw material demanded (in tons) is also presented. Assuming a conversion factor of 0.8 tons/m^3 of pine, in this illustrative example a weekly supply of more than 1,800 m^3 of raw material to a sawmill is considered. According to the last National Census of Sawmills [10], the volume of transported raw material in this illustrative example is equivalent to that handled in large companies (annual production level greater than 23,585 m^3, which means 471.7 m^3 per week approximately).

There are 3 open harvest areas with different stem availabilities to obtain the required logs: 8,000 stems in f_1, 7,000 stems in f_2 and 6,500 stems in f_3. In each harvest area there are 2 assigned harvesting crews, which can use the bucking patterns presented in Table 2, where the percentage of material lost after applying a certain pattern is also presented (this loss of material can be seen in Fig. 1). Since each harvesting crew can process a maximum of 50 stems per hour, the maximum harvesting capacity is set at 100 stems per hour. In a similar fashion, if the harvesting crew performs the cutting activity, then it must work at least 50% of its capacity.

Table 1.　Minimum and total sawmill demands.

Type of logs	Minimum demand per period					Total demand	Total weight (tons)
	t_1	t_2	t_3	t_4	t_5		
l_1d_1	50	100	75	50	50	450	8.55
l_1d_2	100	80	50	50	20	600	36.6
l_1d_3	0	150	150	150	0	900	81.9
l_1d_4	100	100	80	75	50	600	102
l_2d_1	0	100	180	150	90	1050	23.1
l_2d_2	200	120	100	80	120	900	63.9
l_2d_3	150	300	150	200	0	900	95.4
l_2d_4	100	150	250	150	150	1200	237.6
l_3d_1	100	100	250	50	0	750	20.25
l_3d_2	0	0	80	270	350	900	77.4
l_3d_3	200	200	200	80	0	1050	135.45
l_3d_4	0	0	80	250	220	900	216
l_4d_1	100	300	100	0	0	600	18.6
l_4d_2	0	0	150	250	100	600	61.2
l_4d_3	80	80	80	100	200	750	114
l_4d_4	120	120	120	120	0	600	169.8

There is a truck fleet composed of 100 trucks with minimum and maximum capacity of 18 and 27 tons each, respectively. The fleet is distributed in 3 regional bases.

Each truck can perform at most 3 loaded trips in its route, as long as the working time limit is respected (8 h). Both loading and unloading times of each truck are 30 min. The distances between each node of the network are given by Table 3. The rest of the parameters used are omitted due to space reasons.

Table 2. Bucking patterns applicable to each harvest area.

Type of logs	Bucking pattern (harvest area)									
	$b_1(f_1)$	$b_2(f_1)$	$b_3(f_1)$	$b_4(f_2)$	$b_5(f_2)$	$b_6(f_2)$	$b_7(f_2)$	$b_8(f_3)$	$b_9(f_3)$	$b_{10}(f_3)$
l_1d_1	2				1		1			
l_1d_2							1			
l_1d_3	2						1			
l_1d_4	5							2	3	
l_2d_1		1						1		
l_2d_2						1				
l_2d_3		2				1		1		
l_2d_4		4				1	3	2	3	
l_3d_1			1							
l_3d_2					1				1	
l_3d_3			1		1				1	
l_3d_4			4		1	3			1	3
l_4d_1				1						
l_4d_2										1
l_4d_3			1							1
l_4d_4				3	2					1
losses (%)	19	10	13	11	5	8	10	13	12	11

The model is implemented and solved in GAMS [11] 24.7.3 version, with the CPLEX 12.6.3 solver, in an Intel(R) Core(TM) i7-8700, 3.20 GHz.

The size of the considered case study is 30,565 binary variables, 73,461 positive variables and 114,122 equations. Following, the main results obtained after 15 min, reaching 0.87% of optimality gap, are detailed. It is worth highlighting that the model gives a lot of detailed information about the harvesting and log transportation planning, but due to reduced space in this presentation they cannot be shown and it is available for interested readers.

Table 3. Distances between the nodes (in kilometers).

	f_1	f_2	f_3	i_1
p_1	100	80	95	35
p_2	80	30	150	55
p_3	45	105	65	40
i_1	50	110	85	

Regarding to the harvest stage, the number of stems cut in harvest area f_1 is 610, 570 and 770 in periods t_1, t_2 and t_3, respectively, while for the same periods the number of stems cut in f_2 is 800, 664 and 636, respectively. In the harvest area f_3 the number of cut stems in periods t_1 and t_2 is 700 and 800, respectively, while in period t_3 no tree is harvested. There are no harvest activities in periods t_4 and t_5 in any harvest area. The bucking patterns used (and the number of times they are applied) in each period are detailed in Table 4.

Table 4. Bucking patterns used.

Period	Bucking pattern (harvest area)									
	$b_1(f_1)$	$b_2(f_1)$	$b_3(f_1)$	$b_4(f_2)$	$b_5(f_2)$	$b_6(f_2)$	$b_7(f_2)$	$b_8(f_3)$	$b_9(f_3)$	$b_{10}(f_3)$
t_1	150	100	360	147		509	144		700	
t_2		180	390	453		31	180		200	600
t_3		770				360	276			

In Table 5 the evolution of demand levels by period and by diameter for a particular log length (l_2) is shown. In this table the level of minimum committed demand per period ("$dmin_{l,d,i,t}$"), the level of covered demand above the committed demand per period ("$Q^{PL}_{l,d,i,t}$") and the accumulated covered demand ("ACD") are detailed.

Regarding the use of trucks, 13 vehicles are used in period t_1 (9 from p_2 and 4 from p_3), 7 in period t_2 (4 from p_2 and 3 from p_3), 9 in period t_3 (5 from p_2 and 4 from p_3), 5 in period t_4 (all from p_3) and 8 trucks in period t_5 (1 from p_2 and 7 from p_3).

As an example, the load of two of the trucks used in one of the planning periods (t_1) is presented in Table 6. In this case, both trucks perform 2 trips between harvest area f_1 and the sawmill. In this table, the composition of each load can be seen: number of loaded logs of each diameter (of the same length) and total weight of the load.

Logs inventory levels (length l_2) in both the harvest areas and sawmill are presented in Fig. 2 and Fig. 3, respectively. In Fig. 2, the blue, red, and green bars represent the number of logs on the roadside in harvest areas f_1, f_2, and f_3, respectively.

In Fig. 3 it can be seen that, since the logs stored in the plant at the end of the planning horizon are penalized, the model proposes a solution in which there are no logs (of any type) stored in period t_5.

Table 5. Demand satisfaction.

Diameter	Period	$dmin_{l,d,i,t}$	$Q^{PL}_{l,d,i,t}$	ACD
d_1	t_1	0	0	0
	t_2	100	0	100
	t_3	180	0	280
	t_4	150	0	430
	t_5	90	530	1050
d_2	t_1	200	160	360
	t_2	120	0	480
	t_3	100	0	580
	t_4	80	120	780
	t_5	120	0	900
d_3	t_1	150	0	150
	t_2	300	0	450
	t_3	150	100	700
	t_4	200	0	900
	t_5	0	0	900
d_4	t_1	100	0	100
	t_2	150	0	250
	t_3	250	0	500
	t_4	150	260	910
	t_5	150	140	1200

Table 6. Truck loads.

Truck/trip	Log length	Diameter/#Logs			Total weight (tons)
$c_{71}v_1$	l_2	d_2: 137			26.95
$c_{71}v_2$	l_3	d_1: 180	d_4: 92		26.94
$c_{72}v_1$	l_2	d_1: 100	d_3: 50	d_4: 98	26.90
$c_{72}v_2$	l_1	d_3: 180	d_4: 62		26.92

Since the harvesting crews perform the logging activities during the first 3 periods in the harvest areas, in the last 2 periods the demand is satisfied with the logs stored (both in forests and plants). From the analysis of Fig. 2 and Fig. 3, it can be seen that in the last period, only logs that are required to fulfill total demand are transported, leaving the logs generated by excess on the roadside in harvest areas. This is mainly because, since

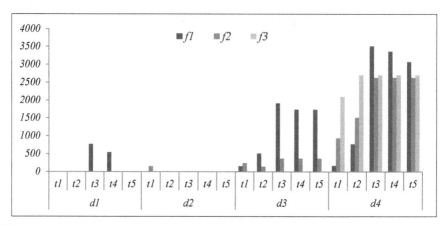

Fig. 2. Remaining logs of length l_2 on the roadside in harvest areas, by diameter. (Color figure online)

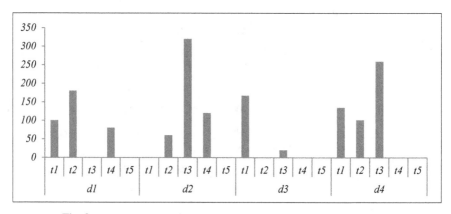

Fig. 3. Inventory level of logs of length l_2 in the sawmill, by diameter.

more trucks will be required (at a higher cost), the model tends to use as few trucks as possible. Another factor influencing these decisions is that the cost of maintaining a log in the sawmill storage yard is slightly higher than the cost of maintaining the same log at the roadside in the harvest area.

5 Final Remarks

In the forest industry, harvesting decisions are strongly related to plant production planning decisions. These plants, based on the commitments assumed with customers, plan detailed raw material supply programs to guarantee production. The harvesting crews receive these supply programs and harvest the raw material accordingly. Therefore, it is necessary to coordinate these efforts to avoid the generation of logs that will not be required by the plants and keep them in the storage yard at their respective cost. Transportation activities, in general, are not taken into account in this structure and have a great influence not only in the efficiency of the system but also in the cost structure.

The works found in the literature usually address these topics separately, which can lead to inefficient solutions. These inefficiencies are related to unnecessary logs inventory levels in both harvest areas and plants, and underutilization of truck loading capacity due to inefficient coordination, generating a high number of trips and a corresponding increase in the number of used trucks.

In this work, the proposed MILP model includes topics related to: vehicle routing, considering heterogeneous truck fleet, multiple periods, multiple plants and multiple depots; truck load composition; harvesting planning; and inventory management, both in harvest areas and plants.

The characteristics of the problem addressed and the solution time for the proposed example demonstrate the capabilities of the model to solve a problem in a weekly time horizon. Although the analyzed example was developed for illustrative purposes, it represents a situation in which more than $1,800$ m^3 of raw material are transported, which in practice represents a situation related to large companies with a large supply flow. The model was solved with a CPU time limit of 15 min reaching a solution with an optimality gap equal to 0.87%.

Regarding the practical utility of the developed model, it can be used as a tool to support the planners' decisions, providing answers related to:: the periods in which the harvest teams must harvest, the level of forest production in each harvest area (number of generated logs of each type), the logs inventory levels, the number of trucks required to transport the harvested and the composition of each load, the route each truck perform in each period, the proportion of weekly demand covered in each period, among others. This solution capacity shows the potential of the developed MILP model when planning forestry activities.

Although the aim of this work is to present a mathematical model that jointly addresses issues that were not addressed before in the literature, key aspects emerge for improvement in future work, such as truck scheduling decisions to avoid or reduce waiting times that could be generated.

References

1. D'Amours, S., Rönnqvist, M., Weintraub, A.: Using operational research for supply chain planning in the forest products industry. Inf. Syst. Oper. Res. **46**(4), 47–64 (2008)
2. Carlsson, D., D'Amours, S., Martel, A., Rönnqvist, M.: Supply chain planning models in the pulp and paper industry. Inf. Syst. Oper. Res. **47**(3), 167–183 (2009)
3. Borges, J.G., Diaz-Balteiro, L., McDill, M.E., Rodriguez, L.C.E. (eds.): The Management of Industrial Forest Plantations. MFE, vol. 33. Springer, Dordrecht (2014). https://doi.org/10. 1007/978-94-017-8899-1
4. El Hachemi, N., El Hallaoui, I., Gendreau, M., Rousseau, L.-M.: Flow-based integer linear programs to solve the weekly log-truck scheduling problem. Ann. Oper. Res. **232**(1), 87–97 (2014). https://doi.org/10.1007/s10479-014-1527-4
5. Bordón, M.R., Montagna, J.M., Corsano, G.: Mixed integer linear programming approaches for solving the raw material allocation, routing and scheduling problems in the forest industry. Int. J. Ind. Eng. Comput. **11**(4), 525–548 (2020)

6. Karlsson, J., Rönnqvist, M., Bergström, J.: An optimization model for annual harvest planning. Canad. J. Forest Res. **34**(8), 1747–1754 (2004)
7. Dems, A., Rousseau, L.M., Frayret, J.M.: Annual timber procurement planning with bucking decisions. Eur. J. Oper. Res. **259**(2), 713–720 (2017)
8. Vanzetti, N., Broz, D., Montagna, J.M., Corsano, G.: Integrated approach for the bucking and production planning problems in forest industry. Comput. Chem. Eng. **125**, 155–163 (2019)
9. Bordón, M.R., Montagna, J.M., Corsano, G.: An exact mathematical formulation for the optimal log transportation. Forest Policy Econ. **95**, 115–122 (2018)
10. Ministry of Agriculture, Livestock and Fisheries, 2015. National Census of Sawmills. https://www.magyp.gob.ar/sitio/areas/ss_desarrollo_foresto_industrial/censos_inventario/_archivos/censo//000000_Informe%20Nacional%20de%20Aserraderos%202015.pdf, Accessed 06 July 2020
11. Rosenthal, R.E.: A GAMS tutorial. https://www.gams.com/31/docs/index.html, Accessed 06 July 2020

An Operational Planning Model to Support First Mile Logistics for Small Fresh-Produce Growers

Álvaro M. Majluf-Manzur[1], Rosa G. González-Ramirez[1(✉)],
Raimundo A. Velasco-Paredes[1], and J. Rene Villalobos[2]

[1] Universidad de los Andes, 12455 Santiago, RM, Chile
ammajluf@miuandes.cl, rgonzalez@uandes.cl
[2] Arizona State University, Tempe, AZ 85281, USA
rene.villalobos@asu.edu

Abstract. In this work, we propose a hierarchical optimization model to support harvest and local routing decisions of fresh agricultural products. The proposed model has been motivated by the fact that small growers of fresh produce very often face difficulties in terms of availability of local (or first-mile) logistics at reasonable prices which results in lower profit margins and more food waste. So, in order to reduce the cost of first-mile logistics they can co-ordinate the consolidation of the harvests from different growers to send their products to the packing facility. Hence, we integrate harvesting and routing operations for a set of small growers with the aim to reduce their operational costs. The proposed framework has been implemented and preliminary evaluated with some instances considering as case study farmers of fresh produce in the central region of Chile.

Keywords: First mile logistics · Small fresh-produce growers · Harvesting and distribution planning

1 Introduction

Fresh fruit and vegetables consumption have been growing in recent years world-wide. The combination of the growth of global population, nutrition concerns and more open borders will sustain, if not accelerate, the current growth of international trade of fresh fruits and vegetables. As an example, Fig. 1 presents the imports of fresh fruits and vegetables into the US for consumption, from 2015 until 2019 expressed in thousands of US dollars. We can observe that in both cases, the graph shows a remarkable increasing trend.

This expanded demand of fresh fruits and vegetables represents a challenge since it requires the development of efficient supply chains that can meet this increasing demand considering all the complexities derived from the perishability of the product. Hence, the transportation, storage and distribution of these products need to be carefully planned to minimize food waste due to inefficient logistics practices. This opens the question of how to reengineer traditional supply chains for fresh produce into flexible and rapid response supply chains to best serve the rapidly changing market.

© Springer Nature Switzerland AG 2021
D. A. Rossit et al. (Eds.): ICPR-Americas 2020, CCIS 1408, pp. 205–219, 2021.
https://doi.org/10.1007/978-3-030-76310-7_17

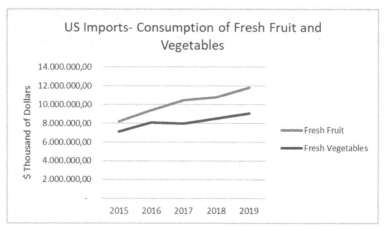

Fig. 1. US Imports for consumption of fresh fruit and vegetables (Thousand of Dollars). Source: FAS- USDA database, accessed on August 13th, 2020.

The covid-19 pandemic has caused serious problems in the supply chains of agricultural products. For instance, the media has reported cases of lack of food at the supermarket shelves and crops being wasted at the farms because of lack of traditional markets, such as food service outlets, for the farmers[1,2]. Furthermore, there has been shift of consumption to supermarket chains and direct-to-consumer channels. This highlights the need to develop tools to make rapid logistics connections between farmers and the evolving markets. This is even more relevant in the case of small growers who very often face additional challenges. For instance, even prior to the pandemic, small growers were experiencing difficulties accessing some markets, due among other factors to poor logistical channels, lack of negotiation power and lack of access to financial resources. Furthermore, due to some food safety regulations and first-mile logistics challenges, usually supermarket chains do not procure directly from small growers. Some of the changes caused by covid-19will be permanent and hence, we need to prepare for the new conditions.

Currently, very often two or more intermediaries take possession of the product between the farm's gate and the final consumer, resulting in multiple locations holding inventory throughout the supply chain. In this case, customers and growers are not directly connected to the final market, and are limited to operate under the market signals transmitted by those intermediaries which results in distorted market information and, capturing a lower margin of the value chain than if they would sell directly to the final consumer. The long distribution chain also contributes to the high levels of waste observed in the fresh produce industry. Accordingly, improvements in the fresh food supply chains, may provide social benefits that range from higher income for small and regional farmers, to more abundant healthy food for the consumer [1].

---

[1] https://www.magazine.bayer.com/en/feeding-the-world-in-a-pandemic-can-food-supply-chains-cope-with-covid-19-.aspx.

[2] https://onlinelibrary.wiley.com/doi/epdf/10.1111/cjag.12237.

Figure 2 presents a schematic view of the main processes of a fresh produce SC. The main echelons of a prototypical fresh produce supply chain are indicated, as well as the physical flow of products through it [2].

	Agricultural practices and Harvesting	Consolidation & Cold-Chain Entry	Processing / Packing	Transportation and logistics	Final Distribution Logistics
Strategic Decisions	• Selection of farming technology & equipment • Financial planning • Design of supply networks, • Evaluation of perennial crops, Crop rotation strategies	• Selection of storage equipment & technology • Financial planning • Design of supply networks • Selection of cold technology • Location of cold chain facilities	• Selection of storage equipment & technology • Financial planning • Design of supply networks	• Design of the global supply chain network • Design of the domestic supply chain network	• Design of supply network • Market selection and contracts with customers (incoterms and other conditions) • Pricing analysis
Tactical and Operational Decisions	• Cropping planning • Harvesting planning • Planting policies • Equipment scheduling • Water allocation • Land preparation • Labor planning	• Storage planning • Cold chain scheduling and planning • Transportation planning and scheduling • Plan opening chambers	• Storage planning • Cold chain scheduling and planning • Transportation planning and scheduling • Batch production planning • Load assignments • Packing	• Domestic transportation planning and scheduling • International transportation planning and scheduling • Packing and unpacking • Inventory decisions and cargo handling • Empty container logistics • Clearance and other procedures management	• Transportation planning and scheduling • Packing and unpacking • Inventory decisions and cargo handling

Fig. 2. Schematic view of the main processes of a fresh produce supply chain. Source: Villalobos et al. [2].

In this paper, we aim to provide planning tools for small growers based on an integrated planning and coordination environment for supply chains of fresh fruits and vegetables that includes information acquisition and coordination supply-demand tools to close the logistics and information gap between small growers and existing and emerging markets, especially those created by direct sales to consumers through ecommerce. The aim is to increase the benefits obtained by the small growers by means of a coordination scheme in their first mile operations. We focus on the operational planning decisions related to the harvesting and transportation of produce to the packing facilities. In particular, we propose a hierarchical planning decision model to determine joint harvest and truck scheduling for a group growers. We apply the proposed model to a case study that consists of small growers that produce fruits and vegetables in the central region of Chile. The underlying problem is to coordinate harvest and truck routing to achieve cost savings in first mile logistics, which it is usually one of the main hurdles for the integration of small farmers have to solve to insert themselves into the main marketing channels.

The remainder of this paper is organized as follows. Section 2 presents a literature review regarding operational harvesting and transportation decisions of a fresh produce supply chain. Section 3 presents the proposed mathematical formulation. Section 4 presents the experimental design and numerical results. Conclusions and recommendations for future research are given in Sect. 5.

2 Literature Review

Research on agricultural supply chains have gained the attention of academics in several aspects. There are many planning decisions in a fresh-produce agricultural supply chain at the strategic, tactical, and operational level. We can find multiple contributions addressing different planning aspects [3–17]. Comprehensive reviews of operational research tools to support planning decisions can be found in [18, 19] and Villalobos et al. [1] present a review of technologies to support real-time decisions for fresh produce logistics.

Among strategic level decisions, we can refer to Flores and Villalobos [8] who focused on the development of market intelligence tools. They developed an opportunistic shipment policy that increases a farmer´s commercialization reach within a secondary market with minimal or no capital investment once a base market has been established. From the perspective of the first mile decisions, we can refer to Ahumada et al. [4] that present an integrated tactical planning model for the production and distribution of fresh agricultural problems under uncertainty. The proposed model aims to maximize the revenues of a single and large producer that has some control over the logistics decisions associated with the distribution of the crop.

Ahumada and Villalobos [3] on the other hand, propose an operational model for planning the harvest and distribution of perishable agricultural products considering a single large grower. The aim of the work is to optimize the savings provided by the trade-off between the freshness with which the products are delivered and the transportation costs. The model generates short-term decisions related to harvesting, packing and transportation. They consider time constraints, available labor, and the effects of harvesting decisions on product quality; their modeling strategy is based on harvest patterns and one of the decisions considered is which pattern should be used for harvesting each plot of the planted area. The model takes the perspective of a single grower who produces tomato and bell peppers.

Ferrer et al. [7] propose an optimization approach for scheduling wine-grapes harvest operations. They propose a mathematical model that schedule the harvesting operations considering the quality loss due to harvesting before or after the optimal date. They consider penalizations for harvesting before or after the optimal date. This is an important difference with respect to the work to be presented in this paper. Rather than modeling the initial and final harvesting times, in this paper we consider patterns that determine the frequency of the harvesting along the days of the planning horizon. For instance, one harvesting pattern corresponds to harvesting every single day, another corresponds to harvesting every other day, and so on… Ferrer et al. [7] consider routing aspects into their model, as they assume that equipment and personnel should be moved from one block of the planted area to another. They model the routing problem as a TSP (Traveling

Salesman Problem). This is another difference with respect to our work in which we consider more than one vehicle to transport the fresh produce from the small farms to a packing center.

In terms of previous works in the literature that specifically consider small growers, we can refer to Flores et al. [9]. They propose a framework for the strategic design of local agricultural systems for small growers. In terms of contributions focused on the transportation of harvested products, we can refer to Alarcon-Gerbier et al., [11]. They address the combined problem of fresh produce harvest-decisions and the scheduling of truck arrivals to processing plants of tomato. González-Araya et al. [12] consider the same problem and propose a GRASP methodology for solving the model. Although it has certain similitudes to our work, they consider a problem with different objectives than those considered in the models to be presented in this paper. For instance, their objective function is based on the minimization of waiting times of trucks, while we aim to reduce total costs.

Another related contribution is presented by Yantong et al. [13] who propose an inventory routing planning model for perishable food with quality considerations. They consider the case of sending products to different retailers from a single depot, focused on the last mile logistics, that differs from our case that focuses on first mile activities. Furthermore, they do not consider labor restrictions which is an important aspect include in our model.

Lamsal et al. [14] address the integrated harvest and delivery scheduling problem for planning the movement of the crop from the farm to a processing plant. They consider the case of multiple and independent producers and assume that no on-farm storage is allowed. They consider a time-frame of 24 h. In contrast, in our case we assume a longer planning horizon (for instance, 2 weeks) to coincide with the planning of harvesting operations observed in practice. Furthermore, while in our case, we are maximizing profits, their model penalizes positive and negative deviations from the mill´s unloading capacity.

Gvozdenovic and Brcanov [15] address the problem of transporting agricultural goods from the fields to the factory using a fleet of vehicles. They consider the transportation of sugar beets used in sugar production. The transport between the fields and the factory is modeled by a time-span network. They do not consider harvesting decisions, which is an important difference with respect to our work.

Another related work is presented by Plessen [16] that addresses a tactical decision problem for harvest-planning based on the coupling of crop assignment with vehicle routing. They determine which crop to assign to each field, the sequence to serve the fields during the harvest and the dispatching of multiple harvesters located at multiple depots to the multiple fields. One important difference is that in our case we consider an operational planning decision model and assume that crops have been already planted. Furthermore, we are not considering the dispatching of harvesters but consider a limited labor available.

Jiang et al. [17] propose an integrated harvest and distribution scheduling model considering time windows for perishable products. The distribution of products is done directly from the farms to the consumers. They consider a single farm and several consumers located in different places, which is the main difference with respect to our work. In our case we assume that products are transported to a packing facility from which they will be transported to a distant location. They consider a single type of product while in our case we allow multiple products. Another important difference is that they model the scheduling problem based on the initial and final time of harvesting at the farm, while in our case, the timing is determined by the harvesting patterns selected which simplifies the modeling structure.

In this paper, we aim to extend the operational planning model proposed by [3] to consider small growers that are organized by a supply chain coordinating agent who uses planned routes for consolidating cargo instead of using direct shipments. In contrast to the most of the existing related literature, we consider a planning horizon of several days, and harvesting patterns as the planning cornerstone as it is done by [3].

3 Mathematical Formulation and Proposed Solution Approach

We consider the problem of planning the harvesting and distribution of fresh products from several small farmers to a packing facility. We propose a hierarchical approach based on two stages that are iteratively solved, as it is illustrated in Fig. 3. The first step consists of solving the harvest planning model, estimating the distribution costs as a direct shipment. Once the harvesting decisions have been made, a routing model is solved to consolidate shipments from several growers. The solution of the routing problem provides the transportation costs to the harvesting model. The procedure is repeated until both models converge.

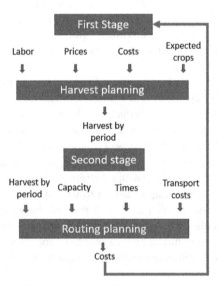

Fig. 3. Hierarchical solution implementation.

Sections 3.1 and 3.2 present the harvesting and routing mathematical formulations are respectively.

3.1 Harvesting Planning Model

We consider the planning problem of harvesting and distributing products by means of consolidated routes serving I farms and delivering the consolidated cargo to a single packing facility. Each farm produce K products. As in Ahumada and Villalobos [3], in order to simplify the model, we define harvesting patterns as the basic planning unit. Each harvesting pattern has different labor requirements and results in different amounts harvested. Then the harvesting decision problem consists of determining the timing and the harvesting pattern to be used to by each grower so that profits are maximized. Thus the problem becomes what harvesting pattern (if any) to use every day to harvest all the fruit that is ripe that day. We assume that only one pattern can be used in a single day for each planted plot. According to the products planted, it is expected certain volume of products to be harvested in each farm. The available labor for each farm is restricted, and for each box of product harvested is necessary certain hours of labor, which is also equivalent to cover a hectare of product in each farm. Each product has a minimum demand to be covered. We aim to maximize profits for the farmers, assuming a certain price for the products to be sent to their market as well as certain price of the product savage. Harvested products are transported to a packing facility, so we also consider the corresponding transportation costs. The notation employed is further described.

Sets and Parameters

$t \in T$: planification periods (days).
$k \in K$: products.
$i \in I$: farms.
$v \in V$: harvest patterns.
EH_{tikv}: expected harvest in period t from farm i of product k by pattern v (in boxes of product).
SH_{tv}: if the pattern v is required to harvest in period t.
VS_{tk}: percentage of product k which is discarded in period t.
LAH_{it}: available labor for the farm i in period t.
LRH_{ik}: hours needed to harvest a box of product k in farm i.
LBH_{ik}: hours of labor necessary to cover one hectare of product k in farm i.
MOP_i: maximum working hours per farm i.
D_{kt}: minimum quantity to send of product k in period t.
P_{kt}: price of the product k in period t.
PS_{kt}: price of the product k salvaged in period t.
CH_k: cost of harvesting a box of product k.
$Clabor$: cost of one hour of work.
CT_i: estimated cost of transporting a box to the consolidation center from farm i.
Cap: truck capacity.

Variables

X_{tikv}: harvested area of product k in the farm i with harvest pattern v in the period t.
QH_{tik}: harvested boxes of product k from farm i in the period t.

SP_{tikr}: shipped boxes of product k from farm i in the period to the consolidation point, using truck r.
QS_{tik}: discarded boxes of product k from farm i in the period t.
Opl_{it}: hours of labor contracted in the farm i in the period t.

The proposed model is formulated as a mixed integer programming (MIP) model as follows:

$$Max(\sum_t \sum_i \sum_k \sum_r SP_{tikr}P_{kt}) + (\sum_t \sum_i \sum_k QS_{tik}PS_{kt} - (\sum_t \sum_i (LAH_{it} + OPL_{it})Clabor) - (\sum_t \sum_i \sum_k QH_{tik}CH_k) - (\sum_t \sum_i \sum_k \sum_r SP_{tikr}CT_l) \tag{1}$$

Subject to:

$$QH_{tik} = \sum_v X_{tikv}EH_{tikv} \qquad \forall\, t, i, k. \tag{2}$$

$$QH_{tik} \leq Cap \qquad \forall\, t, i, k. \tag{3}$$

$$QS_{tik} = \sum_v X_{tikv}EH_{tikv}VS_{tk} \qquad \forall\, t, i, k. \tag{4}$$

$$\sum_k (LRH_{ik}QH_{tik}) + \sum_k \sum_v X_{tikv}SH_{tv}LBH_{ik} \leq LAH_{it} + OPl_{it} \qquad \forall\, t, i. \tag{5}$$

$$\sum_r SP_{tikr} = QH_{tik} \qquad \forall\, t, i, k. \tag{6}$$

$$\sum_i \sum_r SP_{tikr} \geq D_{kt} \qquad \forall\, t, k. \tag{7}$$

$$\sum_v \sum_t X_{tikv} \leq AP_{ik} \qquad \forall\, i, k. \tag{8}$$

$$OPl_{it} \leq MOP_i \qquad \forall\, t.i. \tag{9}$$

$$\sum_i \sum_k SP_{tikr} = Cap \qquad \forall\, t, r. \tag{10}$$

$$X_{tikv}, QH_{tik}, SP_{tikr}, QS_{tik}, Opl_{it} \geq 0\ \forall t, i, k, v, r. \tag{11}$$

The objective function (1) maximizes the benefits of selling the normal and the salvage products minus the cost of labor, harvest and transport. Constraint (2) indicates that the amount harvested depends on the area harvested according to the pattern and the expected harvest. Constraint (3) limits the harvest to the truck capacity. Constraint (4) indicates that the quantity of the product discarded depends on the area harvested according to the pattern, the expected harvest and the estimate of the quantity discarded. Constraint (5) limits the quantity harvest to the labor available. Constraint (6) define that all the quantity harvested must be send. Constraint (7) indicates that the quantity sent to destination must be at least the minimum quantity. Constraint (8) limits the area harvested per pattern to the total area planted. Constraint (9) indicates that the extra labor is limited to a maximum working hours per day. Constraint (10) limits the quantity sent to the truck capacity. Constraint (11) indicates the nature of the variables.

3.2 Routing Problem

The second phase of the hierarchical approach consists of a planning problem for the recollection of the harvests from a set of farms located in different locations. Each one of these farms harvests a given quantity of products in a determined period of time according to a preestablished plan. The aim is to send all the products harvested by the farmers to a packing center, minimizing the costs of transportation. A fleet of trucks with limited load capacity is available to do the transportation of products. Time windows are defined for the collection of these products, as well as the operating times of the packing facility.

Parameters

$r \in R$: trucks.
$I_0 = I \cup \{0\}$, where the node 0 corresponds to the packing facility.
HS_{ti}: amount of harvest to send from the farm i in the period t.
Cap: trucks capacity.
TBF_{ij}: Travel time between farms (i, j).
C_{ij}: Travel cost between farms (i, j).
S_i: Charging time in the farm i or discharge at the consolidation point.
E_i: Earliest time to get to the farm i.
U_i: Later time to get to the farm i.
M: Scale very large.

Variables

X_{tijr}: if the arc (i, j) is visited by the truck r the period t.
Y_{tir}: if the farm i is visited by the truck r the period t.
VT_{tir}: time to visit the farm i by the truck r the period t.
$VTCE_{tr}$: departure time from the consolidation point by the truck r the period t.
$VTCA_{tr}$: time of arrival at consolidation point by the truck r the period t.

Objective Function

$$Min \sum_t \sum_i \sum_j \sum_r C_{ij} X_{tijr} \tag{12}$$

Constraints

$$\sum_i X_{tizr} - \sum_j X_{tzjr} = 0 \quad \forall\, z, r, t. \tag{13}$$

$$\sum_i Y_{tir} \leq \sum_i X_{t0ir} M \quad \forall\, r, t. \tag{14}$$

$$\sum_i Y_{tir} \leq \sum_j X_{tj0r} M \quad \forall\, r, t. \tag{15}$$

$$\sum_i HS_{ti} Y_{tir} \leq Cap \quad \forall\, r, t. \tag{16}$$

$$\sum_{j!=i} X_{tijr} = Y_{tir} \qquad \forall\, i, r, t. \tag{17}$$

$$\sum_{i!=j} X_{tijr} = Y_{tjr} \qquad \forall\, j, r, t. \tag{18}$$

$$\sum_{r} y_{tir} M \geq HS_{ti} \qquad \forall\, i, t \text{ where } i \in I : HS_{ti} \geq 0. \tag{19}$$

$$\sum_{r} y_{tir} = 1 \qquad \forall\, i, t \text{ where } i \in I : HS_{ti} \geq 0. \tag{20}$$

$$VT_{tir} + S_i + TBF_{ij} - (1 - X_{tijr})M \leq VT_{tjr} \qquad \forall\, i, j, r, t. \tag{21}$$

$$VTCE_{tr} + TBT_{0j} - (1 - X_{t0jr})M \leq VT_{tjr} \qquad \forall\, j, r, t. \tag{22}$$

$$VT_{tir} + S_i + TBT_{ij} - (1 - X_{ti0r}) \leq VTCA_{tr} \qquad \forall\, i, r, t. \tag{23}$$

$$E_i\, y_{tir} \leq VT_{tir} \leq U_i\, y_{tir} \qquad \forall\, i, r, t. \tag{24}$$

$$E_o \leq VTCE_{tr} \qquad \forall\, r, t. \tag{25}$$

$$VTCA_{tr} \leq U_o \qquad \forall\, r, t. \tag{26}$$

$$VT_{tir}, VTCA_{tr}, VTCE_{tr} \geq 0 \qquad \forall\, i, r, t. \tag{27}$$

$$X_{tijr}, Y_{tir} \in \{0, 1\} \qquad \forall\, i, j, r, t. \tag{28}$$

The objective function (12) minimize the transportation cost from all farms to the consolidation center. The constraint (13) defines the correct flow of the trucks. The constraint (14) and (15) indicates that each route must start and finish in the consolidation center. The constraint (16) limits the amount sent to the truck capacity. The constraints (17) and (18) defines the relation between variables X_{tijr} and Y_{tir}. The constraint (19) indicates that if there is a harvest to be removed, the truck must visit the farm. The constraint (20) restrict to visit once each farm that has harvest. The constraints (21), (22) and (23) defines times of arrival between farms, to a farm from the consolidation center and to the consolidation center. The constraint (24) restrict the time of arrival to each farm. The constraints (25) and (26) restrict the time of departure and arrival to the consolidation center. The constraints (27) and (28) define the nature of the variables.

4 Experimental Design

To evaluate the performance of the proposed framework numerical experiments were carried out. The instances were generated considering six factors that were varied using two levels (low and high): the number of farms (I), the cost of harvesting a box (CH_k),

Table 1. Parameters used in the experiment

Parameter	Low level (-1)	high level (1)
I (Farms)	4	6
CH_1 (USD)	$1.55	$1.705
CH_2 (USD)	$0.43	$0.473
CH_3 (USD)	$1.5	$1.65
MOP (hours)	120	150
EH_{ti1v} (boxes)	9,000	11,000
EH_{ti2v} (boxes)	5,400	6,600
EH_{ti3v} (boxes)	540	600
CAP (boxes)	1,000	1,800
I (Farms)	$3.28125	$3.9375

the available extra working hours (MOP_i), the harvest yield (EH_{tikv}), the capacity of the trucks (Cap) and the labor cost ($Clabor$). This resulted in an experimental design of 2^6, that is, 64 different instances. The values of the parameters were estimated according to information provided by INDAP (Chilean Agricultural Development Institute) website and interviews with technical personnel of this Institute. Table 1 summarizes the values considered for each level and parameter.

For each of the instances, 6 replicates were created, in which the yield of the harvest varies 10% of the expected yield for each level and product. The location of the farms was set considering the central region of Chile. The location of the farms was defined by using google maps and the areas were the products grow in the central region of Chile (the Metropolitan Region of Santiago and the Region of Valparaiso).

The proposed optimization framework was implemented in Phyton version 3.7.6 with the AMPL 2.0 API. Gurobi 5.0.1 solver is used to solve the mathematical models. To solve the experiments, an MSI computer with an Intel Core i7-7700HQ CPU 2.80 GHz, with 8 GB of installed memory and Windows 10 with a 64-bit system is used.

Due to the computationally difficulty of solving the routing model, the number of farms was limited to 4 and 6 farms (in the low and high-level values). Table 2 shows the minimum, maximum and average computational times to solve the instances (Table 2).

Table 2. Computational times (seconds).

	Minimum	Maximum	Average
Harvesting time (s)	0.15	0.80	0.18
Routing time (s)	0.17	22.80	0.95
Total time (s)	1.70	25.39	3.10

Table 3 presents the benefits obtained when consolidating products for the transportation from the farms to the packing facility, considering the values (low and high) of each parameter, as well as the general results. The table presents the increase (%) of profits with respect to the direct shipment strategy where each farm transports their products without any coordination with other growers.

Table 3. Numerical results in terms of profits increase.

	Profit increase (%)		
	Minimum	Maximum	Average
Overall results	9.20%	26.02%	16.81%
Profit increase by factor			
Farms			
−1 (low level)	9.20%	11.22%	10.00%
1 (high level)	21.63%	26.02%	23.63%
Box cost			
−1	9.21%	25.81%	16.81%
1	9.20%	26.02%	16.82%
Labor hours			
−1	9.20%	26.02%	16.92%
1	9.24%	25.62%	16.71%
Yield			
−1	9.47%	26.02%	17.52%
1	9.20%	23.79%	16.11%
Capacity truck			
−1	9.20%	16.28%	16.28%
1	9.87%	17.35%	17.35%
Labor cost			
−1	9.23%	26.02%	16.82%
1	9.20%	25.74%	16.81%

We performed an analysis to determine the factors that are more relevant in increasing the profits of the small growers. We found that the most relevant factors were the number of farms, the harvest yield, and the truck capacity (see Fig. 4). As the number of farms increases, the profits of the farmers increase due to the reduction of logistics costs.

As an example, we present the routes found in one of the instances solved by the proposed optimization framework. This instance corresponds to 4 farms that produces strawberries and cauliflower. The percentage of profit increase in the final solution with respect to a direct shipment strategy is 10%. Figure 5 presents the routes found in the optimal solution. Figure 5a presents the route of truck 1 which starts in the packing

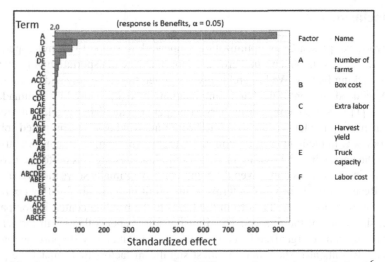

Fig. 4. Pareto chart of the Standardized Effects on the experimental design 2^6.

center (blue point D), then goes to farm 2 (red dot B), continue to farm 1 (red dot C) and finally return to the packing center. Figure 5b presents the route of truck 2 that initiates the route with the farm 4 (yellow dot B), departing from the packing center (blue dot D). Then it visits the farm 3 (yellow dot C), to finally return to the packing center.

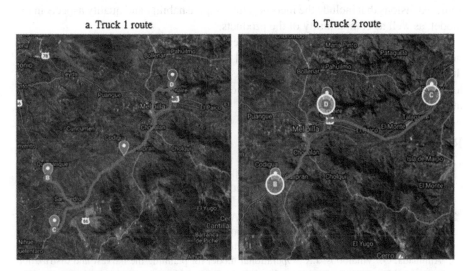

a. Truck 1 route   b. Truck 2 route

Fig. 5. Illustration of the routes found in one of the instances solved. (Color figure online)

5 Conclusions

In this work, we presented an optimization framework to support first mile operations considering the case in which small growers coordinate the transportation of fresh products to a packing facility. We hierarchically integrated the harvesting and routing decisions. A reduced version of the harvesting planning model proposed by Ahumada and Villalobos [1] was used, considering harvesting patterns and labor availability restrictions. The routing model is a classical VRP with time windows. The proposed solution framework was implemented in Phyton and Gurobi. An iterative scheme was used to integrate the harvesting and routing models. Preliminary testing was performed considering small instance sizes only, given the complexity to optimally solve the large routing models. Results provide insights in terms of the profit increase obtained by coordinating the transportation of products from the farms to the packing center, in comparison to direct shipments of the farmers. This illustrates the benefits that small farmers can achieve if they can be organized in a cooperative scheme that may support the logistics operations. Results also show that the most significant factors that impact the profits increase are the number of farms, and then, the capacity of the trucks.

As ongoing research, we are considering several aspects to improve the routing model to apply to more realistic aspects. For instance, we plan to allow cargo splitting to better use the capacity of the trucks. We are also evaluating alternatives that can reduce the computational time of the routing model, such that providing an initial solution to the solver based on a heuristic algorithm. In terms of the harvesting model, as we have simplified the model proposed by Ahumada and Villalobos [3], we are evaluating some extensions that include the incorporation of perishability and quality aspects in the model, as well as compatibility of the products.

Further research should consider the integration of a tactical planning model as the one proposed by Ahumada and Villalobos [4] with the operational planning model so that the planting and harvesting decisions can be related to the coordination with first mile logistics. Cooperative schemes and negotiations among the small farmers to incorporate into the models should also be explored.

References

1. Beitzen-Heineke, E.F., Balta-Ozkan, N., Reefke, H.: The prospects of zero-packaging grocery stores to improve the social and environmental impacts of the food supply chain. J. Cleaner Prod. **140**, 1528–1541 (2017)
2. Villalobos, J.R., Soto-Silva, W.E., González-Araya, M.C., González-Ramírez, R.G.: Research directions in technology development to support real-time decisions of fresh produce logistics: a review and research agenda. Comput. Electron. Agric. **167**, 105092 (2019)
3. Ahumada, O., Villalobos, J.R.: Operational model for planning the harvest and distribution of perishable agricultural products. Int. J. Prod. Econ. **133**, 677–687 (2011)
4. Ahumada, O., Villalobos, J.R., Mason, A.N.: Tactical planning of the production and distribution of fresh agricultural products under uncertainty. Agric. Syst. **112**, 17–26 (2012)
5. Amorim, P., Günther, H.-O., Almada-Lobo, B.: Multi-objective integrated production and distribution planning of perishable products. Int. J. Prod. Econ. **138**, 89–101 (2012)
6. Catalá, L., Durand, G., Blanco, A., Alberto Bandoni, J.: Mathematical model for strategic planning optimization in the pome fruit industry. Agric. Syst. **115**, 63–71 (2013)

7. Ferrer, J.-C., Mac Cawley, A., Maturana, S., Toloza, S., Vera, J.: An optimization approach for scheduling wine grape harvest operations. Int. J. Prod. Econ. **112**, 985–999 (2008)

8. Flores, H., Villalobos, J.R.: Using market intelligence for the opportunistic shipping of fresh produce. Int. J. Prod. Econ. **142**, 89–97 (2013)

9. Flores, H., Villalobos, J.R.: A modeling framework for the strategic design of local fresh-food systems. Agric. Syst. **161**, 1–15 (2018)

10. Flores, H., Villalobos, J.R., Ahumada, O., Uchanski, M., Meneses, C., Sanchez, O.: Use of supply chain planning tools for efficiently placing small farmers into high value vegetable markets. Comput. Electron. Agric. **157**, 205–217 (2019)

11. Alarcón-Gerbier, E.A., Gonzalez-Araya, M.C., Moraga, M.M.R.: Supporting harvest planning decisions in the tomato industry. In ICORES Proceeding, pp. 353–359 (2017)

12. González-Araya, M.C., Soto-Silva, W.E., Díaz-Miranda E.A.: GRASP metaheuristic for planning fresh produce harvest and truck arrivals to agro-industrial processing plants. Working Paper (2018)

13. Yantong, L.I., Feng, C.H.U., Zhen, Y.A.N.G., Calvo, R.W.: A production inventory routing planning for perishable food with quality consideration. Ifac-Papersonline **49**(3), 407–412 (2016)

14. Lamsal, K., Jones, P.C., Thomas, B.W.: Harvest logistics in agricultural systems with multiple, independent producers and no on-farm storage. Comput. Ind. Eng. **91**, 129–138 (2016)

15. Gvozdenović, N., Brcanov, D.: Vehicle scheduling in a harvest season. Econ. Agric. **65**(2), 633–642 (2018)

16. Plessen, M.G.: Coupling of crop assignment and vehicle routing for harvest planning in agriculture. Artif. Intell. Agric. **2**, 99–109 (2019)

17. Jiang, Y., Chen, L., Fang, Y.: Integrated harvest and distribution scheduling with time windows of perishable agri-products in one-belt and one-road context. Sustainability **10**(5), 1570 (2018)

18. Ahumada, O., Villalobos, J.R.: Application of planning models in the agri-food supply chain: a review. Eur. J. Oper. Res. **196**, 1–20 (2009)

19. Soto-Silva, W.E., Nadal-Roig, E., González-Araya, M.C., Pla-Aragones, L.M.: Operational research models applied to the fresh fruit supply chain. Eur. J. Oper. Res. **251**, 345–355 (2016)

Simulation

Impact of Labor Productivity and Multiskilling on Staff Management: A Retail Industry Case

Silvana Vergara, Jairo Del Villar, James Masson, Natalia Pérez,
César Augusto Henao$^{(\boxtimes)}$ (iD), and Virginia I. González (iD)

Universidad del Norte, Barranquilla, Colombia
{scvergara,jadelvillar,jmasson,aristizabaln,cahenao,
vvirginia}@uninorte.edu.co

Abstract. This research seeks to evaluate the effect of the learning and forgetting phenomenon on multiskilling decisions. This phenomenon occurs when multi-skilled employees lose productivity in those tasks they perform less frequently. The literature recommends considering this decrease in productivity in multiskilling decisions, in order to find personnel planning alternatives that improve workforce performance. The methodology proposes a mixed integer programming model that explicitly incorporates the learning and forgetting phenomenon and, also, the use of multiskilled workforce through a k-chaining approach. This formulation allows to determine how many employees should be multiskilled, in how many additional departments they will be trained, and what amount of productivity is expected in each trained department. The solution to the problem minimizes the costs of understaffing, training, and productivity loss. The case study is applied in the retail industry, in which two experiments were evaluated that faced three different levels of demand variability. The first experiment generated lower costs because it did not incorporate the productivity loss associated with the learning and forgetting phenomenon. The second experiment, which incorporated the learning and forgetting phenomenon, generated a higher total cost but its solution is better adjusted to the real operation of a retail store. Finally, given the incorporation of the learning and forgetting phenomenon, the second experiment required higher levels of multiskilling to minimize understaffing costs and, in turn, compensate for the productivity loss.

Keywords: Learning and forgetting · Chaining · Multiskilling · Personnel scheduling · Retail

1 Introduction

The manufacturing and service sectors are fast-growing, highly competitive, and highly demanding of human resources (HR). These sectors include industries such as: automotive, textile, manufacturing, call centers, healthcare, retail, among others. Henao [1] states that HR are generally the costliest productive resources. Therefore, their efficient management allows the industries in these sectors to improve the level of service offered to their customers and also to reduce personnel costs. In particular, the retail industry

© Springer Nature Switzerland AG 2021
D. A. Rossit et al. (Eds.): ICPR-Americas 2020, CCIS 1408, pp. 223–237, 2021.
https://doi.org/10.1007/978-3-030-76310-7_18

must respond not only to predictable phenomena such as seasonal demand, but also to unpredictable such as demand uncertainty and unscheduled staff absenteeism. Such phenomena produce periods of overstaffing and understaffing, which tend to increase personnel costs and deteriorate the level of service offered to customers. [2, 3].

For the retail industry, several authors have addressed the overstaffing and understaffing problems through the use of multiskilled workforce. [4–7]. Note that, multiskilled employees are those who can work in k departments of a store (with $k \geq 2$), and the flexibility they provide allows to transfer employees from over-staffed departments to under-staffed ones. Therefore, their efficient use makes it possible to minimize the mismatch between staffing and personnel demand. However, multiskilling comes with a price, especially in the case of full multiskilling, so hiring and training employees in multiple departments can be very expensive.

In addition, as a result of the phenomenon known in the literature as "learning and forgetting", a multiskilled employee will have a productivity loss in those departments where he/she is less frequently assigned. Consequently, this process of forgetting and loss of productivity will generate a cost for companies, since lower productivity will require the allocation of more man-hours to cover a given level of demand. Additionally, Azzi and Liang [8] explain that there are decreasing performances as the number of skills increases in an employee. Therefore, very high levels of multiskilling may be impractical and unnecessary and, consequently, it is also necessary to establish clear policies to define the most cost-effective levels of multiskilling. So far, literature in retail has focused mainly on modeling the benefits of multiskilling; however, it is also necessary to develop mathematical models that incorporate the costs associated with the productivity loss caused by the learning and forgetting phenomenon.

Considering the need described above, the contribution that this article intends to make to human resources management in the retail industry is to develop a model that determines the levels of multiskilling that minimize the costs of training and understaffing, but also the costs associated with the productivity loss. Therefore, for an annual staff assignment problem, a mixed integer programming model was formulated to explicitly incorporates the learning and forgetting phenomenon. In addition, the model will ensure the implementation of a workforce flexibility policy known as k-chaining. This policy is widely recommended in the literature to face an uncertain demand [5, 6, 9–12]. Under a k-chaining policy, a certain number of employees will be trained in k departments (with $k \geq 2$), such that the assignment decisions are configured through a bipartite graph involving different types of closed chains.

Finally, this methodology will be applied to a Chilean retail store by using CPLEX software. The performance of the optimization approach will be evaluated for different demand variability scenarios, this will allow us to measure the benefits of multiskilling in the face of uncertain demand, but also to evaluate the impact of the learning and forgetting phenomenon.

2 Literature Review

Table 1 summarizes a number of articles that have studied the effect of productivity loss on personnel scheduling problems with multiskilled personnel. The elements present in Table 1 are explained below:

1. *Human resources decision level* (HR): indicated the type of personnel scheduling problem addressed. It can be: (i) staffing (S); (ii) shift scheduling (SS); and (iii) assignment (A). This last problem refers to the staff assignment problem, which is addressed in the present study.
2. *Multiskilling* (MS): indicates whether multiskilling was modeled as a parameter (Par), or a decision variable (Var).
3. *Chaining* (CH): indicated whether or not the study evaluated the benefits of multiskilling through a k-chaining policy.
4. *Workforce* (WF): it can be: (i) homogeneous (Hom), where employee's productivity is the same regardless of the number of task types or departments to which he/she is trained; and (ii) heterogeneous (Het), where employee's productivity varies when he/she is trained to work in multiple task types or departments.
5. *Productivity modeling* (PM): it indicates how productivity is modeled to consider the loss/gain effect caused by the use of multiskilled personnel. According to Henao et al. [5] three types of approaches are usually considered: (i) learning and forgetting (L/F), where the employee's productivity is modeled as a natural process of learning, forgetting, and relearning. (ii) productivity matrix (M), where each employee is assumed to have different productivity levels in the trained departments, which is presented matrixially; and (iii) constant productivity (CP), where all employees have equal productivity in all trained departments.
6. *Cost of productivity loss (CPL):* indicates whether the study considered the monetary cost of the productivity loss generated by multiskilling.
7. *Solution method* (SM): It can be: (i) heuristic/metaheuristics (H); (ii) optimization (OPT), (iii) simulation (S), y (iv) analytics (AN).
8. *Planning horizon (PH):* indicates the planning horizon covered by the study.
9. *Application* (AP): represents the industrial sector in which the study was applied. It can be: (i) Manufacturing (M); (ii) Call Centers (CC); and (iii) Retail (R).

Table 1 reports several findings in the literature that are worth highlighting. First, few studies evaluated for the retail industry the benefits of incorporating multiskilling in human resource management [4–7]. Second, few studies used a k-chaining policy to structure the employee training plans through closed chains [5–7]. These studies were applied to the retail industry, and they reported that a k-chaining policy with $k = 2$ was highly effective in minimizing the costs of over/understaffing in the face of uncertain demand. However, unlike the present paper, such studies considered a homogeneous workforce and, therefore, they not incorporated the learning and forgetting effect in the mathematical models. In addition, they used a k-chaining policy with $k = 2$, which meant that employees could only be trained in a maximum of two departments. Instead, this paper will use a k-chaining policy con $k \geq 2$ and will also incorporate the process of learning and forgetting into the mathematical modeling.

Third, most of the previous studies were applied to the manufacturing sector, and in turn, most of these modeled the multiskilling considering the effects of the learning and forgetting phenomenon. This demonstrates the importance of studying the learning and forgetting phenomenon, and exploring its impact on the service sector industry, particularly in the retail industry, where it is also observed. Fourth, most of the proposed models do not consider the monetary impact that arises when an employee's productivity

Table 1. Studies that address how to incorporate productivity loss into multiskilling decisions.

References	HR	MS	CH	WF	PM	CPL	SM	PH	AP
Attia et al. [13]	S + SS	Par	No	Het	M	Yes	OPT + S	4 wks	M
Attia et al. [14]	SS	Par	No	Het	L/F	Yes	H + S	–	M
Bentefouet and Nembhard [15]	A	Par	No	Het	L/F	No	OPT	–	M
Kim et al. [16]	A	Par	No	Het	L/F	No	OPT	48 h	M
Azizi and Liang [8]	A	Par	No	Het	L/F	Yes	H	17 wks	M
Sawhney [17]	SS	Par	No	Hom	CP	No	OPT	–	M
Bentefouet [18]	SS	Var	No	Het	L/F	No	S	16 wks	M
Nembhard [19]	S + A	Par	No	Het	L/F	No	S	–	M
Attia et al. [20]	S + A	Par	No	Het	M + L/F	Yes	H + S	6 wks	M
Henao et al. [4]	SS	Var	No	Hom	CP	No	OPT	–	R
Henao et al. [5]	A	Var	Yes	Hom	CP	No	OPT + H	1 wks	R
Malachowski & Korytkowski [21]	A	Var	No	Het	L/F	No	S	4 wks	M
Mossa et al. [22]	SS	Var	No	Hom	CP	No	OPT + H	–	M
Munoz and Bastian [23]	SS	Var	No	Het	L/F	Yes	S	–	CC
Valeva et al. [24]	A	Var	No	Het	L/F	No	OPT + S	–	M
López and Nembhard [25]	SS	Par	No	Het	M + L/F	No	OPT + H	–	M
Korytkowski [26]	S + A	Par	No	Het	L/F	No	H	–	M
Kiassat and Safaei [27]	A	Var	No	Het	L/F	Yes	S	–	M
Biel & Glock [28]	SS	Var	No	Het	L/F	Yes	S	–	M
Henao et al. [6]	A	Var	Yes	Hom	CP	No	AN + H	1 wks	R
Porto et al. [7]	S + SS	Var	Yes	Hom	CP	No	OPT	1 wks	R
This paper	*A*	*Var*	*Yes*	*Het*	*L/F*	*Yes*	*OPT*	*52 wks*	*R*

decreases as his/her multiskilling level increases. Attia et al. [20] highlight the importance of incorporating this cost in the problem objective function, since this way the costs and benefits associated with the use of multiskilled staff are better quantified. Consequently, this study will consider this cost in the objective function.

3 Problem Description

The problem is to design a suitable training plan for a known workforce and, in turn, to obtain their weekly working hours assignment for a planning horizon of 52 weeks (i.e., one year). In a novel way, our mathematical formulation will explicitly incorporate terms in the objective function and constraints that allow to model a multiskilling k-chaining policy (with $k \geq 2$) and the learning and forgetting phenomenon. In specific terms, the challenge will be to solve an annual staff assignment problem for a retail store, which simultaneously decides: (i) How many employees will be multiskilled? (ii) In how many additional departments each multiskilled employee will be trained? (iii) How many weekly working hours must work each employee in each assigned department? and (iv) What will be the productivity levels that each employee will have in the departments where he/she has been trained? In other words, although a multiskilled employee may work in two or more departments, he/she will not have the same productivity in all of them.

The proposed model considers the following assumptions: (i) understaffing costs are the same for all departments, and represent the costs per unsatisfied demand. (ii) Training costs are the same for all departments. These first two assumptions are common in the retail industry [5–7, 29]. (iii) Unscheduled absenteeism of personnel is not considered. (iv) All employees must work exactly 45 h per week. (v) Each employee is initially trained to work in a single department (i.e., single-skilled employees), but they could be trained in any department of the store. (vi) Each employee has an associated productivity equal to 1 (i.e., 100%) in the department in which he/she is initially trained (i.e., primary department). (vii) Three employee types will be considered, which are classified according to their experience degree: Senior, Standard, and Junior. Each employee type will have a different initial productivity level in the departments where he/she is additionally trained (i.e., secondary departments). Finally, Table 2 shows the used notation in our problem formulation.

Table 2. Sets, parameters, and variables of the addressed problem.

Sets	
I	Store employees, indexed by i
L	Store departments, indexed by l
S	Weeks, indexed by s
I_l	Employees under contract in the department $l, I_l \subset I$
General parameters	
r_{ls}	Number of hours required in department l and week $s, \forall l \in L, s \in S$
h	Number of weekly hours an employee must work under his/her contract
m_i	Department for which employee i is initially trained, $\forall i \in I$
u	Understaffing cost per hour
c	Cost of training an employee to work in any department
Parameters associated with the learning and forgetting phenomenon	
σ_{il}	Learning rate of employee i in department l, $\forall i \in I, l \in L$
b_{il}	Exponential factor for employee i in department l, $\forall i \in I, l \in L$
n	Number of work repetitions
f_{il}	Slope of the forgetting curve for employee i in department l, $\forall i \in I, l \in L$
TA	Uninterrupted period of practice of the skill
TB	Interruption period after which productivity has decreased to its initial level
ξ	Represents the ratio between periods TA y TB. Such that $\xi = \frac{TB}{TA}$
η_{eq}	Equivalent number of work repetitions on a specified date
U	Monetary cost of developing employee productivity
λ_{il}	Number of work repetitions by employee i in department l that should be performed if the interruption had not occurred, $\forall i \in I, l \in L$

(continued)

Table 2. (*continued*)

Parameters associated with the learning and forgetting phenomenon	
θ_{il}^{ini}	Initial productivity of employee i in department l, when he/she is assigned for first time, $\forall i \in I, l \in L$. It varies according to employee type: Senior, Standard, Junior

Variables	
x_{il}	Equal to 1 if employee i is trained in department l, otherwise 0, $\forall i \in I, l \in L$
v_i	Number of additional skills of the multiskilled employee i, $\forall i \in I$
ω_{ils}	Number of hours per week assigned to employee i in department l, week s, $\forall i \in I, l \in L, s \in S$
θ_{il}^{SP}	Employee productivity i in department l, at the beginning of the planning horizon, $\forall i \in I, l \in L$
θ_{il}^{FP}	Employee productivity i in department l, at the end of the planning horizon, $\forall i \in I, l \in L$
k_{ls}	Understaff hours in department l and week s, $\forall l \in L, s \in S$

4   Methodology

The methodology is presented in two subsections. Based on the article by Attia et al. [20], Subsect. 4.1 presents a set of formulas to model the learning and forgetting phenomenon. Then, using this set of formulas, Subsect. 4.2 presents the mixed integer programming model to solve the problem described above.

4.1   Modeling the Learning and Forgetting Phenomenon

Learning Phenomenon. According to Attia et al. [20], as a result of the learning phenomenon, the time required for an employee to perform a task is reduced each time the employee repeats such task. This learning phenomenon represents an evolution of productivity, which can be measured in terms of learning rate σ_{il}, their initial productivity θ_{il}^{ini} and the uninterrupted time the activity was carried out expressed in weeks, n. This evolution in productivity due to learning can be expressed by Eq (1). Where the exponential factor b_{il} can be calculated using Eq (2).

$$\theta_{il}^{SP} = \frac{1}{\left[1 + \left(\frac{1}{\theta_{il}^{ini}} - 1 \right)(n)^{b_{il}} \right]} \tag{1}$$

$$b_{il} = \frac{\log(\sigma_{il})}{\log(2)} \tag{2}$$

Forgetting Phenomenon. The productivity loss generated by the interruption of work and transfer of employees to other departments must also be considered. Constant switching between departments can cause employees to forget those skills associated with less frequently assigned departments. This is known as the forgetting phenomenon, which causes a productivity loss. In this case the evolution of productivity can be expressed by Eq. (3). This equation is in function of the slope of the forgetting curve, f_{il}, the number of work repetitions of the skill before the interruption expressed in weeks, η_{eq}, and the interruption period λ_{il}. Where the slope of the forgetting curve f_{il} is calculated using Eq. (4).

$$\theta_{il}^{FP} = \frac{1}{\left[1 + \left(\frac{1}{\theta_{il}^{ini}} - 1\right)(\eta_{eq})^{b_{il} - f_{il}}(\eta_{eq} - \lambda_{il})^{f_{il}}\right]} \tag{3}$$

$$f_{il} = -b_{il}(b_{il} + 1)\left[\frac{\log(\eta_{eq})}{\log(\xi + 1)}\right] \tag{4}$$

Therefore, due to the phenomenon of learning and forgetting, it seeks to minimize not only the costs of training and understaffing, but also the cost of productivity loss. This will better guide decision making on which additional departments should be trained on each employee.

4.2 Mathematical Model Incorporating k-chaining and the Learning and Forgetting Phenomenon

The optimization model can now be formulated as follows:

$$Min \sum_{i \in I} \sum_{l \in L: l \neq m_i} cx_{il} + \sum_{l \in L} \sum_{s \in S} uk_{ls} + \sum_{l \in L: l \neq m_i} U\left[\frac{\sum_{i \in I}(\theta_{il}^{FP} - \theta_{il}^{SP})}{\sum_{i \in I}\theta_{il}^{SP}}\right]x_{il} \tag{5}$$

s.t.

$$\sum_{i \in I} \theta_{il}^{FP}\omega_{ils} + k_{ls} \geq r_{ls} \qquad \forall l \in L, \forall s \in S \tag{6}$$

$$\sum_{l \in L} \omega_{ils} = h \qquad \forall i \in I, \forall s \in S \tag{7}$$

$$\omega_{ils} \leq hx_{il} \qquad \forall i \in I, \forall l \in L, \forall s \in S \tag{8}$$

$$x_{il} = 1 \qquad \forall i \in I, \forall l \in L : l = m_i \tag{9}$$

$$v_i = \sum_{i \in L: l \neq m_i} x_{il} \qquad \forall i \in I \tag{10}$$

$$\sum_{i \in l_l} v_i = \sum_{i \in \{I - I_l\}} x_{il} \qquad \forall l \in L \tag{11}$$

$$\theta_{il}^{SP} = \frac{1}{\left[1 + \left(\frac{1}{\theta_{il}^{ini}} - 1\right)(n)^{b_{il}}\right]} \qquad \forall i \in I, \forall l \in L \qquad (12)$$

$$\theta_{il}^{FP} = \frac{1}{\left[1 + \left(\frac{1}{\theta_{il}^{ini}} - 1\right)\left(\eta_{eq}\right)^{b_{il}-f_{il}}\left(\eta_{eq} - \lambda_{il}\right)^{f_{il}}\right]} \qquad \forall i \in I, \forall l \in L \qquad (13)$$

$$0 \le \theta_{il}^{SP} \le 1 \qquad \forall i \in I, \forall l \in L \qquad (14)$$

$$0 \le \theta_{il}^{FP} \le 1 \qquad \forall i \in I, \forall l \in L \qquad (15)$$

$$x_{il} \in \{0, 1\} \qquad \forall i \in I, \forall l \in L \qquad (16)$$

$$v_i \ge 0 \qquad \forall i \in I, \forall l \in L \qquad (17)$$

$$\omega_{ils} \ge 0 \qquad \forall i \in I, \forall l \in L \qquad (18)$$

$$k_{ls} \ge 0 \qquad \forall i \in I, \forall l \in L \qquad (19)$$

The objective function (5) minimizes the following costs: (i) training; (ii) under-staffing; and (iii) productivity loss. Constraints (6) and (19) produce the (non-negative) understaffing level associated with each department. Constraints (7) ensure that each employee works exactly the number of hours required by his/her contract in each week. Constraints (8) impose that each employee is assigned to work only in the departments for which he/she was trained. Constraints (9) determine the department in which each employee is initially trained. Constraints (10) indicate, for each employee, how many additional departments he/she was trained in. He/she could continue as a single-skilled employee or become multiskilled employee. The constraints (11) guarantee that for each department l, the number of multiskilled employees belonging to it will be equal to the number of employees belonging to other departments but who were trained to work in department l as well. Constraints (10) and (11) guarantee the formation of closed chains through a k-chaining approach. Constraints (12) represent the evolution of productivity under the learning phenomenon, while constraints (13) represent the evolution of productivity under the forgetting phenomenon. Finally, constraints (14)–(19) define the domain of each variable of the problem.

5   Case Study, Results and Discussion

This section is presented in three subsections. First, the used data to solve our case study are presented. Second, we present the experiments and performance metrics that will allow us to evaluate the impact of productivity evolution on multiskilling decisions. Third, we present the results of the methodology as well as an analysis of the results.

5.1 Case Study

This case study considers real, processed and simulated data associated with a Chilean retail store.

Real Data. The store considers 6 departments and 30 hired employees, i.e., $|L| = 6$ and $|I| = 30$. Such that, in each department there is an initial number of single-skilled employees, I_l. In addition, all employees must work 45 h per week, i.e., $h = 45$, and the planning horizon is 52 weeks, $|S| = 52$. With respect to labor costs, it is considered a minimum training cost, $c = US\$1$ wk/employee [5], the understaffing cost is equal to $u = US\$15/h$, and the cost associated with the productivity loss is equal to $P = US\$2$ wk/employee.

Processed Data. These data are required to model the learning and forgetting phenomenon. This case study considers three employee types according to their initial productivity, these are: Senior, Standard and Junior. These have an initial productivity (θ_{il}^{ini}) equal to $0.8, 0.7$ y 0.6 respectively, in the additional departments in which they will be trained. At the same time, all employees have an initial productivity, $\theta_{il}^{ini} = 1$, in the department in which they are initially trained. In addition, in this study, the learning rate was considered constant for each employee and department, and its value is $\sigma_{il} = 0.8$. Likewise, it was assumed that the number of work repetitions prior to the beginning of the study (n) is equal to 1, because the employees are not yet trained in additional skills [20]. The exponential factor b_{il} and the slope of the forgetting curve f_{il} do not differ between departments. Finally, in order to model the evolution of productivity considering the effect of learning and forgetting, it was considered $\xi = 3$ y $n_{eq} = 2$. All values indicated are typically reported in the literature.

Simulated Data. Historical data show that the staff demand in hours (r_{ls}) by department l and week s follows a normal distribution. In order to evaluate the impact of demand variability on this methodology results, three variability scenarios were considered; represented by the following coefficients of variation (CV): 10%, 20% y 30%. Then, for each CV and following a normal probability distribution truncated at zero (to avoid negative values), a Monte Carlo simulation was performed to obtain the demand values for each of the 52 weeks of the planning horizon.

5.2 Experiments and Performance Metrics

Experiments. In order to compare the impact of productivity evolution on multiskilling decisions, the following two experiments were defined:

1. *Experiment 1*: Employees may be trained in one or more additional departments, but they will be modeled as a homogeneous workforce. That is, all employees have a productivity equal to 1 in all departments where they are trained.
2. *Experiment 2*: Employees may be trained in one or more additional departments, but they will be modeled as a heterogeneous workforce. Such that the effect of the learning and forgetting phenomenon is incorporated.

Two things are worth mentioning. First, Experiment 2 is equivalent to the methodology proposed in this paper. Second, for each experiment, the three demand variability scenarios previously defined are considered ($CV = 10\%, 20\%$ y 30%).

Performance Metrics. In order to evaluate the performance of the solutions, two metrics will be used to measure the multiskilling requirements.

1. *Percentage of multiskilled employees:* calculates the number of multiskilled employees required, in relation to the total number of employees in the store.

$$\%ME = \frac{No.\,Multiskilled\;Employees}{|I|}.100 \tag{20}$$

2. *Percentage of total multiskilling:* is calculated as the number of additional skills trained, in relation to the maximum number of skills possible to train.

$$\%TM = \frac{\sum\limits_{i \in I}\sum\limits_{l \in L: l \neq mi} x_{il}}{|I|(|L| - 1)}.100 \tag{21}$$

5.3   Results and Discussion

The proposed formulation was written in AMPL and solved using ILOG CPLEX 12.9.0 software. A PC with an Intel (R) Core (TM) i5-7200U CPU @ 2.50 GHz 2.71 GHz and 4GB of RAM was used. For all scenarios, optimality was reached in less than 60 s. In addition, Table 3 shows the number of variables and constraints generated in each of the tested scenarios. Note that, since the structure of the problem does not change from one scenario to another, the size of the model is the same in all of them.

Table 3. Model features.

Experiment	%CV	Binary variables	Continuous variables	Constraints
1	10	150	10 341	10 035
	20	150	10 341	10 035
	30	150	10 341	10 035
2	10	150	10 342	10 036
	20	150	10 342	10 036
	30	150	10 342	10 036

The learning and forgetting phenomenon has an impact on multiskilling decisions. Therefore, based on a cost analysis and the %ME and %TM performance metrics, the results obtained in Experiments 1 and 2 were evaluated and compared. First, for Experiment 2, it was reported that the final productivity (θ_{il}^{FP}) obtained for the three employee types (in the secondary departments) was as follows: Senior, 0.9; Standard, 0.8; and Junior, 0.7. This indicates that those models that assume a homogeneous workforce as described in Experiment 1 may provide solutions that do not fit the reality of the companies.

To continue with the analysis, Table 4 presents the results obtained for each experiment, from which the following analyses can be derived. First, for both experiments, it is observed that as the coefficient of variation (%CV) increases, the multiskilling requirements (i.e., %ME and %TM) increase. This result is intuitive, since the higher the levels of uncertainty in demand, greater investment in labor flexibility (i.e., multiskilling) will allow to minimize the mismatch between staffing and staff demand. Second, for Experiment 2, we observe that the required multiskilling levels are higher than in Experiment 1. This result is interesting, as it indicates that the incorporation of the learning and forgetting phenomenon requires higher levels of labor flexibility to compensate for the fact that the final productivity levels of multiskilled employees (in secondary departments) are no longer the maximum, i.e., they are less than 1.

Table 4. Results and performance measures associated with Experiments 1 and 2.

Exp	%CV	N° Multiskilled employees					%ME	%TM	Annual cost (US$)			
		$v = 1$	$v = 2$	$v = 3$	$v = 4$	Total			TC	UC	PL	Total
1	10	8	1	–	–	9	30	7	10	57 360	–	57 370
	20	13	2	–	–	15	50	13	19	140 760	–	140 779
	30	14	5	1	–	20	67	18	27	163 320	–	163 347
2	10	5	3	2	1	11	37	14	21	62 124	5	62 150
	20	4	7	4	–	15	50	20	30	151 854	8	151 892
	30	5	6	4	2	17	57	25	37	176 256	10	176 303

Third, it should be noted that even when two scenarios have the same %ME value, the required %TM not be equal. For example, for %CV = 20, it is observed that in both experiments %ME = 50 was obtained; however, the %TM is higher in Experiment 2. This means that in both experiments the same number of multiskilled employees was required, but in Experiment 2 the multiskilled employees were trained in a greater number of additional departments. In general, the results reflected in Table 4 show that in Experiment 2 the maximum number of additional skills trained in a multiskilled employee was $v = 4$, while for Experiment 1 this number was $v = 3$.

Table 4 also shows a fourth analysis associated with the costs incurred in each experiment and CV. It is observed that in Experiment 2 there is a greater investment in multiskilling and, therefore, training costs (UC) are higher. In addition, in both experiments, it is observed that the understaffing costs increase as the CV increases. However, Experiment 1 records lower understaffing costs than Experiment 2. This result occurs because Experiment 1 assumes that multiskilled employees are homogeneous and have a maximum productivity level in each of the trained departments (i.e., $\theta_{il}^{SP} = \theta_{il}^{FP} = 1$). That is, in Experiment 1 the supply of man-hours is 100% available, which is not the case in Experiment 2, where the forgetting phenomenon generates a productivity loss.

Consequently, Experiment 2 provides more realistic required multiskilling levels and, at the same time, does not underestimate the costs of understaffing. Finally, as a result of the higher training (TC) and understaffing (UC) costs, and the incorporation of the virtual cost associated with the productivity loss (PL) generated by the learning and forgetting phenomenon, Experiment 2 always reports a higher total cost than Experiment 1.

6 Conclusions

In the present study, a mixed integer programming model was proposed to solve an annual assignment problem that explicitly incorporates the learning and forgetting phenomenon and, also, the use of multiskilled personnel through a k-chaining approach. The objective was to measure and evaluate the impact of the learning and forgetting phenomenon on multiskilling decisions. The case study was applied in a Chilean retail store, for a planning horizon equal to one year, in which two experiments facing three different levels of demand variability were evaluated. The first experiment did not incorporate the learning and forgetting phenomenon and, in turn, assumed a homogeneous workforce with maximum productivity. The second experiment did incorporate the learning and forgetting phenomenon and therefore modeled a heterogeneous workforce.

This case study shows several interesting results. First, Experiment 1 always reported lower costs compared to Experiment 2. This is because Experiment 1 underestimates the costs of training and understaffing by assuming (unlikely) that all multiskilled employees always have maximum productivity in all additional departments where they were trained. Moreover, Experiment 2 required higher levels of multiskilling to compensate for the productivity loss caused by the forgetting phenomenon. In turn, given that the multiskilled employees reported final productivities lower than 1 in the additional departments where they were trained, the reported levels of understaffing were higher relative to Experiment 1. These results are interesting, as they demonstrate the importance of incorporating the learning and forgetting phenomenon in multiskilling decisions, which leads to the achievement of solutions that are more in line with reality.

In addition, it was found that greater the variability in demand, greater the levels of multiskilling are required. Additionally, for Experiment 2, the number of additional skills trained on a multiskilled employee took a maximum of four, while for Experiment 1 this maximum was equal to three.

Finally, for future research, it would be interesting to add the following analyses: (1) simultaneously incorporating staff hiring decisions, which would allow to better adjust to scenarios with high variability in demand; (2) explicitly considering stochastic demand in the mathematical formulation, which would allow to obtain robust and feasible solutions to changes in that parameter; and (3) performing a sensitivity analysis on the used cost parameters, which would allow to explore changes in the solution structure of the problem.

Acknowledgements. This research was supported by *"Fundación para la Promoción de la Investigación y la Tecnología (FPIT)"* under Grant 4.523.

References

1. Henao, C.A.: Diseño de una fuerza laboral polifuncional para el sector servicios: caso aplicado a la industria del retail (Tesis Doctoral, Pontificia Universidad Católica de Chile, Santiago, Chile) (2015). https://repositorio.uc.cl/handle/11534/11764
2. Mac-Vicar, M., Ferrer, J.C., Muñoz, J.C., Henao, C.A.: Real-time recovering strategies on personnel scheduling in the retail industry. Comput. Ind. Eng. **113**, 589–601 (2017)
3. Álvarez, E., Ferrer, J.C., Muñoz, J.C., Henao, C.A.: Efficient shift scheduling with multiple breaks for full-time employees: a retail industry case. Comput. Ind. Eng. **150**, 106884 (2020)
4. Henao, C.A., Muñoz, J.C., Ferrer, J.C.: The impact of multi-skilling on personnel scheduling in the service sector: a retail industry case. J. Oper. Res. Soc. **66**(12), 1949–1959 (2015)
5. Henao, C.A., Ferrer, J.C., Muñoz, J.C., Vera, J.: Multiskilling with closed chains in a service industry: a robust optimization approach. Int. J. Prod. Econ. **179**, 166–178 (2016)
6. Henao, C.A., Muñoz, J.C., Ferrer, J.C.: Multiskilled workforce management by utilizing closed chains under uncertain demand: a retail industry case. Comput. Ind. Eng. **127**, 74–88 (2019)
7. Porto, A.F., Henao, C.A., López-Ospina, H., González, E.R.: Hybrid flexibility strategic on personnel scheduling: retail case study. Comput. Ind. Eng. **133**, 220–230 (2019)
8. Azizi, N., Liang, M.: An integrated approach to worker assignment, workforce flexibility acquisition, and task rotation. J. Oper. Res. Soc. **64**(2), 260–275 (2013)
9. Henao, C.A., Batista, A., Pozo, D., Porto, A.F., González, V.I.: Multiskilled personnel assignment problem under uncertain demand: a benchmarking analysis. In: Submitted to Computers & Operations Research, Under 1st review (2021)
10. Fontalvo Echavez, O., Fuentes Quintero, L., Henao, C.A., González, V.I.: Two-stage stochastic optimization model for personnel days-off scheduling using closed-chained multiskilling structures. In: Rossit, D.A., Tohmé, F., Mejía, G. (eds.) Production Research. ICPR-Americas 2020. Communications in Computer and Information Science, vol. 1407, Springer Nature Switzerland AG (2021)
11. Abello, M.A., Ospina, N.M., De la Ossa, J.M., Henao, C.A., González, V.I.: Using the k-chaining approach to solve a stochastic days-off-scheduling problem in a retail store. In: Rossit, D.A., Tohmé, F., Mejía, G. (eds.) Production Research. ICPR-Americas 2020. Communications in Computer and Information Science, vol. 1407, Springer Nature Switzerland AG (2021)
12. Mercado, Y.A., Henao, C.A.: Benefits of multiskilling in the retail industry: k-chaining approach with uncertain demand. In: Rossit, D.A., Tohmé, F., Mejía, G. (eds.) Production Research. ICPR-Americas 2020. Communications in Computer and Information Science, vol. 1407, Springer Nature Switzerland AG (2021)
13. Attia, E.A., Edi, K.H., Duquenne, P.: Flexible resources allocation techniques: characteristics and modelling. Int. J. Oper. Res. **14**(2), 221–254 (2012)
14. Attia, E.A., Dumbrava, V., Duquenne, P.: Factors affecting the development of workforce versatility. IFAC Proc. Vol. **45**(6), 1221–1226 (2012)
15. Bentefouet, F., Nembhard, D.A.: Optimal flow-line conditions with worker variability. Int. J. Prod. Econ. **141**(2), 675–684 (2013)
16. Kim, S., Nembhard, D.A.: Rule mining for scheduling cross training with a het-erogeneous workforce. Int. J. Prod. Res. **51**(8), 2281–2300 (2013)
17. Sawhney, R.: Implementing labor flexibility: a missing link between acquired labor flexibility and plant performance. J. Oper. Manag. **31**(1–2), 98–108 (2013)
18. Bentefouet, F.: Workforce schedulling in the context of human perfromance variability: the worker approach. Doctoral dissertation, The Pennsylvania State University, Pennsylvania, U.S.A. (2013). https://etda.libraries.psu.edu/catalog/16925.

19. Nembhard, D.A.: Cross training efficiency and flexibility with process change. Int. J. Oper. Prod. Manag. **34**(11), 1417–1439 (2014)
20. Attia, E.A., Duquenne, P., Le-Lann, J.M.: Considering skills evolutions in multi-skilled workforce allocation with flexible working hours. Int. J. Prod. Res. **52**(15), 4548–4573 (2014)
21. Malachowski, B., Korytkowski, P.: Competences-based performance model of multi-skilled workers. Comput. Ind. Eng. **91**, 165–177 (2016)
22. Mossa, G., Boenzi, F., Digiesi, S., Mummolo, G., Romano, V.A.: Productivity and ergonomic risk in human based production sys-tems: a job-rotation scheduling model. Int. J. Prod. Econ. **171**, 471–477 (2016)
23. Munoz, D., Bastian, N.: Estimating cross-training call center capacity through simulation. J. Syst. Sci. Syst. Eng. **25**(4), 448–468 (2016). https://doi.org/10.1007/s11518-015-5286-9
24. Valeva, S., Hewitt, M., Thomas, B.W., Brown, K.G.: Balancing flexibility and inventory in workforce planning with learning. Int. J. Prod. Econ. **183**, 194–207 (2017)
25. López, C.E., Nembhard, D.: Cooperative workforce planning heuristic with worker learning and forgetting and demand constraints. In: Coperich, K., Cudney, E., Nembhard, H. (eds.) IEE Annual Conference. Proceedings, Pittsburgh, Pennsylvania, USA. (2017)
26. Korytkowski, P.: Competences-based performance model of multi-skilled workers with learning and forgetting. Expert Syst. Appl. **77**, 226–235 (2017)
27. Kiassat, C., Safaei, N.: Effect of imprecise skill level on workforce rotation in a dynamic market. Comput. Ind. Eng. **131**, 464–476 (2018)
28. Biel, K., Glock, C.H.: Governing the dynamics of multi-stage production systems subject to learning and forgetting effects: a simulation study. Int. J. Prod. Res. **56**(10), 3439–3461 (2018)
29. Porto, A.F., Henao, C.A., López-Ospina, H., González, E.R., González, V.I.: Dataset for solving a hybrid flexibility strategy on personnel scheduling problem in the retail industry. Data Brief **32**, 106066 (2020)

Modelling the Dynamics of a Digital Twin

Marisa Analía Sánchez[1]([✉]) [iD], Daniel Rossit[2] [iD], and Fernando Tohmé[3] [iD]

[1] Dpto. de Ciencias de la Administración, Universidad Nacional del Sur,
Bahía Blanca, Argentina
mas@uns.edu.ar

[2] Dpto. de Ingeniería, Universidad Nacional del Sur, INMABB-UNS-CONICET,
Bahía Blanca, Argentina
daniel.rossit@uns.edu.ar

[3] Dpto. de Economía, Universidad Nacional del Sur, INMABB-UNS-CONICET,
Bahía Blanca, Argentina
ftohme@criba.edu.ar

Abstract. "Digital twining" is one the main ways of establishing data channels in cyber-physical systems using both the outputs of running a virtual model and real time data collected by sensors. The purpose to this paper is to outline the digital twin of a cyber-physical *production* system. We apply the System Dynamics paradigm to the case of a shop-floor factory devoted to cloud manufacturing. The digital twin uses data from the real production line, providing assistance to maintenance procedures triggered by inconsistencies between the real and the virtual processes.

Keywords: Digital twin · Cyber-physical systems · System Dynamics · Simulation

1 Introduction

One of the main components of the Industry 4.0 infrastructure are "smart factories", able to adapt their production processes to demand, assigning their resources more efficiently than traditional factories. The key technologies enabling smart factories are the Internet of Things [2], Cyber-physical Systems (CPS) [3, 4], cloud computing [4] and big data [5, 6].

These technologies allow the efficient integration of different functionalities of the control specification ISA-95 into a single production system. In terms of its five levels, all the processes that can be digitalized can be thus absorbed in Industry 4.0 systems [7, 8]. This integration can be both at horizontal and vertical levels. The vertical integration extends from the physical process level (level 0 of ISA-95) to Manufacturing Execution Systems (level 3 of ISA-95) [9]. These levels, handled by CPS, translate the physical events into digital data, creating a digital twin (DT) of the production system. The digital twin provides useful information about the real workload of the production system, which can be used in planning and business management processes [10, 11]. Managers can simulate the inner workings of the plant under different scenarios, gaining further

© Springer Nature Switzerland AG 2021
D. A. Rossit et al. (Eds.): ICPR-Americas 2020, CCIS 1408, pp. 238–252, 2021.
https://doi.org/10.1007/978-3-030-76310-7_19

information contributing to make better decisions. The ultimate objective of digital twins is to improve the operation and efficiency of manufacturing assets, reducing costs by forecasting future states and supporting advanced decision-making throughout the entire manufacturing lifecycle [12].

A DT of a production system is a type of simulation that allows the real-time control of production. Ding et al. [1] use the term "digital twining" to refer to the process of building a digital twin in the cyber world of physical objects and systems, establishing data channels for cyber-physical connection and synchronization. The implication of this definition is that a DT should use both simulated values and real time data collected by sensors. The simulation has the ability to compress time running ahead the real world process since the digital twin simulation clock is set at one relevant period ahead of real time. If at time t_r a sensor reports a machine failure that was not generated in the simulation when its internal clock was at t_r, the latter should be reset at t_r updating its state with this new information. This feature enables a DT to keep the simulation running at a par with the real production line.

The aim of this work is to describe a simulation model for a digital twin cyber-physical system. In order to define the guidelines for a smart plant environment, the modelling methodology of System Dynamics can be used to handle the uncertainties and nonlinear relationships among interacting system components [13, 14]. The complexity in the definition of the model may ensue from structural or dynamic aspects. Structural complexity refers to the number of components in a system, or the number of combinations involved in making a decision. Dynamic complexities result from the non-linear and history-dependent nature of self-organizing and adaptive systems [15]. System Dynamics addresses these two kinds of complexity by postulating that the behavior of complex systems results from an underlying structure of flows, delays, and feedback loops [16]. The emphasis in System Dynamics is not on forecasting the future, but on learning how the actions in a period of time can trigger reactions in the future [17, 18].

The paper is organized as follows. Section 2 presents concepts drawn from the literature on manufacturing processes, providing information about the system to be modelled. Section 3 presents the running example that will illustrate the main notions presented in this article. Section 4 introduces Causal Loop diagrams, the basic graphic modelling tool needed to sketch the description of the system. Section 5 presents Stocks-and-Flow diagrams, which allow developing a full model of the system. Section 6 shows how this model provides the grounds to simulate different scenarios using the digital twin to support decision-making. Finally, Sect. 7 concludes.

2 Conceptual Model of the Smart Factory

The digital twin concept was first devised as a rich digital representation of actual devices, being widely used in the aerospace field [19]. In 2015, the scope of the original definition was expanded, opening the possibility of using it in other fields [20]. Virtual factories were characterized as the digitalization of plants integrated with the real systems assisting the production process along the lifecycle of each asset [21]. This development led to two different concepts of what a digital twin is: some researchers believe that it can be identified with its virtual component (i.e. the simulation process) while others emphasize on the connections between the virtual and the physical aspects [22]. Tao *et al.* [23] present a design for a smart factory including four main components: a physical shop-floor (PS), a virtual shop-floor (VS), the data generated by the digital twin (D), and a shop-floor service system (SSS).

The physical part includes entities such as human, machines, and materials. The virtual part refers to the digital twin of the physical shop-floor which supports the decision-making and control of the physical part [23]. It provides control orders for the PS and optimization strategies for the SSS. The data component integrates physical, virtual, and service data with the aim of providing consistent information. The service component embodies services like the definition of production and resource allocation plans as well as guidelines for data fusion. These services, in turn, provide information for both the physical and the virtual shop-floor. The authors propose a three stage operation of the digital twin:

- Before production, the SSS defines a production plan based on data collected from different sources (customer orders, real-time data read by sensors, outputs from the simulation, as well as business data) and transmits it to the VS for verification.
- The VS simulates the production plan and if it works correctly, it sends control orders to the PS to start production. The real-time data generated by the PS is then recorded by the VS.
- Once the production process finishes, its recorded history is analysed to extract knowledge that can be used by the VS to simulate future production plans.

In the digital twin shop-floor framework, physical and simulated data are synchronized. The operational implication is that the simulation is fed with real-time sensor data, sending back controls to the PS in a continuous interaction between the physical and virtual components. Tao et al. [23] framework is the basis for implementing the simulation model of a *digital twin cyber-physical system*. We, in turn, add on top of that the process of updating in real-time the production plan in the context of a cloud manufacturing environment. The ensuing digital twin embodies many specialized digital twins: parts DT, machines DT, and a shop-floor DT. The model assumes the use of real-time production data (shop-floor and cloud-shop floor) to support the factory operations.

While simulation has been traditionally used to identify bottlenecks in production plan, it can now be incorporated into smart machines or tools equipped with devices able to capture, process and transmit data. A simulation fed with such information becomes able to check whether the behavior of the physical factory is consistent with the "simulated" behavior. In particular, when (real) data indicates an unforeseen failure, the simulation synchronizes accordingly (data fusion), updating its behavior, and transmitting new control orders to smart devices. Hence, such simulation of the digital twin cyber-physical system helps to handle failures. Figure 1 depicts our simulation model of a digital twin cyber-physical system.

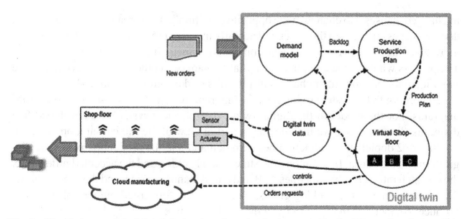

Fig. 1. Simulation model of the digital twin cyber-physical system. Dashed lines: data flow, solid lines: control flow.

3 Running Example

Let us consider a smart factory consisting of two production lines A and B. Orders are scheduled on line A. If A is operating above its capacity or it is being repaired, the orders are sent to line B. If line B is operating above its capacity, then the remaining orders are assigned to cloud manufacturing suppliers. For simplicity, we assume that only line A suffers failures and the time to repair is of 3 time units. The cycle times of production lines A, B as well as that of the cloud suppliers are represented by the probability distributions detailed in Table 1. These normal distributions are assumed to have been fitted with historical data of demand.

Table 1. Probability distributions describing cycle time of production facilities.

Production facilities	Cycle time (in days)
A	Normal distribution (mean = 2, st. deviation = 2%)
B	Normal distribution (mean = 4, st. deviation = 2%)
Cloud manufacturing suppliers	Normal distribution (mean = 2, st. deviation = 2%)

4 Causal Loop Diagrams

The first step towards creating a dynamic model of the digital twin CPS is to draw its Causal Loop diagram [18]. In a Causal Loop diagram variables are connected by arrows denoting the causal influences among them: $x \rightarrow y$ means that the input variable x has some causal influence on the output variable y. A positive influence, denoted by $+$, means "a change in x, being the rest of variables unchanged, causes y to change in the same direction". In turn, a negative influence means "a change in x, being the rest of variables unchanged, causes *y* to change in the opposite direction". An increase in a cause variable does not necessarily mean the effect will actually increase [15]. Since a variable may have more than one input, a change depends on the combined effect of all its input variables. The process wherein one component *x* initiates changes in other components, and those modifications lead to further changes in *x* itself is said a feedback loop [24].

According to the system of interest, the diagram may be decomposed in several sub-diagrams. For the model of the digital twin cyber-physical system, we developed five diagrams. To save space we include only a partial view of the full causal loop diagram.

Figure 2 depicts the interaction between demand and the production plan. The diagram shows the causal links among the state of the backlog and the accepted orders rate, the dispatch rate of production lines A, B and cloud manufacturing, and the sales volume. As orders increase, the backlog grows, increasing the forecasted number of orders to produce. An increase in the capacity of the production line A adds up the number of units it produces but decreases the shop floor production of line B as well as the production in cloud manufacturing. In turn, increases in failures of production line A decreases its running time and thus its available capacity. As shop-floor production increases, the work in progress and the required dispatch rate grow in time. Finally, increases in the dispatch rate impacts positively on the rate of fulfilled orders and on sales.

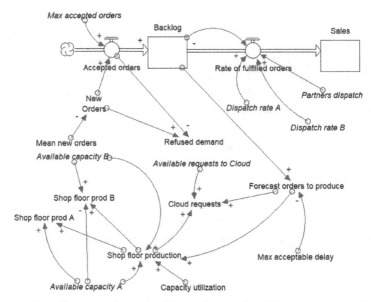

Fig. 2. Causal loop diagram of demand and the production plan. This represents a partial view of the causal loop diagram of the complete system (Generated with Stella™).

The accumulation of resources in a system (stocks), their rates of change (flows) and the feedback loops are the main components of the syntax of dynamic systems models. Stocks are represented by rectangles, inflows and outflows represented by pipes pointing into or out of stocks; valves that control the flows indicate the rates of change; and clouds represent the sources and sinks for the flows originating outside the boundary of the model. Converters representing processes that convert inputs into outputs are depicted as circles.

5 Stocks and Flows Diagrams

We structured the smart shop-floor simulation model in five interacting sectors. Cloud manufacturing requests are incorporated in addition to the physical shop-floor production in such a way that production is not limited by plant capacity. The overall structure of interactions between customer demand and flows of orders and products is based on the supply chain models in Sterman [15].

Demand Sector. The stock Backlog is the accumulation of accepted orders less the satisfied orders (Fig. 3). It is assumed that orders cannot be changed or cancelled.

$$\text{Backlog(t)} = \text{Backlog(t-dt)} + (\text{Accepted orders-Rate of fulfilled orders}) * dt \quad (1)$$

There is a cap on the amount of production based on the capacity of the plant and the number of manufacturing orders that can be requested to other manufacturers in the cloud. This cap sets the theoretical maximum production rate that may depend on constraints such as the availability of materials, labor, equipment in the plant and equipment of partners in the cloud. The fulfilled orders rate is given by Eq. (2).

$$\text{Rate of fulfilled orders} = \text{Dispatch rate A} + \text{Dispatch B} + \text{Partners dispatch} \quad (2)$$

Figure 3 depicts the Stock and flow diagram for the Demand sector. There are converters and flows defined in such a way as to capture the logic of the synchronization when the behavior of the simulation is not consistent with sensor data. The *Real time gap* is set to 10 meaning that the simulation clock is 10 time units ahead of real time (this value is arbitrarily defined only for illustration). Let the simulation clock be t_s. *Sync*

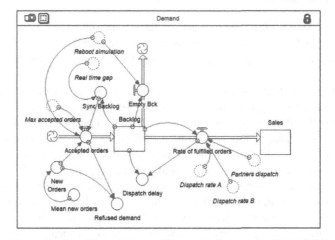

Fig. 3. Stock and flow diagram of Demand (Generated with Stella™).

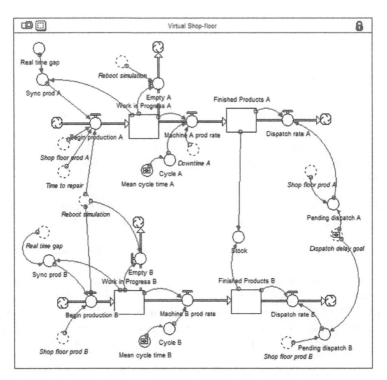

Fig. 4. Stock and flow diagram of the Virtual Shop-floor (Generated with Stella™).

backlog is a variable recording the value of *Backlog* at period t_s − *Real time gap* − 1. The initial value of converter *Reboot simulation* is 0 and is set to 1 when simulation synchronization is required. Then, if *Reboot simulation* is 1 at simulation time t_s, then the *Backlog* stock is emptied (the flow *Empty Bck* is habilitated) and set to the historical value at t_s − *Real time gap* − 1.

Virtual Shop-Floor Sector. Production is captured with a chain of two stocks: *Work in progress* and *Finished products* (Fig. 4). For the purpose of this model, all stages of the production process are aggregated together into the *Work in progress* stock as prescribed in Sterman [15].

There are two manufacturing facilities in the plant (A and B) and if necessary, requests are made to cloud manufacturing partners. As mentioned in the description of the Demand sector, there are stocks and flows defined to support synchronization.

Orders are scheduled on production line A. If A is operating above its capacity or it is being repaired, the orders are sent to line B. If also B is operating above its capacity, the required orders are placed to cloud manufacturing providers (Figs. 5, 6, 7 and 8). There exist a maximum of orders that can be requested from the cloud depending on previous agreements with partners. The cycle time is dependent on process-technology and product design (converters *Mean cycle time A*, *Mean cycle time B*). Also, plant A is subject to failure events and the rate of production is thus dependent on the availability of the plant. The stocks, inflows and outflows are formulated as

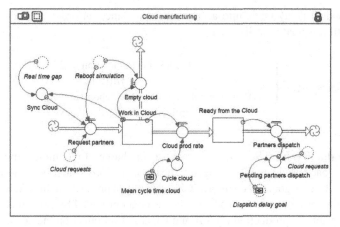

Fig. 5. Stock and flow diagram of Cloud manufacturing (Generated with Stella™).

$$\text{Work in Progress A}(t) = \text{Work in Progress A}(t - dt)$$
$$+ (\text{Begin production A} - \text{Machine A prod rate} - \text{Empty A})*dt \qquad (3)$$

$$\text{Empty A }(t) = \begin{cases} \text{Work in Progress A}(t), \text{ for Reboot simulation } = 1 \\ 0, \text{ for Reboot simulation } = 0 \end{cases} \qquad (4)$$

$$\text{Machine A prod rate} = \begin{cases} (\text{Work in Progress A}/\text{Cycle A}), \text{ for Downtime A} = 0 \\ 0, \text{ for Downtime A} = 1 \end{cases} \qquad (5)$$

Most of the time, the inflow *Begin production A* assumes the value *Shop floor production A*. However, when simulation synchronization is required at instant t_s, the simulation should reflect machine A failure at time t_s, and a downtime interval of length $t_s + $ *Time to repair*, resetting stock *Work in Progress A* with the value at period $t_s - $ *Real time gap* $- $ *Time to repair* (value recorded by converter *Sync prod A*). Hence, the inflow *Begin production A* is defined as

$$\text{Begin production A}(t) = \begin{cases} \text{Sync prod A}(t - \text{Time to repair}), \text{ if}(\text{Reboot simulation}(t) = 0 \\ \text{AND Reboot simulation}(t - \text{Time to repair}) = 1) \\ \text{Shop floor production A}(t), \text{ otherwise} \end{cases}$$
$$(6)$$

We assume that products are dispatched taking into account the dispatch delay goal. For example, if the dispatch delay goal is set to be 3 days and current time is t_s, then finished products which orders were placed at time $t_s - 3$ become dispatched.

$$\text{Finished products A }(t) = \text{Finished products A }(t - dt) + (\text{Machine A prod rate} - \text{Dispatch rate A}) * dt \qquad (7)$$

$$\text{Dispatch rate A} = \text{Min (Pending dispatch A, Finished products A)} \qquad (8)$$

$$\text{Pending dispatch A}(t) = \text{Shop floor production A}(\text{Max}(0, t - \text{Dispatch delay goal})) \qquad (9)$$

The dispatch delay takes into account the time since an order is included in the backlog until it is ready for dispatch.

$$
\text{Begin production}\,B = \left\{ \begin{array}{c} 0,\,\text{for Reboot simulation} = 1 \\ \text{Sync prod}\,B(t\text{ - }1), \\ \text{for (Reboot simulation} = 0\,\text{AND Reboot simulation}\,(t\text{ - }1) = 1) \\ \text{Shop floor production}\,B,\,\text{otherwise} \end{array} \right\} \tag{10}
$$

Service Production Plan Sector. This sector is in charge of recommending orders of production to each production line (Fig. 6). The number of orders to produce is dependent on the dispatch delay goal and on the available capacity of production lines A, B and of the cloud manufacturing providers. Calculations use data provided by (a) simulations (the available capacity of production lines A and B), and (b) production managers (capacity utilization and the dispatch delay goal).

$$
\text{Max acceptable delay} = \text{Dispatch delay goal} \tag{11}
$$

$$
\text{Forecast orders to produce} = \text{Backlog}/\text{Max acceptable delay} \tag{12}
$$

The following formulas represent the scheduling criteria for production lines A, B and cloud manufacturing. The production volume is based on forecasts, assuming that no more than the available capacity is allowed.

Shop floor production =
Min ((Available capacity A + Available capacity B) * Capacity utilization),
Forescast orders to produce) $\tag{13}$

$$
\text{Shop floor production A} = \text{Min (Available capacity A, Shop floor production)} \tag{14}
$$

$$
\text{Shop floor production B} = \left\{ \begin{array}{c} \text{Min(Available capacity B, Shop floor production - Available capacity A),} \\ \text{for (Shop floor production} > \text{Available capacity A)} \\ 0,\,\text{otherwise} \end{array} \right\} \tag{15}
$$

Cloud requests = Min(Available requests to cloud,Max (Forecast orders to produce - Shop floor production, 0)) $\tag{16}$

Digital Twin Data Sector. The digital twin data sector emulates (Fig. 7) the digital twin data component introduced by Tao [23]. This component is in charge of calculating measures used by the Service Production Plan and the Virtual Shop-floor. These measures are ratios between the capacity utilization of plants A and B, the number of orders that can be requested to cloud partners, and the maximum number of order that can be accepted considering the current level of utilization of plants A and B and cloud services. The dispatch performance rate compares the current dispatch delay and the dispatch goal defined by the firm's goals. The time to repair machine A is defined as constant, but it can be also represented with a probability distribution function or can be a simulation input entered by the operator.

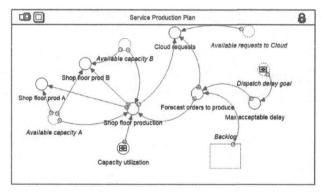

Fig. 6. Stock and flow diagram of Service Production Plan (Generated with Stella™).

The most relevant part of the digital twin sector is the comparison between data retrieved by sensors and simulated data. For simplicity, we only consider a sensor for machine A (converter *Sensor A*). The sensor detects machine A failures. If this information does not agree with simulated data (converter *Failure A*) at period t_r, then the behavior of the simulation after t_r should be updated to take into account this (the converter *Reboot simulation* is set to 1). In a more realistic setting, the presence of failures should be checked on a time interval (not a point interval).

$$\text{Reboot simulation(t)} = \left\{ \begin{array}{c} 1, \text{ for Sensor A AND(Failure A(t - Real time gap)} = 0 \\ 0, \text{ otherwise} \end{array} \right\}$$

(17)

The current available capacity of machine A is restricted to its maximum (physical) capacity. In addition, the downtime in case of failure and the reboot of the simulation are also taken into account.

$$\text{Available capacity A} = \left\{ \begin{array}{c} 0, \text{ for Downtime A} = 0 \text{ AND} \left(\begin{array}{c} \text{Reboot simulation} = 0 \text{ AND} \\ \text{Reboot simulation(t - Time to repair)} = 1 \end{array} \right) \\ \text{Max(Initial capacity A - Work in Progress A, 0), otherwise} \end{array} \right\}$$

(18)

$$\text{Available capacity B} = \left\{ \begin{array}{c} 0, \text{ for Reboot simulation} = 1 \\ \text{Max(Initial capacity B - Work in Progress B, 0), otherwise} \end{array} \right\}$$

(19)

6 Scenario Modelling

By using the System Dynamics diagrams as a modeling tool and its derived equations we are in conditions of computing numerical solutions. Tools such as Stella™ allow automatizing this computation. There are many techniques of numerical integration of differential equations that can be used to solve the resulting system. The most popular are the Runge-Kutta methods, although their use of finite steps and the approximation to average rates over the interval introduce errors, affecting the results (Table 2).

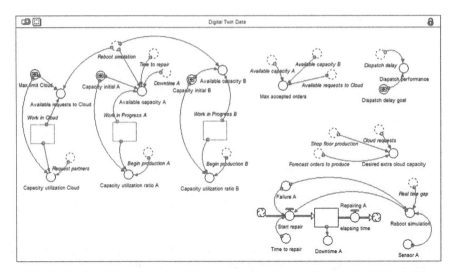

Fig. 7. Stock and flow diagram of Digital twin data. Source (Generated with Stella™).

Table 2. Parameters for the scenario simulation

Parameter description	Value
Demand (number of orders)	Normal distribution (mean = 500, st. deviation = 10%)
Dispatch delay goal	3 days
Initial capacity of line A	500 units
Initial capacity of line B	400 units
Available cloud capacity	1000 units

Running our system we find that during the first five days, the system is unstable and dominated by the details of the initial conditions. So this period is discarded for analytical purposes. There is a simulated failure of production line A at time 10 and the line is down for 3 days. There is an increase in line B and cloud production (Fig. 9).

There is a failure of line A (detected by sensor data) at real time 20 (the simulation clock is set 10 units ahead of real time). Since the simulation behavior did not consider a failure at time instant 20, then production lines A, B and cloud are set to 0 and re-initialized with real values recorded at simulation time 20 (Fig. 9). In addition, line A is down for 3 days. Simulation forecasts are "corrected" by incorporating real time data when available. Hence, a synchronization of the physical and virtual factory takes place. The available cloud capacity is not enough (notice that *Desired cloud capacity* in Fig. 10 increases after the simulated failure at period 10 and after synchronization due to a real failure of A). This suggests that during downtime plant periods (or an abrupt increase in demand), improving the access to more cloud providers would be recommended. The maximum acquired cloud capacity is not enough to satisfy the desired dispatch delay of 3 days.

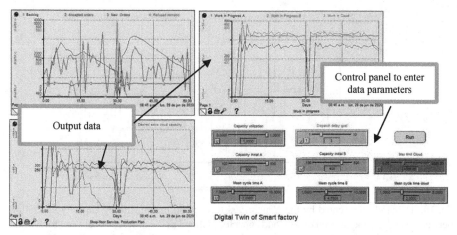

Fig. 8. Digital twin interface including graphics depicting the state of the smart factory and slider input devices to set values of capacity utilization, dispatch delay goal, initial capacity of production lines A and B, maximum limit of requests to cloud manufacturers, and mean cycle times (Generated with Stella™).

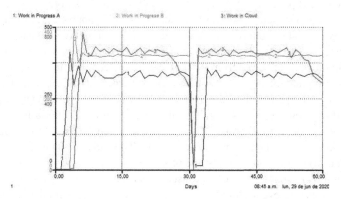

Fig. 9. Production of lines A, B and cloud partners (Generated with Stella™).

In [25] the authors use sensitivity analysis to evaluate the way that behavior patterns vary with different parameters values. We do this to assess the impact of different levels of cloud manufacturing capacity. An increase in that variable would reduce the number of rejected orders. At the same time, such increment entails higher costs incurred in partner agreements.

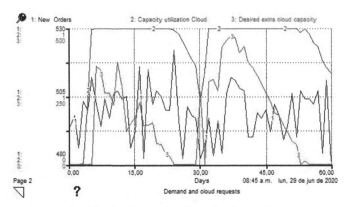

Fig. 10. Demand represented by the *New orders* line; cloud manufacturing capacity utilization (between 0 and 1); and desired extra cloud capacity (defined as the difference between forecasted production orders and available shop-floor and cloud capacity) (Generated with Stella™).

7 Conclusions

The purpose to this paper was to lay down the guidelines for a digital twin of a cyber-physical production system based on a System Dynamics approach. Although we did not intend to cover all systems that may interact with the digital twin, we described how to synchronize the simulation when an inconsistency is detected when data comes from different sources (data retrieved from sensors and simulated data). It is possible to extend the model to integrate the digital twin with different databases of the organization.

Due to the anticipatory nature of simulation runs, different production plans can be examined to choose the most convenient one according to criteria like optimizing cost, dispatch rate or rejected orders. This information may be useful to support human decision-making or to control a physical machine. This enhances the value of a digital twin (simulation) by supporting real-time production control.

Other simulation paradigms can be used to model and implement a digital twin. For example, discrete event simulation [26] [1] is adequate to model the shop-floor production lines. However, a digital twin of a cyber-physical production system is a complex system (many components or subsystems interact and the knowledge of the impact of a change in a variable over another is incomplete). To model problems with such levels of uncertainty an iterative learning process and modeling tools able to deal with incomplete information are required [18]. The model presented in this work serves as a basis to add interactions with other subsystems belonging to different tiers of the supply chain.

The system dynamics paradigm is adequate to model uncertainty and study the evolution of the system over time (for example, the interactions among demand, logistic disruptions, and machine fatigue). System dynamics shows its potential in forecasting or understanding the far future. Hence, a promising approach is to use an hybrid method based on discrete event simulation to handle short term phenomena (machine failure) and system dynamics comprising all subsystems and assess the far future.

References

1. Ding, K., Chan, F., Zhang, X., Zhou, G., Zhang, F.: Defining a digital twin-based cyber-physical production system for autonomous manufacturing in smart shop floors. Int. J. Prod. Res. **57**(20), 6315–6334 (2019)
2. Yao, X., Zhou, J., Lin, Y., Li, Y., Yu, H, Liu, Y.: Smart manufacturing based on cyber-physical systems and beyond. J. Intell. Manuf. **30**(8), 2805–2817 (2017). https://doi.org/10.1007/s10 845-017-1384-5
3. Lee, E.: Cyber physical systems: Design challenges. In: 11th IEEE International Symposium Object Oriented Real-time Distributed Computing, Florida (2008)
4. Wang, L., Wang, X.: Cloud-Based Cyber-Physical Systems in Manufacturing. Springer, London (2018)
5. Li, J., Tao, F., Cheng, Y., Zhao, L.: Big data in product lifecycle management. Int. J. Adv. Manuf. Technol. **81**(4), 667–684 (2015)
6. Tao, F., Qi, Q., Liu, A., Kusiak, A.: Data-driven smart manufacturing. J. Manuf. Syst. **48**, 157–169 (2018)
7. Monostori, L.: Cyber-physical production systems: Roots, expectations and R&D challenges. Procedia Cirp **17**, 9–13 (2014)
8. Rossit, D., Tohmé, F.: Scheduling research contributions to Smart manufacturing. Manuf. Lett. **15**, 111–114 (2018)
9. Rossit, D., Tohmé, F., Frutos, M.: Industry 4.0: smart scheduling. Int. J. Prod. Res. 1–12 (2018)
10. Parsanejad, M., Matsukawa, H.: Work-in-process analysis in a production system using a control engineering approach. J. Jpn. Indus. Manage. Assoc. **67**(2), 106–113 (2016)
11. Rossit, D., Tohmé, F., Frutos, M.: An Industry 4.0 approach to assembly line resequencing. Int. J. Adv. Manuf. Technol. 1–12 (2019)
12. Damjanovic-Behrendt, V., Behrendt, W.: An open source approach to the design and implementation of digital twins for smart manufacturing. Int. J. Comput. Integr. Manuf. 1–19 (2019)
13. Barlas, Y.: Formal aspects of model validity and validation in system dynamics. Syst. Dyn. Rev. **12**(3), 183–210 (1996)
14. Morecroft, J.: Strategic Modelling and Business Dynamics, A feedback systems approach. Wiley, Chichester (2007)
15. Sterman, J.: Business Dynamics: Systems Thinking and Modeling for a Complex World. McGraw-Hill, New York (2000)
16. Forrester, J.: Industrial Dynamics. Pegasus Communications, Massachutses (1961)
17. Senge, P.: The fifth discipline: the art and practice of the learning organization. Doubleday/Curency, New York (1990)
18. Sánchez, M.: Modeling for System´s Understanding," in Formal Languages for Computer Simulation: Transdisciplinary Models and Applications, Fonseca i Casas, P. (ed.) Hershey, IGI Global, pp. 38–61 (2013)
19. Grieves, M.: Digital twin: Manufacturing excellence through virtual factory replication. White Paper **1**, 1–7 (2014)
20. Rios, J., Oliva, M., Mas, F.: Product Avatar as Digital Counterpart of a Physical Individual Product: Literature Review and Implications in an Aircraft. In: 22nd ISPE Inc. International Conference on Concurrent Engineering, Delft (2015)
21. Sacco, M., Pedrazzoli, P., Terkaj, W.: VFF: Virtual Factory Framework. In: 2010 IEEE International Technology Management Conference, Lugano (2010)
22. Tao, F., Zhang, H., Liu, A., Nee, A.: Digital Twin in Industry: State-of-the-Art. IEEE Trans. Indus. In-f. **15**(4), 2405–2415 (2019)

23. Tao, F., Zhang, M.: Digital twin shop-floor: a new shop-floor paradigm towards smart manufacturing. IEEE Access **5**, 20418–20427 (2017)
24. McGarvey, B., Hannon, B.: Modeling Dynamic Systems, Springer, New York (2004)
25. Hekimoglu, M., Barlas, Y.: Sensitivity Analysis of System Dynamics Models by behavior Pattern Measures. In: Proceedings of the 28th International Conference of the System Dynamics Society, Seul (2010)
26. Banks, J., Carson, J., Nelson, B., Nicol, D.: Discrete-Event System Simulation. Prentice Hall, New Jersey (2001)

A Simulation-Optimization Approach for the Household Energy Planning Problem Considering Uncertainty in Users Preferences

Diego Gabriel Rossit[1]([⊠])(iD), Sergio Nesmachnow[2](iD), Jamal Toutouh[3](iD), and Francisco Luna[4](iD)

[1] INMABB, Department of Engineering, Universidad Nacional del Sur (UNS)–CONICET, Bahía Blanca, Argentina
diego.rossit@uns.edu.ar
[2] Universidad de la República, Montevideo, Uruguay
sergion@fing.edu.uy
[3] Massachusetts Institute of Technology, Cambridge, USA
toutouh@mit.edu
[4] Universidad de Málaga, Málaga, Spain
flv@lcc.uma.es

Abstract. Power supply is one of the basic needs in modern *smart homes*. Computer-aid tools help optimizing energy utilization, contributing to sustainable goals of modern societies. For this purpose, this article presents a mathematical formulation to the household energy planning problem and a specific resolution method to build schedules for using deferrable electric that can reduce the cost of the electricity bill while keeping user satisfaction at a satisfactory level. User satisfaction have a great variability, since it is based on human preferences, thus a stochastic simulation-optimization approach is applied for handling uncertainty in the optimization process. Results over instances based on real-world data show the competitiveness of the proposed approach, which is able to compute different compromise solution accounting for the trade-off between these two conflicting optimization criteria.

Keywords: Smart cities · Energy planning · Mixed-integer programming · Simulation · Multiobjective optimization

1 Introduction

The paradigm of smart cities aims at increasing resource efficiency in several daily activities that citizens perform in urban environments. In the case of energy management, this aim is not only related to the amount of energy consumed, but also to the infrastructure required to distribute the energy [3]. The capacity of this infrastructure is often conditioned by peak consumption, as it should be

D. A. Rossit et al. (Eds.): ICPR-Americas 2020, CCIS 1408, pp. 253–267, 2021.
https://doi.org/10.1007/978-3-030-76310-7_20

able to distribute the energy during the periods of high demand without producing power outages. However, if consumption of a certain area is remarkably unbalanced (having important variations along the day), this would required a large investment in infrastructure that will be idle the most of the time [10].

The introduction of time-of-use pricing in electricity bills for households is a major contributor to the overall efficiency of the electrical system, since it incentives citizens to have a smoother consumption patron, displacing the usage of electric devices from expensive peak hours to relatively cheaper off-peak hours. This behavior reduces the maximal instant power consumption of an urban area and, therefore, cuts back the required infrastructure investment to handle the peak and the risk of power outages [10]. However, usually off-peak hours, in which electricity is cheaper, are not preferred by users for using their appliances. This effect, which is known as inconvenience due to timing [1], can affect the well-being of the users. Therefore, there is a trade-off between both criteria, i.e., electricity cost and users satisfaction. Intelligent computer-aid tools may help users in the decision-making process of scheduling their deferrable devices [19].

This article proposes a novel mixed integer programming model for scheduling the deferrable electric appliances usage in households, which simultaneously considers minimizing the electricity cost and maximizing the users satisfaction. Users satisfaction measures to what extend the starting time and duration for appliances usage scheduled by the model match the users preferences–which is estimated through the analysis of historical data [4,5,22]. However, since this parameter can show certain variability between different days, a simulation-optimization resolution approach that considers this stochastic behaviour is devised. Therefore, the main contributions of the research reported in this work include: i) a compact mathematical formulation for the energy household problem, ii) the application of a stochastic resolution approach to consider uncertain users preferences, and iii) experimental evaluation over instances based on real-world data and analysis of the results.

The article is structured as follows. Section 2 presents the analysis of the main related works. The proposed mathematical formulation is outlined in Sect. 3. Section 4 describes the proposed simulation-optimization resolution approach. Section 5 describes the computational experimentation conducted to evaluate its effectiveness, and reports the numerical results for realistic problem instances. Finally, Sect. 6 formulates the conclusions and describes the main lines of future research.

2   Related Work

Household energy planning has been considered as a complex problem in the related literature. The deterministic version (scheduling non-interruptible electric appliances) is associated with bin packing [16], a well-known NP-hard problem. Moreover, including uncertainty increases the complexity of the problem [17]. Several articles have considered uncertainty in this kind of problem. For a recent review, we refer to Lu et al. [18] for a comprehensive analysis of the topic and Liang and Zhuang [17], who focused on stochastic applications.

Uncertainty in energy household planning problems has been considered in several aspects. Chen et al. [6] considered uncertainties in the power consumed by the appliances and the renewable solar energy gathered by a photovoltaic array. A three-stages resolution process was proposed. First, a deterministic linear programming optimization model considering mean values for the appliances consumption and maximum solar power generation was solved. A stochastic procedure based on Monte Carlo simulation was applied to the resulting solution. The simulation considers different energy consumption rates of appliances and selects the consumption rate that minimizes the probability of shortcuts, which occurs when the overall consumption of electricity surpass a certain threshold value. Finally, an online adjustment of the previous (offline) solution was applied, which monitors the instant solar power generation and the consumption of appliances in real-time, compensating the household electric balance of the offline solution with a larger power storage in the battery or purchase from the grid. Hemmati and Saboori [12] proposed a particle swarm optimization algorithm to deal with uncertainty of photovoltaic panels in a similar problem. Assuming that the energy generated in the panels has a Gaussian probabilistic distribution, a Monte Carlo simulation was used each time the stochastic the stochastic function has to be evaluated to obtain a sample of the generation values.

Other researchers have used robust optimization, which aims at minimizing the impact of the worst-case scenario, considering that aleatory parameters have a bounded probabilistic distribution [1]. Jacomino and Le [13] presented a robust optimization approach to simultaneously minimize energy cost and maximize the comfort of users. They considered uncertainty in two aspects: the outdoor temperature and the solar radiation (related to weather forecast), that affect the energy to be consumed to satisfy the required indoor temperature, and users decisions related to not programmable services, i.e., despite the scheduled starting time and duration of the appliances the user can modified these conditions when actually using them. For handling uncertainty on users behaviour, a decomposition approach based on estimating the probability of occurrence of each scenario was used. Wang et al. [28] proposed a robust optimization approach for dealing with photovoltaic energy generation in household planning by using a mixed integer quadratic programming model, and Wang et al. [29] for dealing with uncertainty in hot water utilization and outdoor temperature that influences the usage of heating and air conditioning systems.

Other authors, although they have not consider uncertainty in their models, they have explored the trade-off that usually exists between electricity cost and users satisfaction through linear mathematical programming approaches -as it is performed in this work-. Among them, Yahia et al. [30] modeled a bi-objective problem considering these two objectives, which were combined by means of a linear weighted sum to form a unique objective function. They solved two single-household instances, i.e., a real South African case study and an artificial large instance, using LINGO. Additionally, they performed an extensive analysis of the sensitivity of the results to the modifications of certain parameters. The same authors expanded their work in [31] by considering as a third objective the reduction of the peak load. Moreover, in this last work they solved an instance considering several households simultaneously. They applied and compared three different multiobjective approaches: lexicographic optimization, normalized weighted sum and compromise programming.

This work contributes to the literature in several aspects. Firstly, a novel linear mathematical formulation of the household planning energy problem that explicitly considers users satisfaction as an objective function is presented. Approaches as such are not common in the related work [30]. Moreover, this is an improved mathematical formulation compared to the one presented in our previous work [20] for a similar conceptual model, having a smaller number of variables and restrictions. Secondly, this work considers stochastic users preferences which differentiates it to other linear programming applications in the related work [30,31]. This leads to the final aspect that differentiates this work that is the application of the simulation-optimization Sample Average Approximation method to handle the uncertainty which has not been applied to this specific problem before.

3   Mathematical Formulation

The household energy planning problem addressed in this article aims at reducing expenses of electricity in households while enhancing users satisfaction. This last objective was estimated by considering in which part of the day users prefer to use the appliances (inferred from historical data). Then, the mathematical formulation considers the following elements:

Sets:

- a set of users $U = (u_1 \dots u_{|U|})$, each user represents a household;
- a set of time slots $T = (t_1 \dots t_{|T|})$ in the planning period;
- sets of domestic appliances $L^u = \left(l_1^u \dots l_{|L|}^u\right)$ for each user u;

Parameters:

- a penalty term ρ^u applied to those users that surpass the maximum electric power contracted;
- a parameter D_l^u that indicates the average time of utilization for user u of appliance $l \in L^u$;
- a parameter C_t that indicates the utilization cost (per kW) of the energy in time slot t;
- a parameter P_l^u that indicates the power (in kW) consumed by appliance l;
- a binary parameter UP_{lt}^u that is 1 if user u prefers to use the appliance $l \in L^u$ at time slot t, 0 in other case;
- a parameter E^u that indicates the maximum electric power contracted by user u;
- a parameter E^{joint} that indicates the maximum electric power that the (whole) set of users U are allowed to consume;

Variables:

- a binary variable x_{lt}^u that indicates if user u has appliance $l \in L^u$ turn on at time slot t;
- a binary variable δ_{lt}^u that indicates if the appliance $l \in L^u$ of user u is turn on from time slot t up to a period of time that its at least equal to D_l^u;
- a binary variable ψ_t^u that indicates if the user is using more power than the maximum power contracted E^u.
- a binary variable Ψ_t^u that indicates if the user is using more power than 130% of the maximum power contracted E^u.

The problem aims at finding a planning function $X = \{x_{lt}^u\}$ for the use of each household appliance that simultaneously maximizes the user satisfaction (given the users preference functions) and minimize the total cost of the energy consumed. The mathematical formulation is outlined in Eqs. (1)–(11).

$$\max F = \sum_{u \in U} \sum_{l \in L^u} \sum_{\substack{t_1 \in T \\ t \leq |T| - D_l^u}} \left(\delta_{lt_1}^u \left(\sum_{\substack{t_2 \in T \\ t_1 \leq t_2 < t_1 + D_l^u}} U P_{lt_2}^u \right) \right) \tag{1}$$

$$\min G = \sum_{t \in T} \sum_{u \in U} \left(\sum_{l \in L^u} x_{lt}^u P_l^u C_t + \rho^u \left(0.3 \psi_t^u + 0.7 \Psi_t^u \right) \right) \tag{2}$$

Subject to

$$\delta_{lt}^u \leq 1 - \frac{D_l^u - \left(\sum_{\substack{t_2 \in T \\ t \leq t_1 < t + D_l^u}} x_{lt_1}^u \right)}{D_l^u}, \ \forall \, u \in U, l \in L^u, t \in T \tag{3}$$

$$\sum_{t \in T} \delta_{lt}^u = n_l^u, \ \forall \, u \in U, l \in L^u \tag{4}$$

$$\psi_t^u \geq \frac{\sum_{l \in L^u} P_l^u x_{lt}^u - E^u}{\sum_{l \in L^u} P_l^u}, \ \forall \, t \in T \tag{5}$$

$$\Psi_t^u \geq \frac{\sum_{l \in L^u} P_l^u x_{lt}^u - 1.3 E^u}{\sum_{l \in L^u} P_l^u}, \ \forall \, t \in T \tag{6}$$

$$\sum_{\substack{u \in U \\ l \in L^u}} P_l^u x_{lt}^u \leq E_{joint}, \ \forall \, t \in T \tag{7}$$

$$\psi_t^u \in \{0, 1\}, \ u \in U \forall \, t \in T \tag{8}$$

$$\Psi_t^u \in \{0, 1\}, \ u \in U \forall \, t \in T \tag{9}$$

$$\delta_{lt}^u \in \{0, 1\}, \ \forall \, u \in U, l \in L^u, t \in T \tag{10}$$

$$x_{lt}^u \in \{0, 1\}, \ \forall \, u \in U, l \in L^u, t \in T \tag{11}$$

Equation (1) aims at maximizing the users satisfaction according to their preferences. Equation (2) aims at minimizing the energy expense budget, which include the charge for energy consumption and the penalization for exceeding the maximum power contracted. Equation (3) enforces δ_{lt}^i to be one when the length of time an appliance will be on is equal or larger than the required by the user. Equation (5) enforces ψ_j^i to be one if the user exceeds the maximum power contracted. Equation (6) enforces Ψ_j^i to be one if the user exceeds the maximum power contracted for more than 30%. Equation (7) enforces that the joint electric consumption by the set of users do not surpass a certain threshold maximum power. This equation is included when users are part of the same housing unit, e.g., an apartment building. Equations (8)–(11) establishes the binary nature of the variables.

4   The Proposed Simulation-Optimization Resolution Approach for the Stochastic Household Energy Planning

Real-world data shows that considering users preferences (UP) as a deterministic parameter does not adjust to reality [15]. Users satisfaction can be modelled more accurately if uncertainty is taken into account for preferences in the model. Therefore, this article develops a resolution approach that considers this stochastic behaviour.

4.1   Bi-objective Optimization

In order to handle the biobjective nature of the optimization problem presented in Sect. 3, a weighted sum optimization approach is applied. The weighted sum is a traditional method in the multiobjective optimization literature which has extensively been used in many applications, including for the energy household related problems [1]. Applying this approach, Eqs. (1) and (2) are jointly optimized with Eq. (12), where w_F and w_G are the relative weights given to each criteria by the decision-maker.

$$\max H = w_F \frac{F - F^{best}}{F^{best} - F^{worst}} - w_G \frac{G - G^{best}}{G^{worst} - G^{best}} \tag{12}$$

One of the main drawbacks of this method is to know the actual best and worst values of each objective within the set of non-dominated solutions which are used for normalization (i.e., F^{best} and G^{best}, F^{worst} and G^{worst} in Eq. (12), respectively). In this work, for addressing this issue, the procedure proposed in Rossit [24] and applied in Rossit et al. [25] is used. This is a two step procedure. In the first step, the best and worst values of each objective are approximated by solving the single objective problem of each of the criteria involved. These values, which are likely to be dominated [2], are improved in the second step of the procedure. In this second phase, these best and worst values are used in the weighted sum formula (Eq. (12)) along with a biased combination of weights. This is, two different problems are solved, one problem using $w_F >> w_G > 0$ and the other problem using $w_G >> w_F > 0$. Finally, from the solutions of these last two multiobjective problems, the new best and worst values are obtained.

4.2 Sample Average Approximation Method for Considering Stochastic Users Preferences

Formally, in a stochastic optimization problem with a probabilistic objective function, the expected value of this function should be optimized. In the case of the formulation described in Sect. 3, if parameters UP are considered stochastic, Eq. (1) should be replaced by Eq. (13).

$$e = \mathbb{E}_{\mathbf{P}} \left[F \left(\mathbf{\Delta}, \mathbf{UP} \right) \right]. \tag{13}$$

In Eq. (13), $\mathbf{UP}$ is the random vector of the stochastic users preferences and $\mathbf{\Delta}$ is the vector of decisional variables δ described in Sect. 3. In order to optimize Eq. (13), all the possible realizations of vector $\mathbf{UP}$ with its corresponding probability should be considered. Taking into account that the model of Sect. 3 uses a finite set of time slots, the set of possible realizations of $\mathbf{UP}$ is also finite. Particularly, there are $|T|^{\sum_{u \in U} |L^u|}$ realizations of this vector, each one constituting a possible scenario for the stochastic problem. For example, consider an instance in which the day is split in intervals of 30 min, i.e., $|T| = 48$, there two users (households) and each user has only two appliances ($|L^{u_1}| = |L^{u_2}| = 2$). Then, the number of possible scenarios would be $48^4 = 5,308,416$.

For the cases in which the large number of scenarios of real-world instances makes impractical to compute the exact expected value of Eq. (13), the expected value can be approximated with an independently and identically distributed (i.i.d.) random sample. This technique is called the "sample-path optimization" [23] or "sample average approximation" [26]. Thus, Eq. (14) is an estimator of the expected value of Eq. (13).

$$\hat{e} = \frac{1}{N} \sum_{j=1}^{N} F \left(\mathbf{\Delta}, \mathbf{UP^j} \right) \tag{14}$$

As aforementioned, the set of values $UP^1, ..., UP^N$, is an i.i.d. random sample of N realizations of the stochastic vector parameter $\mathbf{UP}$. The optimization problem obtained when Eq. (14) is used instead of Eq. (13), is the sample average approximation optimization problem (hereafter SAA) and can be solved deterministically with commercial solvers. Clearly, the solution of the SAA problem depends on the realizations $\mathbf{UP}$ that are included in the random sample. Moreover, the larger the size of the sample (N), the smaller is the difference between Eq. (13) and its estimator Eq. (14). Particularly when $N \to \infty$, $\hat{e} \to e$ [14].

Different samples of size N (i.e., different set of realizations of the stochastic vector parameter $\mathbf{UP}$) will shape different forms of Eq. (14). Therefore, the algorithms based on sample average solve the SAA problem several times with different samples [14,27] and then select the most promising solution according to some predefined criteria as the final solution.

In this article the procedure proposed in Norkin et al. [21] and implemented in Verweij et al. [27] is applied. This is described as follows. Let $\hat{e}_N^1, \hat{e}_N^2, ..., \hat{e}_N^M$

be the values of Eq. (14) when solving M SAA problems, each one with a different sample of size N. Moreover, considered that $\hat{s}_N^1, \hat{s}_N^2, ..., \hat{s}_N^M$ are the solution (values of decision values) obtained when each of the aforementioned M SAA problems. An intuitive criteria for selecting the best solution among the M possibilities, would be to pick the solution with the best $\hat{e}_N$ value. A more sophisticated idea is to build an independent sample of size N', with $N' >> N$, and evaluate the solutions using this sample. Then, select the solution with the best value as it is expressed in Eq. (15) for a maximization problem.

$$\hat{s}_N^* \in \arg\max\{\hat{e}_{N'}(\hat{s}_N) : \hat{s}_N \in \hat{s}_N^1, \hat{s}_N^2, ..., \hat{s}_N^M\} \tag{15}$$

This idea takes advantage from the fact that even though using the large sample size N' for the optimization phase can be very time consuming (specially in NP-hard problems as the one addressed in this paper), using it for just for evaluation of the objective function Eq. (14) can be achievable in reasonable computing time [14].

5   Computational Experiments

This section describes the instances and methodology used for the evaluation of the proposed approach, and reports quantitative and qualitative results.

5.1   Problem Instances

The instances construction is based on information from the REDD dataset [15]. As performed in Colacurcio et al. [7], instances with different sizes were considered. One of the key parameters to estimate was the users preferences. For estimating this parameter, information about the power consumption of the selected appliances on each household was analyzed. This involved cleaning the data from comparatively very small power consumption which are related to stand-by consumption of each appliance, for example, small screen leds. After this, for each combination of user and appliance, a probability of usage for each time slot was estimated (p_t^u). With this probability, M instances were constructed for each sample size N as is described in Sect. 5.2. Additionally, from REDD dataset the mean power consumption of each appliance in KW (P_l^u) and the duration of the average time of utilization of each appliance (D_l^u) were estimated. When performing this noticeable differences were identified during the weekend, a behaviour that is usual for household users [8]. Therefore the instances were classified in weekdays and weekends.

Parameters E^u (maximum electric power contracted for each household) and C_t were obtained from the website of the Electric Company of Montevideo, Uruguay (https://portal.ute.com.uy/). Two instances size were considering, each one with two variations: weekdays (wd) and weekend (we):

– small ($s.wd$ and $s.we$). It has one household with seven deferrable appliances.

– large ($l.wd$ and $l.we$). It has two households with six and seven deferrable appliances respectively.

In the instances that corresponds to the small size Eq. (7) is not used since there is only one household and, thus, Eqs. (5) and (6) are enough for limiting the maximum consumed energy.

5.2   Experiment Design

After some preliminary experimentation, the following sample sizes were chosen $N = 50, 200, 500, 1000, 2000, 3000$ and 10000. Within each sample size, the number of independent samples (M) was set to 100. The evaluation sample size (N') was set to 100000. The estimation of the ideal and nadir value for the weighted sum function were estimated with the procedure introduced in Sect. 4.1. This is performed within each N value. Additionally, five different weights configurations are used for exploring different trade-off between energy cost and user satisfaction (w_f, w_g): (0.99,0.01), (0.25, 0.75), (0.5, 0.5), (0.75, 0.25) and (0.01, 0.99). The SAA problems were solved with Gurobi [9] through Pyomo as modelling language [11].

Algorithm 1. Schema of a the Simulation-optimization (SO) approach.

1: **initialize** $SO(p_{lt}^u, N, M, \alpha, \beta)$
2: **initialize** list S of size M
3: **for** $m \leftarrow 0, m + +, m \leq M$ **do**
4:     **for** $n \leftarrow 0, n + +, n \leq N$ **do**
5:         **for all** $u \in U$ **do**
6:             **for all** $l \in L^u$ **do**
7:                 **for all** $t \in T$ **do**
8:                     **initialize** $t \leftarrow random(0, 1)$
9:                     **if** $t \leq p_{lt}^u$ **then** $UP_{lt}^u = 1$
10:                    **else** $UP_{lt}^u = 0$
11:                    **end if**
12:                **end for**
13:            **end for**
14:        **end for**
15:    **end for**
16:    $S\,[m] \leftarrow$ **Solve MDR**$(\alpha, \beta, \mathbf{UP})$)
17: **end for**
18: **return** S

5.3   Experimental Results

The experimental results are reported in Tables 1–2, for the small and large instances considering weekday and weekend usage conditions, respectively. These

Table 1. Results of small instance.

N	(α, β)	Time (sec)		$F^{N'}$		$G^{N'}$		$F(H_{best}^{N'})$	$G(H_{best}^{N'})$
		Avg	Std	Avg	Std	Avg	Std		
Small instance weekday (s.wd)									
50	(0.99,0.01)	0.04	0.07	0.5065	0.0405	115.91	8.65	0.5761	103.73
	(0.01,0.99)	0.13	0.09	0.2245	0.0192	89.39	0.00	0.2468	89.39
	(0.50,0.50)	0.08	0.08	0.4128	0.0464	96.84	2.36	0.4592	93.37
	(0.75,0.25)	0.05	0.07	0.4849	0.0476	103.63	4.40	0.5062	98.75
	(0.25,0.75)	0.11	0.08	0.3104	0.0478	90.99	0.95	0.2455	89.39
200	(0.99,0.01)	0.05	0.09	0.5445	0.0302	115.10	6.94	0.5877	104.83
	(0.01,0.99)	0.13	0.09	0.2342	0.0151	89.39	0.00	0.2462	89.39
	(0.50,0.50)	0.09	0.07	0.4270	0.0401	94.38	2.05	0.4608	93.37
	(0.75,0.25)	0.08	0.11	0.4835	0.0400	99.14	1.58	0.5170	98.20
	(0.25,0.75)	0.14	0.12	0.3036	0.0355	90.50	0.37	0.3048	89.74
500	(0.99,0.01)	0.06	0.11	0.5590	0.0251	113.08	2.86	0.5869	104.84
	(0.01,0.99)	0.14	0.10	0.2404	0.0088	89.39	0.00	0.2463	89.39
	(0.50,0.50)	0.11	0.11	0.4854	0.0266	98.07	0.72	0.4503	93.37
	(0.75,0.25)	0.10	0.13	0.5138	0.0432	100.75	2.82	0.5866	102.63
	(0.25,0.75)	0.14	0.13	0.3380	0.0301	90.86	0.44	0.3034	89.95
1000	(0.99,0.01)	0.02	0.01	0.5699	0.0229	112.93	1.11	0.5318	104.28
	(0.01,0.99)	0.10	0.01	0.2419	0.0073	89.39	0.00	0.2497	89.39
	(0.50,0.50)	0.07	0.01	0.5016	0.0235	98.15	0.50	0.4353	93.37
	(0.75,0.25)	0.05	0.07	0.5192	0.0324	99.54	2.00	0.5150	98.20
	(0.25,0.75)	0.09	0.01	0.3402	0.0236	90.80	0.12	0.3432	90.57
Small instance weekend (s.we)									
50	(0.99,0.01)	0.03	0.08	0.5271	0.0446	41.49	11.05	0.6095	29.66
	(0.01,0.99)	0.05	0.08	0.2725	0.0305	23.18	0.00	0.3160	23.18
	(0.50,0.50)	0.05	0.09	0.4864	0.0448	26.53	1.69	0.5298	25.41
	(0.75,0.25)	0.02	0.01	0.5348	0.0349	30.88	3.60	0.6003	28.81
	(0.25,0.75)	0.03	0.01	0.4163	0.0499	24.36	0.65	0.3334	22.05
200	(0.99,0.01)	0.03	0.08	0.5748	0.0358	38.37	9.51	0.5907	28.81
	(0.01,0.99)	0.05	0.09	0.2955	0.0152	23.18	0.00	0.3157	23.18
	(0.50,0.50)	0.08	0.12	0.4842	0.0280	25.08	0.48	0.5412	25.41
	(0.75,0.25)	0.04	0.08	0.5612	0.0242	28.95	0.88	0.5984	28.81
	(0.25,0.75)	0.08	0.11	0.3976	0.0346	23.67	0.28	0.3013	21.76
500	(0.99,0.01)	0.04	0.08	0.6048	0.0234	37.03	3.77	0.6005	28.81
	(0.01,0.99)	0.08	0.13	0.3011	0.0053	23.18	0.00	0.3157	23.18
	(0.50,0.50)	0.09	0.14	0.4783	0.0203	24.73	0.19	0.4809	23.33
	(0.75,0.25)	0.04	0.09	0.5611	0.0275	28.22	1.18	0.5316	25.41
	(0.25,0.75)	0.07	0.10	0.3174	0.0300	23.23	0.14	0.2948	22.76
1000	(0.99,0.01)	0.01	0.00	0.6176	0.0155	36.92	2.03	0.5901	28.82
	(0.01,0.99)	0.03	0.00	0.3027	0.0024	23.18	0.00	0.3054	23.18
	(0.50,0.50)	0.03	0.00	0.5467	0.0368	26.94	1.78	0.5321	25.41
	(0.75,0.25)	0.02	0.03	0.5845	0.0092	28.89	0.11	0.6084	28.81
	(0.25,0.75)	0.04	0.00	0.4665	0.0298	24.38	0.48	0.4355	23.76

Table 2. Results of large instance.

N	(α, β)	Time (sec)		$F^{N'}$		$G^{N'}$		$F(H_{best}^{N'})$	$G(H_{best}^{N'})$
		Avg	Std	Avg	Std	Avg	Std		
Large instance weekday (l.wd)									
50	(0.99,0.01)	0.07	0.07	0.9626	0.0606	200.60	15.23	1.0071	170.06
	(0.01,0.99)	0.40	0.15	0.3721	0.0350	131.41	0.00	0.4312	131.41
	(0.50,0.50)	0.21	0.10	0.8650	0.0622	144.47	2.86	0.7668	136.91
	(0.75,0.25)	0.19	0.15	0.9439	0.0621	160.94	9.52	1.0695	151.33
	(0.25,0.75)	0.24	0.08	0.7078	0.0886	135.99	1.62	0.6876	133.36
200	(0.99,0.01)	0.09	0.12	1.0262	0.0470	197.83	11.94	0.9115	165.68
	(0.01,0.99)	0.33	0.11	0.3890	0.0295	131.41	0.00	0.4322	131.41
	(0.50,0.50)	0.19	0.09	0.9230	0.0384	144.78	1.39	0.9230	140.15
	(0.75,0.25)	0.17	0.11	0.9901	0.0508	152.26	5.89	1.0213	145.69
	(0.25,0.75)	0.28	0.10	0.7248	0.0428	134.90	0.81	0.7450	134.03
500	(0.99,0.01)	0.08	0.12	1.0361	0.0373	195.71	8.13	1.0370	180.96
	(0.01,0.99)	0.31	0.09	0.3883	0.0277	131.41	0.00	0.4332	131.41
	(0.50,0.50)	0.22	0.14	0.9557	0.0299	145.26	0.33	1.0302	145.69
	(0.75,0.25)	0.16	0.11	1.0133	0.0411	151.98	5.51	1.0880	150.11
	(0.25,0.75)	0.29	0.12	0.7860	0.0441	136.49	1.26	0.7368	134.19
1000	(0.99,0.01)	0.08	0.07	1.0336	0.0277	195.49	8.24	1.0852	191.66
	(0.01,0.99)	0.32	0.08	0.3892	0.0263	131.41	0.00	0.4281	131.41
	(0.50,0.50)	0.19	0.08	0.9581	0.0245	145.21	0.22	0.9714	144.67
	(0.75,0.25)	0.13	0.05	1.0195	0.0398	151.37	5.50	1.0582	149.76
	(0.25,0.75)	0.27	0.09	0.7879	0.0364	136.46	0.96	0.7436	134.62
Large instance weekend (l.wd)									
50	(0.99,0.01)	0.10	0.13	1.0007	0.0620	287.69	13.86	1.1515	277.16
	(0.01,0.99)	0.33	0.12	0.4020	0.0190	197.55	0.16	0.3935	195.93
	(0.50,0.50)	0.17	0.12	0.8944	0.0620	209.69	2.72	0.9664	207.08
	(0.75,0.25)	0.15	0.11	0.9592	0.0693	222.08	9.62	1.0140	211.18
	(0.25,0.75)	0.22	0.12	0.7009	0.0834	201.07	1.49	0.6986	199.33
200	(0.99,0.01)	0.07	0.10	1.0614	0.0417	280.86	7.46	1.1160	271.33
	(0.01,0.99)	0.32	0.13	0.4091	0.0174	197.57	0.00	0.4347	197.57
	(0.50,0.50)	0.15	0.11	0.9796	0.0470	212.28	0.98	1.0626	211.73
	(0.75,0.25)	0.15	0.11	1.0018	0.0458	251.5656	13.567	1.1007	216.66
	(0.25,0.75)	0.22	0.13	0.8277	0.0563	204.74	1.33	0.7558	200.77
500	(0.99,0.01)	0.10	0.12	1.0944	0.0333	278.59	3.21	1.1403	268.53
	(0.01,0.99)	0.32	0.10	0.4139	0.0152	197.57	0.00	0.4345	197.57
	(0.50,0.50)	0.17	0.12	1.0065	0.0356	211.98	0.43	1.0541	211.73
	(0.75,0.25)	0.16	0.11	1.0171	0.0419	249.10	11.78	1.1226	216.16
	(0.25,0.75)	0.21	0.11	0.8301	0.0464	204.57	1.25	0.8179	202.28
1000	(0.99,0.01)	0.05	0.00	1.1060	0.0279	278.80	2.26	1.1365	271.58
	(0.01,0.99)	0.29	0.01	0.4129	0.0136	197.57	0.00	0.4319	197.57
	(0.50,0.50)	0.12	0.01	1.0155	0.0289	211.84	0.23	1.0491	211.73
	(0.75,0.25)	0.11	0.01	1.0347	0.0365	251.14	7.83	1.1169	216.16
	(0.25,0.75)	0.18	0.03	0.8362	0.0330	204.48	1.21	0.8124	202.19

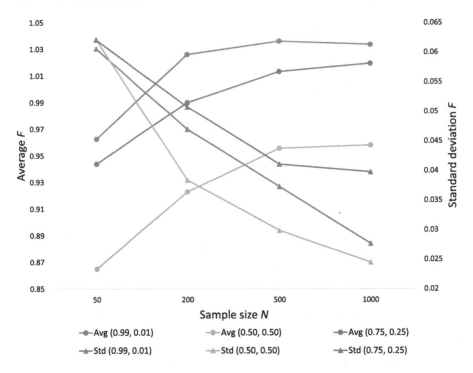

Fig. 1. Sensitivity analysis of average and standard deviation of F in l.wd instance.

Tables report for each instance, sample size (N) and weight vector values the following results: the mean and standard deviation of the runtime, the user satisfaction function F evaluated over N', the mean and standard deviation of the cost function G evaluated over N' (for the M different runs), and the values of F and G of the best compromising solution, i.e., the solution that has the minimal value of function H (Eq. (12)). It should be highlighted that since the cost function G (Eq. (2)) is deterministic, it is not affected by the sample size after the optimization process. In other words, for evaluation purposes: $G^N = G^{N'}$.

All the solutions of the SAA problems were solved to optimality since Gurobi was able to find solutions with 0% MIPGap for the compact mathematical formulation presented in Sect. 3 in relatively short computing times. The different combinations of weights were able to effectively explore the trade-off of the problem. Something interesting is that schedules that are biased towards minimizing the cost objective (with higher values of β) are more difficult to solve for Gurobi (computing times are as much of three times higher).

As expected, in general the larger the sample size N the higher the average the user satisfaction function value, since the expected value is better approximated by Eq. (14). Additionally, another important feature is that increasing sample size N led to a remarkable reduction of the variability of the results (measured through the standard deviation). Figure 1 exemplifies this behavior, reporting the average and standard deviation of function F for the M independent samples for the three combinations with larger α, i.e., in which F has preponderance over the cost, in the large instance for the weekday patron.

6   Conclusions

This article studied the household energy planning problem, a relevant optimization problem that arises in the context of modern smart cities.

A novel bi-objective mathematical formulation of the problem was presented, accounting for uncertainty in the preferences of using each appliance. In this formulation, the aim is to schedule on a daily basis the usage of deferrable appliances while optimizing two conflicting objectives: the cost of the electricity–based on time-of-use tariff–and the users satisfaction. The users satisfaction is estimated through historical data of when (which part of the day) users prefer to use each appliance. However, since there is considerable variation of these preferences, a stochastic resolution approach was used to include the randomness of this parameter. The proposed problem was solved using the Sample Average Approximation method, which is a simulation-optimization approach that combines Monte Carlo simulation and deterministic mixed integer programming. Different real-world instances were considered and solved with different parametric combinations. The approach was able to propose different solutions that explore the trade-off between the two criteria in reasonable computing times. Particularly, for larger values of sample size the standard deviation of the results given by the method was significantly reduced.

Additionally, the initial tests performed in this work were execute in relatively small computing times even for the large instance. This shows the validity of the proposed integer mathematical formulation and the simulation-optimization approach as useful tools to perform practical load scheduling in smart homes. However, for enhancing performance of the model in a particular household, it is important to first gather specific data of that household to accurately estimate the key parameters of the model, such as the stochastic user preferences.

For future work, a crucial research line is to expand the computational experimentation. This should include using larger sample sizes and number of independent samples within each size (M) for exploring whether the accuracy of the solutions can be enhanced. Also, to study the competitiveness of the approach for larger instances in which buildings with more than two households are considered. Another way the instances can be enlarged is to include other kinds of appliances, such as non-deferrable loads, and renewable power generators within the household, e.g., solar or wind power generators.

Acknowledgements. J. Toutouh research was partially funded by European Union's Horizon 2020 research and innovation program under the Marie Skłodowska-Curie grant agreement No 799078, by the Junta de Andalucía UMA18-FEDERJA-003, European Union H2020-ICT-2019-3, and the Systems that Learn Initiative at MIT CSAIL.

References

1. Beaudin, M., Zareipour, H.: Home energy management systems: a review of modelling and complexity. Renew. Sustain. Energy Rev. **45**, 318–335 (2015)
2. Beeson, R.: Optimization with respect to multiple criteria. Ph.D. thesis, University of Southern California, Los Ángeles, United States of America, June 1972
3. Calvillo, C., Sánchez-Miralles, A., Villar, J.: Energy management and planning in smart cities. Renew. Sustain. Energy Rev. **55**, 273–287 (2016)
4. Chavat, J., Graneri, J., Nesmachnow, S.: Household energy disaggregation based on pattern consumption similarities. In: Nesmachnow, S., Hernández Callejo, L. (eds.) ICSC-CITIES 2019. CCIS, vol. 1152, pp. 54–69. Springer, Cham (2020). https://doi.org/10.1007/978-3-030-38889-8_5
5. Chavat, J., Nesmachnow, S., Graneri, J.: Non-intrusive energy disaggregation by detecting similarities in consumption patterns. Revista Facultad de Ingeniería Universidad de Antioquia (2020)
6. Chen, X., Wei, T., Hu, S.: Uncertainty-aware household appliance scheduling considering dynamic electricity pricing in smart home. IEEE Trans. Smart Grid **4**(2), 932–941 (2013)
7. Colacurcio, G., Nesmachnow, S., Toutouh, J., Luna, F., Rossit, D.: Multiobjective household energy planning using evolutionary algorithms. In: Nesmachnow, S., Hernández Callejo, L. (eds.) ICSC-CITIES 2019. CCIS, vol. 1152, pp. 269–284. Springer, Cham (2020). https://doi.org/10.1007/978-3-030-38889-8_21
8. Goldberg, M.: Measure twice, cut once. IEEE Power Energ. Mag. **8**(3), 46–54 (2010)
9. Gurobi Optimization, LLC: Gurobi Optimizer Reference Manual (2020). http://www.gurobi.com
10. Harding, M., Lamarche, C.: Empowering consumers through data and smart technology: experimental evidence on the consequences of time-of-use electricity pricing policies. J. Policy Anal. Manage. **35**(4), 906–931 (2016)
11. Hart, W., et al.: Pyomo-optimization modeling in Python, vol. 67. Springer Science & Business Media, 2 edn. (2017). https://doi.org/10.1007/978-3-319-58821-6
12. Hemmati, R., Saboori, H.: Stochastic optimal battery storage sizing and scheduling in home energy management systems equipped with solar photovoltaic panels. Energy Buil. **152**, 290–300 (2017)
13. Jacomino, M., Le, M.: Robust energy planning in buildings with energy and comfort costs. 4OR **10**(1), 81–103 (2012)
14. Kleywegt, A., Shapiro, A., Homem-de Mello, T.: The sample average approximation method for stochastic discrete optimization. SIAM J. Optim. **12**(2), 479–502 (2002)
15. Kolter, J., Johnson, M.: Redd: A public data set for energy disaggregation research. Workshop on data mining applications in sustainability, San Diego, CA. 25, 59–62 (2011)
16. Koutsopoulos, I., Tassiulas, L.: Control and optimization meet the smart power grid: scheduling of power demands for optimal energy management. In: Proceedings of the 2nd International Conference on Energy-efficient Computing and Networking, pp. 41–50 (2011)
17. Liang, H., Zhuang, W.: Stochastic modeling and optimization in a microgrid: a survey. Energies **7**(4), 2027–2050 (2014)
18. Lu, X., Zhou, K., Zhang, X., Yang, S.: A systematic review of supply and demand side optimal load scheduling in a smart grid environment. J. Clean. Prod. **203**, 757–768 (2018)

19. Luján, E., Otero, A.D., Valenzuela, S., Mocskos, E., Steffenel, A., Nesmachnow, S.: An integrated platform for smart energy management: the CC-SEM project. Revista Facultad de Ingeniería, Universidad de Antioquia **97**, 41–55 (2019)
20. Nesmachnow, S., Colacurcio, G., Rossit, D., Toutouh, J., Luna, F.: Optimizing household energy planning in smart cities: a multiobjective approach. Revista Facultad de Ingeniería Universidad de Antioquia (2020). (in press)
21. Norkin, V., Pflug, G., Ruszczyński, A.: A branch and bound method for stochastic global optimization. Math. Program. **83**(1–3), 425–450 (1998)
22. Porteiro, R., Nesmachnow, S., Hernández-Callejo, L.: Short term load forecasting of industrial electricity using machine learning. In: Nesmachnow, S., Hernández Callejo, L. (eds.) ICSC-CITIES 2019. CCIS, vol. 1152, pp. 146–161. Springer, Cham (2020). https://doi.org/10.1007/978-3-030-38889-8_12
23. Robinson, S.: Analysis of sample-path optimization. Math. Oper. Res. **21**(3), 513–528 (1996)
24. Rossit, D.: Desarrollo de modelos y algoritmos para optimizar redes logísticas de residuos sólidos urbanos. Ph.D. thesis, Universidad Nacional del Sur, Bahía Blanca, Argentina, November 2018
25. Rossit, D., Toutouh, J., Nesmachnow, S.: Exact and heuristic approaches for multiobjective garbage accumulation points location in real scenarios. Waste Manage. **105**, 467–481 (2020)
26. Shapiro, A.: Monte Carlo simulation approach to stochastic programming. In: Proceeding of the 2001 Winter Simulation Conference (cat. no. 01CH37304), vol. 1, pp. 428–431. IEEE (2001)
27. Verweij, B., Ahmed, S., Kleywegt, A., Nemhauser, G., Shapiro, A.: The sample average approximation method applied to stochastic routing problems: a computational study. Comput. Optim. Appl. **24**(2–3), 289–333 (2003)
28. Wang, C., Zhou, Y., Jiao, B., Wang, Y., Liu, W., Wang, D.: Robust optimization for load scheduling of a smart home with photovoltaic system. Energy Convers. Manage. **102**, 247–257 (2015)
29. Wang, J., Li, P., Fang, K., Zhou, Y.: Robust optimization for household load scheduling with uncertain parameters. Appl. Sci. **8**(4), 575 (2018)
30. Yahia, Z., Pradhan, A.: Optimal load scheduling of household appliances considering consumer preferences: an experimental analysis. Energy **163**, 15–26 (2018)
31. Yahia, Z., Pradhan, A.: Multi-objective optimization of household appliance scheduling problem considering consumer preference and peak load reduction. Sustain. Urban Areas **55**, 102058 (2020)

Production Capacity Study in Footwear Production Systems Based on Simulation

Florencia Dornes[1]([⊠]), Daniel Alejandro Rossit[1,2], and Nancy B. López[1]

[1] Departamento de Ingeniería, Universidad Nacional del Sur, Bahía Blanca, Argentina
[2] INMABB, Universidad Nacional del Sur and CONICET, Bahía Blanca, Argentina

Abstract. In this work, a case study of the footwear manufacturing industry is approached. In particular, the case of a leather footwear SME in Argentina is analyzed. The study focuses on the variability of performance that involves working with natural raw materials and its consequent impact on production capacity. For this, the production process was modeled respecting all sources of uncertainty of the problem. Discrete event simulations were performed to analyze the system, and the bottleneck was detected. When analyzing the potential alternatives to overcome the capacity limitation, it was decided to enlarge the workforce. The results of the proposed solution showed that incorporating a worker increased productivity more than proportionally to the direct increase in labor.

Keywords: Stochastic production planning · Footwear production · Simulation · Plant simulation · Bottleneck

1 Introduction

Footwear manufacturing is one of the most important productive activities in Coronel Suárez, a town in Buenos Aires Province, Argentina. It consists of a highly complex production process, which involves various specific machines and a large number of specialized operators. The diversity of products that can be obtained entails the need to plan different particular procedures, with adequate operations for each model. Likewise, in general terms, manufacturing involves the following stages: reception of raw material, cutting, stitching, assembly and packaging.

The case at hand is a safety footwear production problem for an Argentine company, whose main clients are the security forces. The main problem lies in the need to work with natural leather. This material, of animal origin, is the critical input of the process since it intervenes in the greatest proportion in the manufacture of the product and, due to its natural characteristics, the performance is non-constant because of the variability of the leather piece size and the high probability of the presence of holes in it. Therefore, the initial operations of the entire manufacturing process, that is, the process of cutting and preparing the leather, have a variable performance depending on the conditions of the input leather. Also, given the mechanical characteristics of leather (much more resistant than standards fabrics), the stitching operation becomes really critical. Therefore, the probability of occurrence of failures (e.g., needle breakage) or faulty operations (e.g.,

© Springer Nature Switzerland AG 2021
D. A. Rossit et al. (Eds.): ICPR-Americas 2020, CCIS 1408, pp. 268–276, 2021.
https://doi.org/10.1007/978-3-030-76310-7_21

poor stitching) is significant. The occurrence of such events (non-standard leathers and operational failures) negatively impacts production planning.

In the literature, there are numerous antecedents of published articles that use discrete events simulation for the study of similar manufacturing cases. Among them, we can cite Malega et al. (2017) [1], Yang et al. (2019) [2], Zhang et al. (2019) [3] and Wang et al. (2020) [4]. Regarding the footwear industry, its study has been approached by Dang & Pham (2016) [5], where the design of an assembly line with 62 tasks, uncertain operating times and parallel workstations is carried out. The objective followed in [5] is to maximize the performance of the line by determining an optimized setting of operating parameters and an optimal assignment of tasks. On the other hand, Sadegui et al. (2018) [6] optimize the production line, focusing on the stitching area, identified as the bottleneck of the process. The goal is to balance the assembly line, minimizing the number of required workstations and reducing the workload on operators.

On the other hand, there are several cases where the simulation of equipment failures is approached. Ahmadi et al. (2016) [7] assess machine failures in a stochastic environment through simulation, considering that the mean time between failures and the repair follow an exponential distribution. The same distributional considerations are used by Zandieh & Gholami (2009) [8].

In this work, it is addressed a case study of a leather footwear production plant. The production process is subject to different sources of uncertainty that affect production capacity, both due to the variable performance of the leather and the probability of machine breakdown. That is why, in order to study the possibility of increasing production capacity, it is required an approach that allows to analyze production considering the proper uncertainties of the process. Given the high number of operations, the complexity of the process and its own nature (significant influence of random factors), an approach based on the discrete events simulation is appropriate (Banks et al. 2013) [9]. The incorporation of a new worker is proposed as an alternative for increasing production capacity. The results support the proposed alternative, improving productivity more than proportionally.

The rest of the work is organized as follows. In Sect. 2 the problem is presented together with the production process. In Sect. 3, the improvement proposal is analyzed and the tool to be used is introduced. While in Sect. 4, the type of experiment to be carried out is defined together with its parameterizations, and the results obtained are presented. In Sect. 5, the conclusions of the work are presented.

2 Problem Description

In this section, firstly, it is described the complete manufacturing process of the product, and secondly, it is identified and detailed the problem to be solved in this work.

2.1 Manufacturing Process Description

The safety boot is a product made from natural leather, internally lined and suitably reinforced, provided with soles with dielectric properties, resistant to oils and hydrocarbons, cemented and glued with a safety seam in Scátola.

The manufacturing process of the product under analysis is represented in the flowchart of Fig. 1. Although the real process consists of multiple specific stages, for the purposes of the problem that is intended to be studied in this work, the real process information was condensed in a sufficiently descriptive model. This modeling consisted in grouping operations by work center, maintaining a sufficient level of detail to accurately represent the problem of production capacity. In turn, this modeling also respected the stochastic nature of the problem, allowing to represent the variability inherent to work with natural raw materials (the leather performance is variable), as well as its exigent sewing process (machine breakdown).

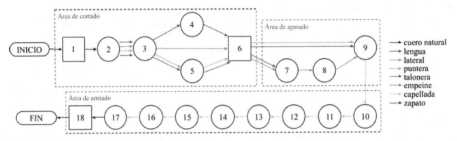

Fig. 1. Flowchart of the production process of the safety boots. References: 1-Reception of raw material; 2-Cut; 3-Lower; 4-Embroider the logo; 5-Paste the reinforcements; 6-Intermediate quality control; 7-Sew the tongue and sides; 8-Eyelet; 9-Sew the upper; 10-Strobel stitching; 11-Moisturize; 12-Put on the last and mark; 13-Scrape, halogenate and cement the sole and cuts; 14-Oven; 15-Flue the sole; 16-Press the sole; 17-Scátola safety stitching; 18-Final quality control, clean and pack.

Safety boots production begins with the reception of raw material (1). In this task, the tactile and visual inspection of the leather is carried out, along with its thickness control and calibration. The materials are transported to the cutting area, where the 10 pieces required for a pair of shoes are obtained: 4 sides, 2 tongues, 2 toecaps and 2 heels. The cut is carried out on a rocker press (2), using special dies or punches for each part, taking care to discard the perforated, wrinkled or variable thickness sections of the leather. It is important to trim the leather (3) in such a way as to reduce the thickness around the perimeter of the cut pieces so that, when sewing between pieces, a constant thickness is obtained and also avoids excessive breakage of sewing machine needles. Subsequently, the safety boot brand logo is embroidered (4) on the tongue, the reinforcements are pasted (5) on the toecaps and heels, and the intermediate quality control of all the components is done (6).

In the stitching area, the different pieces of the shoe are glued, folded and assembled. The operators are in charge of finishing the upper by assembling the pieces by stitching them, using one (9) and two-needle (7) sewing machines. Additionally, the eyelets (8) required for lacing the shoe are made.

Then, in the assembly area, the shoe is finished. This requires, first of all, the assembly process, where the reinforcement insole is sewn using the Strobel machine (10) and the leather is treated with steam (11) to soften it in the machine to moisten cuts. Subsequently, the soling is carried out by the traditional gluing process, which consists of putting the shoe on the last, and marking the sole (12), to finally scrape the sole and glue it (13). To do this, the temperature is increased in an oven (14) and the sole is pressed under pressure for 30 s (15 and 16). Next, the last is removed and a safety seam is made in Scátola (17).

Finally, the product is cleaned and its final quality control is carried out, proceeding to tie with a cord and pack it (18).

Table 1 shows the number of workers currently working in the factory and the details of the activities carried out by each. The operation number matches the operations in the flowchart presented in Fig. 1.

Table 1. Assignment of the productive operations of the workers. Operations are defined from the references in the flowchart. Workers who perform manual parts transport are indicated with an X in the last column.

Worker	Quantity	Operations	Manual parts transport
Worker:1	1	(2)	
Worker:2	1	(3); (4)	X
Worker:3	1	(5); (6)	
Worker:4	1	(7); (8)	
Worker:5	1	(9); (10); (11)	X
Worker:6	1	(12)	
Worker:7	1	(13)	
Worker:8	1	(14); (15)	
Worker:9	1	(16)	
Worker:10	1	(17); (18)	X

2.2 Detail of the Problem to Be Solved

The main problem with this process is that the conditioning of the natural raw materials requires intensive processing, which ends up limiting the actual production capacity of the plant. This limitation occurs in the manufacturing process due to the great variability in the performance of the main raw material, which is natural leather. As it is an input of animal origin, not only the size of the received pieces is variable, but it can also have holes in it, or some burned sections (due to the tanning process), and the thickness is, very often, variable. To minimize the impact of these conditions on the production process, the choice of the raw material supplier is essential, as are the skills of the cutter (operation 2, Fig. 1), who must position the dies in such a way as to minimize the discarding of unsuitable cut pieces and to achieve the desired quality level of the final product.

However, once the cut pieces have been achieved, the greatest requirement is found in the stitching area (operations 7,8 and 9, Fig. 1), where the operators must calibrate the sewing machines according to the leather thickness. If the machine is not adjusted properly to the thickness of the leather, then, the probability of needle breakage increases considerably. That is why the performance of the stitching area has a direct link with the leather variability. The criticality of the machine adjustment to the leather thickness is such that, in a standard working day, 10 needles can be broken per day (making all the necessary adjustments) and, in case of skipping an adjustment, the amount of needle breakage increases. Therefore, this is an area that requires a lot of setup time (not productive time), and even the repair time for breakage is also considerable.

On the other hand, the sewing operations have a direct impact on the quality and aesthetics of the finished boots. Thus, poor performance generates a greater quantity of second-quality shoes, which must be sold at a lower price. Therefore, it is an operation that must be carried out with maximum care, which in operational terms translates into very significant operating times, limiting production capacity. That is why, to increase production capacity, the stitching area must be carefully analyzed. For this, a series of simulations under standard conditions were carried out to evaluate, in relative terms, how the stitching area was performing. Figure 2 represents the use of working hours by each worker. At a first glance, it can be seen that the only worker without idle time is worker 4 (the green part implies "operation" mode and the gray part "waiting"), who turns out to be the worker assigned to operations (7) and (8) (Table 1). Therefore, it is identified that the capacity limitation or bottleneck is found in operations (7) and (8), assigned to worker 4.

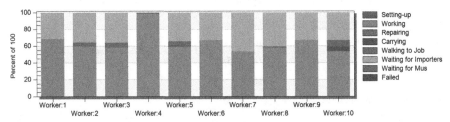

Fig. 2. Bar graph of the time proportions of the different activities carried out by each worker as a function of the total normalized working time, under standard conditions. (Color figure online)

3   Solution Proposal

This section presents the proposed solution for the problem studied. First, the proposal is described, and then, the evaluation and experimentation methods are detailed.

3.1   Solution Proposal Description

The capacity problem is evidenced in operations (7) and (8), which are carried out by the worker 4. These capacity limitations are caused by the fact that it is a very labour-intensive operation due to the permanent adjustment that must be made on the machine to process the leather of variable thickness. Therefore, to increase the production capacity of the system, it would be enough to increase the capacity of the bottleneck (Goldratt

1990) [10]. Now, one way to increase the production capacity of these operations would be to reduce the number of adjustment times that this machine requires, since these are not productive times in the strict sense. However, the need for machine adjustments is because of the variability of the leather thickness, which, as discussed in the previous section, is a characteristic of the raw material used for the process. Therefore, to reduce the time spent on adjustments, it would be necessary to reduce the variability of the leather thickness, and that is not possible.

Therefore, the possibility of increasing the capacity of these operations lies in increasing the productive resources available in that operation, either by increasing the number of machines, or by increasing the number of workers. In this work, we opted for the latter option. What is proposed as a solution for increasing production capacity is the incorporation of a new worker assigned to the same operations as worker 4 in Table 1.

3.2 Solution Implementation Approach

To verify the effectiveness of the alternative chosen for increasing capacity, the Tecnomatix Plant Simulation 15 software (Bangsow, S. 2015) [11] is used. It is a software for simulating material flow in a discrete event system. The software allows the evaluation of complex and dynamic processes, providing support to decision-making related to production. For this, the simulator allows the execution of different experiments representing different alternatives and configurations.

For modeling the production process, the real layout of the company was taken into consideration, with estimated times based on the average daily production. This layout is presented in Fig. 3. The numbers associated with each of the workstations correspond to those presented in the flowchart of Fig. 1.

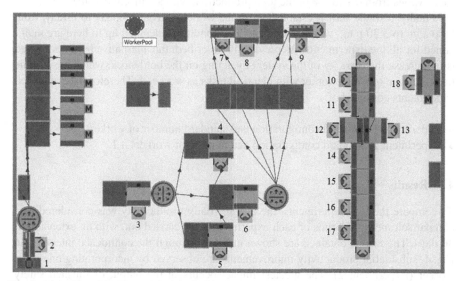

Fig. 3. Plan view of the 3D model of the real footwear production process, using Plant Simulation. The references correspond to the operations in the flowchart of Fig. 1.

4 Experiments and Results

4.1 Experimental Design

To perform the experiments, it is required to set the conditions of the production system to study. Various points of variability were introduced. These are associated with the performance of the raw material, the operating times, the operational failures due to broken needles in the stitching area, and the quality of the finished product.

The natural leather performance was defined considering a supplier of medium-high quality raw material, where the pieces discarded due to material deficiencies vary from 0 to 4%, with a uniform distribution.

Regarding operating times, these were estimated from the information of the real case provided by the SME, taking into consideration the daily production of 100 shoes/day. Since the activities are carried out with the intervention of the workers, the times may differ because of inherent factors in the system. This variability is modeled from a triangular distribution of operating times, considering the minimum and maximum values as a proportion equivalent to 95% and 115% of the mode value, respectively.

Needle break failures of sewing machines in the stitching area are determined from historical needle breakage throughout the factory, which is dependent on the material being processed. For the particular case of the leather, due to its characteristics, the breakage is approximately 10 needles/day and, for both machines, it was defined with a mean time between failures characterized by a negative exponential distribution. On the other hand, the repair duration is defined from a normal distribution, with an expected value of 4 min and a standard deviation of 15 s.

The variability of the final quality control (operation 18 of Fig. 1) was modeled analogously to the leather performance. In this case, the appearance of second-quality shoes varies from 0% to 2% of the total production, with a uniform distribution.

Finally, a single shift is established for the workday, from Monday to Friday, from 6:00 a.m. to 3:30 p.m., with a 1-h break. The conditions detailed up to here are maintained for all the experiments. As previously described, the alternative being evaluated is to increase the capacity of the system by acting on the bottleneck, which implies the incorporation of a new worker with identical tasks as worker 4. Therefore, the analyzed experiments consisted of:

- Experiment 1: standard configuration and standard number of workers.
- Experiment 2: standard configuration and number of workers +1.

4.2 Results

To compare the two experiments, the average daily productivity was considered as a comparison metric. 50 runs of each experiment were carried out, with an extension of 30 days. The results obtained are shown in Fig. 4 through the confidence intervals. In Fig. 4, substantial productivity improvements are observed by incorporating one more worker (Experiment 2). The greatest impact occurs in the output of finished safety boots. Under the initial conditions (Experiment 1), the average production is 2,786.73 pairs/month, with a standard deviation of 286.04 pairs. By implementing the second strategy, incorporating only one new worker in the bottleneck, production is 3,562.47 pairs/month, with a standard deviation of 340.63 pairs. Ergo, productivity is increased by 27.84%. When compared with the percentage increase in the workforce (1 worker out of 10 workers, 10%), this percentage increase in productivity implies more than a proportional increase.

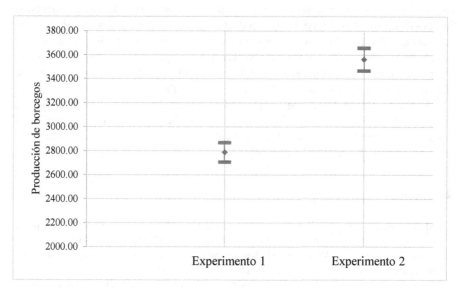

Fig. 4. Confidence interval of the production of safety boots for experiment 1, with the initial conditions, and experiment 2, with the capacity of the bottleneck increased.

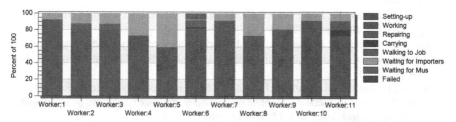

Fig. 5. Bar graph of the times proportions of the different activities carried out by each worker as a function of the total normalized work time, for Experiment 2. The additional worker is referenced as Worker: 5, therefore the other workers are located out of phase from their previous designation in Fig. 2. (Color figure online)

Moreover, when observing the distribution of the proportion of the time worked by each worker (Fig. 5), a general increase is observed, along with a decrease in waiting times, compared to Fig. 2. It is of interest to note that the new incorporated worker is called Worker 5 in Fig. 5, so that Worker 6 to 11 in Fig. 5 correspond to Worker 5 to 10 in Fig. 2. On the other hand, it is observed, in Fig. 5, that the workload is more balanced between the different workers in the process (the green bars tend to be more balanced than in Fig. 2).

5 Conclusions

The work addressed the resolution of the capacity problem of a real footwear production line. For this, a simulation model of the production process was carried out using the Plant Simulation software, which made it possible to detect the main restriction of the system, namely the stitching area, particularly worker 4. Consequently, an alternative solution was proposed to increase the capacity of the bottleneck by adding an extra worker, and experiments were carried out to evaluate the impact of the proposed solution.

The results obtained from the comparison of the initial situation and the alternative situation reflect a significant positive impact on the factory's monthly production of shoes. In other words, the proposed solution enhances the system as a whole, more than proportionally to the increase in capacity.

References

1. Malega, P., Kadarova, J., Kobulnicky, J.: Improvement of production efficiency of tapered roller bearing by using Plant Simulation. Int. J. Simul. Model. **16**(4), 682–693 (2017)
2. Yang, S.L., Xu, Z.G., Wang, J.Y.: Modelling and production configuration optimization for an assembly shop. Int. J. Simul. Model. **18**(2), 366–377 (2019)
3. Zhang, L., Zhou, L., Ren, L., Laili, Y.: Modeling and simulation in intelligent manufacturing. Comput. Indus. **112**, (2019)
4. Wang, X., Lu, J., Chen, R., Xu, M., Xia, L.: Research on design and planning of pulsating aero-engine assembly line based on plant simulation. In: 2020 IEEE 4th Information Technology, Networking, Electronic and Automation Control Conference (ITNEC), vol. 1, pp. 591–595. IEEE (June 2020)
5. Dang, Q.V., Pham, K.: Design of a footwear assembly line using simulation-based ALNS. Procedia CIRP **40**, 596–601 (2016)
6. Sadeghi, P., Rebelo, R.D., Ferreira, J.S.: Balancing mixed-model assembly systems in the footwear industry with a variable neighbourhood descent method. Comput. Indus. Eng. **121**, 161–176 (2018)
7. Ahmadi, E., Zandieh, M., Farrokh, M., Emami, S.M.: A multi objective optimization approach for flexible job shop scheduling problem under random machine breakdown by evolutionary algorithms. Comput. Oper. Res. **73**, 56–66 (2016)
8. Zandieh, M., Gholami, M.: An immune algorithm for scheduling a hybrid flow shop with sequence-dependent setup times and machines with random breakdowns. Int. J. Prod. Res. **47**(24), 6999–7027 (2009)
9. Banks, J., Carson, J.S., Nelson, B.L., Nicol, D.M.: Discrete-Event System Simulation: Pearson new International Edition. Pearson Higher Ed (2013)
10. Goldratt, E.M.: Theory of constraints. North River, Croton-on-Hudson (1990)
11. Bangsow, S.: Tecnomatix Plant Simulation. Springer, Berlin (2015)

Machine Learning and Big Data

A Database Curation for Prediction of the Refractive Index in the Virtual Testing of Polymeric Materials by Using Machine Learning

Santiago A. Schustik[1,2]([✉]), Fiorella Cravero[3]([✉]), Ignacio Ponzoni[3,4]([✉]), and Mónica F. Díaz[1,5]([✉])

[1] Planta Piloto de Ingeniería Química (PLAPIQUI), Universidad Nacional del Sur (UNS), CONICET, Camino La Carrindanga, 8000 Bahía Blanca, Argentina
{sschustik,mdiaz}@plapiqui.edu.ar

[2] Comisión de Investigaciones Científicas de la Provincia de Buenos Aires (CIC-PBA), Buenos Aires, Argentina

[3] Instituto de Ciencias e Ingeniería de la Computación (ICIC) UNS-CONICET, Campus Palihue, 8000 Bahía Blanca, Argentina
{fiorella.cravero,ip}@cs.uns.edu.ar

[4] Departamento de Ciencias e Ingeniería de la Computación (DCIC-UNS), Campus Palihue, 8000 Bahía Blanca, Argentina

[5] Departamento de Ingeniería Química (DIQ-UNS), Av Alem 1253, 8000 Bahía Blanca, Argentina

Abstract. The aim of industry 4.0 is to promote productivity and innovation by incorporating emerging IT technologies, where machine learning is playing a central role in this industrial revolution. In this sense, the production of new materials could take advantage of novel virtual testing approaches based on data science for supporting the design of new polymers. Nevertheless, the lack of data for learning virtual testing models constitutes a hard challenge for progressing in these innovative techniques. Therefore, it is especially important to create reliable databases for polymer study and make them available to the scientific community. In this work, we have focused on the generation of a trustworthy database of Refractive Index (RI) of synthetic polymers. This paper details the different types of errors found in the data source and the corrections made during the curation and cleaning of this database. Additionally, some Quantitative Structure-Property Relationship models for predicting RI, inferred without domain expert intervention, are presented and discussed for illustrating how virtual testing can be applied using this database.

Keywords: Refractive Index · Machine learning · Polymer database

© Springer Nature Switzerland AG 2021
D. A. Rossit et al. (Eds.): ICPR-Americas 2020, CCIS 1408, pp. 279–294, 2021.
https://doi.org/10.1007/978-3-030-76310-7_22

1 Introduction

At present, the world industry is going through a new revolution, known as Industry 4.0, which combines advanced production and operation methods with smart technologies that are integrated into organizations, people and assets. This revolution is driven by emerging new technologies such as artificial intelligence, big data analytics, nanotechnology, robotics and the Internet of Things (IoT), among others. Consequently, organizations must recognise the technologies that best meet their needs to capitalize in them, because companies that do not understand the evolution and opportunities in business that Industry 4.0 brings are at risk of dropping market share. In this context, technologies based on machine learning are playing an important role in the innovation of production processes and, in particular, are having a huge impact in the discovery of new industrial materials [1].

A traditional project for development of new materials follows three main phases: design, synthesis and testing. During the first phase, different alternatives prototypes are designed for the potential new material. This design is guided by domain experts, which pursued some specific characteristics related to required industrial use for the future material. Then, the new materials are synthetized by polymerization reactions. Finally, specimens (physical samples) of them are tried by typical laboratory tests used in polymer science, such as the tensile test, in order to evaluate which polymeric material fulfil the desired industrial profile. Nowadays, this traditional process has been improved by the use of virtual testing models derived from machine learning methods [2]. Figure 1 illustrates how the use of virtual tests is being incorporated in the traditional process. This virtual testing phase makes it possible to predict relevant properties of the alternative prototypes even before synthesizing them, helping to prioritize the best candidate prototypes for the next steps of the polymer development project. Additionally, as a result of virtual testing, it is possible to rethink some design decisions and iteratively reformulate the original prototypes before moving to the more expensive phases of the project, thus saving money, resources and time [3].

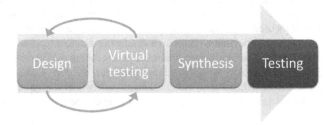

Fig. 1. Incorporating virtual testing in a traditional project for development of new materials.

These virtual tests are generated as Quantitative Structure–Property Relationships (QSPR) models inferred by supervised learning methods. In Polymer Informatics, QSPR models usually are regression models where a subset of molecular descriptors (*features*), which characterize the chemical structure of the molecules, are used for predicting the value of a target property under study. Figure 2 shows the sequence of steps for obtaining a QSPR model. First, the database to be used for learning the model must be defined and revised. In this step, it is important to revise that all required pieces of information be available for each polymer included in the database (polymer structure along with its corresponding Simplified Molecular Input Line Entry Specification (SMILES) code and the experimental value of the target property). If inconsistencies or mistakes are detected during this revision, a curation and cleaning step must be conducted in order to obtain a trusty database. After that, QSPR models can be inferred applying to main procedures: feature selection, where the most relevant molecular descriptors for target property are identified, and models inference, where alternative QSPR models are generated by supervised learning methods. Lastly, an external validation of the models is executed for selecting the best QSPR models trying to avoid overfitting and correlations by chance.

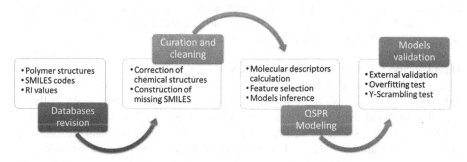

Fig. 2. Steps for generating a QSPR model for virtual testing.

Unfortunately, although this novel proposal can help to speed up the development projects of new polymers by supporting the design stage, the lack of data is slowing the progress and adoption of this innovative technology [3, 4]. Addressing this need, the "Materials Genome Initiative" was created in 2011, involving multiple US agencies collaborating to create policies, resources, and infrastructure that would support US institutions as they worked to rapidly discover and manufacture advanced materials with low cost [5]. Therefore, it is imperative to create reliable databases and make them available to the scientific community for supporting the incorporation of virtual testing in the polymer industry.

In this work we have focused on the generation, curation and cleaning of a database of Refractive Index (RI) of synthetic polymers, which is used in Quantitative Structure-Property Relationship (QSPR) predictive modelling. This property has special interest in the design of optical fibres, which combine two materials with different RI values: high RI (in core) and low RI (cladding). Although some databases that contain several polymers, together with their chemical structures and RI values, can be found in the literature, we have detected several mistakes and inconsistencies that must be corrected [6]. Besides, the repeating unit structure of the polymers and their computational representations in SMILES code are needed for QSPR modelling. Therefore, this report details the different types of errors found in the data source and the corrections made as part of the revision, curation and cleaning of this RI database. Finally, in order to illustrate the use of the

RI database in virtual testing, the results of preliminary QSPR models generated by machine learning, without domain expert intervention during the modelling step, are presented and discussed.

2   Materials and Methods

2.1   Databases: Curation and Cleaning

Database Sources
We have used the database reported by Jabeen et al. [6] for the purpose of making a fair performance comparison of our QSPR models with his reported model. This database contains 127 polymers, with their experimental RI values. However, during the database revision, we found some inconsistencies in terms of structure, property values and chemical names, so we began a process of curing and cleaning. To perform this cure, it was necessary to verify the data with other sources [7–12].

Data Curation and Cleaning Methodology
Data cleaning consists of detecting and correcting or removing incorrect or corrupt data from a database. We did a review of the database reported by Jabeen, collating the RI values and the polymer names with the databases reported by Bicerano [7], Duchowicz et al. [8], Xu et al. [9], Khan et al. [10], Seferis [11] and the chemical structure with PubChem [12]. Additionally, the database reported by Duchowicz was used for obtaining the SMILES codes, which are necessary for computing the molecular descriptors required for training our QSPR models. However, not all polymers were found in the Duchowicz's database, then the rest of the molecule SMILES codes were created by using Molinspiration Galaxy 3D Structure Generator tool [13].

During the review process of the Jabeen's database we found inconsistencies related to the polymer structures. Furthermore, some of them did not correspond to the reported name, and we also found some duplicate polymers. In this sense, we proceeded to make a structure correction due to the inconsistencies that we found and, additionally, we proceeded to develop the missing SMILES codes. Figure 3 shows the steps to curate the database. First, we have made a general revision of the reference database [6], checking polymer names, RI values and chemical structures. The second step was to carry out a detailed verification of each one of the structures, corroborating them against PubChem [12] and Sigma-Aldrich Product Catalog [14]. Since during the verification some errors were found in the structures, the third step was carried out by correcting the structures according to the information collated from the previous sources (Further explanation and examples of this step can be found in Sect. 3.1). Then, in the fourth step, the property values were reviewed according to those reported in the databases of Bicerano [7], Duchowicz et al. [8], Xu et al. [9], Khan et al. [10], and Seferis [11]. After finding some errors, we proceeded to the fifth step, where they were corrected by homogenizing all RI values to four corrected decimal places. In the sixth step, we have made a revision of the SMILES associated with the polymers structure that we need to perform the cure. This revision was carried out by visualizing and verifying that the SMILES code of each polymer was structurally correct. For this, we have used the Molinspiration Galaxy 3D

Structure Generator tool [13], which allows to obtain a visualization of the molecule corresponding to a certain SMILES code. In the seventh step, seeing that not all of the polymers reported by Jabeen et al. [6] were contained in the Duchowicz's database [8], we proceeded to develop the missing SMILES. Finally, in the eighth step, we have made a revision of the already curated database, visualizing and checking that everything - i.e. polymer name, polymer structure, SMILES code, and RI value - is matching for each polymer.

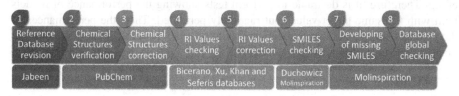

Fig. 3. Steps for the database curation process.

2.2 QSPR Modelling

Feature Selection
The feature selection was made with the WEKA tool [15], using fixed seeds randomization in order to ensure reproducibility and default settings. For this purpose, several well-known methods were used to obtain a total 6 different subsets of molecular descriptors using 75% of the database (training dataset). The methods used were Correlation-based Feature Subset Selection (CFS) and Wrapper Subset Eval (WSE). The last feature selection method evaluates attribute sets by using a learning scheme. In particular, in this work, WSE was defined using different learning techniques: Linear Regression (LR), Neural Networks (NN), Random Forest (RF), Random Committee (RC), and K-Nearest Neighbours (KNN).

QSPR Models Inference
To carry out the learning process and structure-activity modelling 5 different machine learning methods have been used on each subset: Linear Regression (LR), Neural Networks (NN), Random Forest (RF), Random Committee (RC) and K-Nearest Neighbours (KNN). Therefore, a total of 5 putative QSPR models were generated from each subset. The WEKA [15] implementation of each method was used.

We select one QSPR model for each subset for the next instance. The criterion to select the models is to keep the QSPR model with the highest R^2. This metric is used because it is the same reported by Jabeen's model [6] that will be used as reference performance in our work.

QSPR Models Validation
External Validation. The goal of any machine learning algorithm is to learn a generalizable knowledge. Generalizability is the capability of the model to adapt appropriately to

previously unseen data, but similar as that used to create the model. Underfitting occurs when the model does not fit the data well enough, and overfitting occurs when a model captures the noise of the data. External validation was performed in this work using the remaining 25% of the molecules in the database (validation dataset). In addition, we also used Roy's method [16] for classifying the QSPR models according to the quality of their predictions (Good, Moderate or Bad).

Chance Correlations Test. There is a risk of finding random correlations in QSPR modelling. Therefore, it is desirable to perform tests showing the performance of models built with the same target values but randomly permuted. Thus, the performances of the models fitted for random procedures should be much lower than the QSPR models obtained from the original target values [17].

3 Results and Discussions

3.1 Data Curation and Cleaning

The importance of database quality used in supervised learning lies in ensuring that the information is adequately collected and curated. Therefore, a reliable dataset constitutes the key foundations for a meaningful knowledge extraction process. At the end of the learning process, the model is capable of mapping the input data (molecular descriptors) to the output data (target property) in a trustworthy way.

Despite the great effort involved in creating a new database, mistakes are often observed since the creation process is a manual one subject to human error. One of the main problems found in the database reported by Jabeen et al. [6] were the inconsistencies in the structures, and the IDs of the polymers also did not match with the IDs of the existing structures in the supplementary material.

A detailed revision of each polymer was performed. We found mistakes in 21 molecules of Jabeen's database [6], but we will illustrate the curation and cleaning process only explaining a few representative cases. For example, we found that the Poly[oxy(methyl n-octadecylsilylene)] was duplicated but with different RI values, so we decided to solve this inconsistency by keeping the RI value reported in Bicerano's database [7]. Besides, we found that 16 polymers presented in Jabeen's database [6] contained errors in their chemical structure. For example, the Poly(2-ethoxyethyl acrylate), i.e. ID 37 in Table 1, lacked a carbon in the structure, therefore we added the missing carbon and developed the corresponding SMILES code. Figure 4 shows the structural difference.

In all the cases that we had to correct the SMILES code, the Molinspiration Galaxy 3D Structure Generator tool [13] was used. For example, in the case of ID 37, which is poly(2-ethoxyethyl acrylate), we create the SMILES code: CCOCCOC(=O)CC, as shown in Fig. 5.

Figure 6 shows the Poly[oxy(methyl n-octadecylsilylene)], i.e. ID 11 in Table 1. In this case, we verified that a carbon atom was wrongly located in the chemical structure. The carbon indicated by the cross was correctly relocated (highlighted in green). The resulting SMILES code is O[Si](C)(CCCCCCCCCCCCCCCCCC).

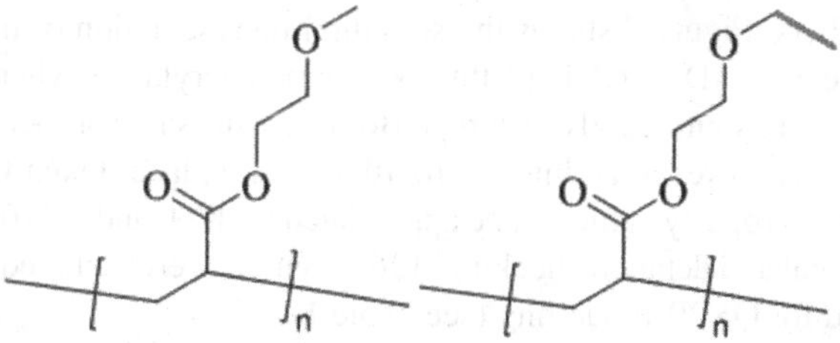

Fig. 4. The original structure is shown (left) and on the right the added carbon in green. (Color figure online)

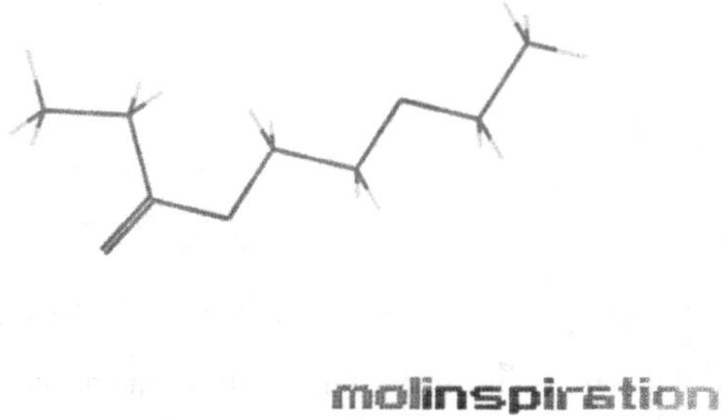

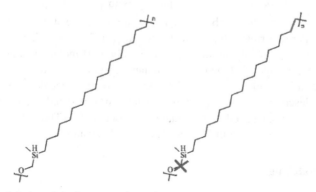

Fig. 5. The Poly(2-ethoxyethyl acrylate) represented by Molinspiration Galaxy 3D Structure Generator. The resulting SMILES code is as follows: CCOCCOC(=O)CC

Fig. 6. On the left the original structure for Poly[oxy(methyl n-octadecylsilylene)] is shown and on the right its correction. (Color figure online)

Regarding the RI values checking, we detected in Jabeen's database that several polymers with different structures have the same property value, which is an undesirable behaviour to QSPR modelling. For this reason, we looked up the RI values in the literature [7, 9–11], and we realized that the target values must be represented with four decimal places in order to denote those structural differences. Therefore, we have updated the property values, so now it is possible to differentiate the RI values for polymers with structural differences such as pairs ID 21–22, ID 25–26, ID 31–32, ID 46–47, ID 52–53,

ID 69–70, and others. Figure 7 shows the structural representation of the polymers ID 46 (Poly (ethylene)) and ID 47 (Poly (2-fluoroethyl methacrylate)), where it can be seen that their structures are significantly different. Both have the same property value (1.476) in Jabeen's database, however, adding the fourth decimal place -taken from literature-, the difference in the property value can be appreciated (1.4760 and 1.4768 respectively). Finally, after a global and detailed checking, 126 polymers were included in the resulting database and used for QSPR modelling (see Table 1).

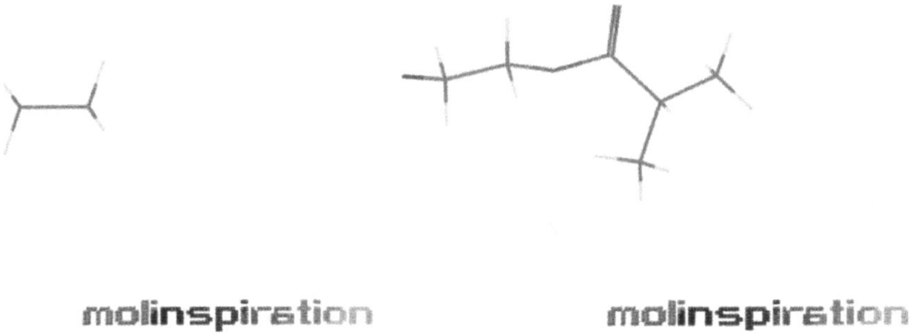

Fig. 7. Structures of two polymers: on the left the ID 46 and on the right the ID 47.

Molecular Descriptors Computation
The DRAGON tool [18] was used to calculate 3839 molecular descriptors for each of the 126 molecules in the database. Therefore, a 95% correlation filter was applied to avoid molecular descriptors that provide redundant information. This filter reduced approximately to 82% of all the descriptors. Finally, 690 molecular descriptors that characterize the molecules and integrate the database were included.

In this way, the definitive database contains 126 curated molecules characterized by 690 molecular descriptors. It was split into two datasets, one saved as a training dataset (75%) and the other one saved as a validation dataset (25%). The value range of the property Refractive Index is from 1.301 to 1.683 for this database.

3.2   QSPR Modelling

Feature Selection
A total of 6 molecular descriptor subsets were selected from the initial pool of 690 molecular descriptors by using 6 different methods. In Table 2, all the subsets, together with used methods and cardinalities are listed. The cardinality for all subsets is higher than the reported by Jabeen et al. [6], which is 4. However, a popular rule suggests that there should be at most one descriptor for every five molecules of the dataset used [19]. It is noteworthy that the 6 selected subsets meet this rule, because they are below to 25 descriptors.

QSPR Models Inference
In total, 30 models were inferred by training each of the 6 subsets with 5 learning

Table 1. Curated database of Refractive Index (RI) for 126 polymers.

ID	Name	RI
1	Poly(hexafluoropropylene oxide)	1.3010
2	Polytetrafluoroethylene	1.3500
3	Poly(trifluorovinyl acetate)	1.3750
4	Poly(pentafluoropropyl acrylate)	1.3850
5	Poly(dimenthyl siloxane)	1.4035
6	Poly(trifluoroethyl acrylate)	1.4070
7	Poly(2,2,2-trifluoro-1-methylethyl methacrylate)	1.4185
8	Poly(vinylidene fluoride)	1.4200
9	Poly[oxy(methyl n-hexylsilylene)]	1.4330
10	Poly(trifluoroethyl methacrylate)	1.4370
11	Poly[oxy(methyl n-octadecylsilylene)]	1.4430
12	Poly[oxy(methyl n-octylsilylene)]	1.4450
13	Poly(vinyl isobutyl ether)	1.4507
14	Poly[oxy(methyl n-hexadecylsilylene)]	1.4510
15	Poly(vinyl ethyl ether)	1.4540
16	Poly[oxy(methyl n-tetradecylsilylene)]	1.4550
17	Poly(oxyethylene)	1.4563
18	Poly(propylene oxide)	1.4570
19	Poly(3-butoxylpropylene oxide)	1.4580
20	Poly(vinyl n-pentyl ether)	1.4590
21	Poly(3-hexoxylpropylene oxide)	1.4590
22	Poly(vinyl hexyl ether)	1.4591
23	Poly(4-fluoro-2-trifluoromethylstyrene)	1.4600
24	Poly(vinyl n-octyl ether)	1.4613
25	Poly(vinyl 2-ethylhexyl ether)	1.4626
26	Poly(vinyl n-decyl ether)	1.4628
27	Poly(2-methoxyethyl acrylate)	1.4630
28	Poly(tert-butyl methacrylate)	1.4638
29	Poly(4-methyl-1-pentene)	1.4650
30	Poly(3-ethoxypropyl acrylate)	1.4650
31	Poly(n-butyl acrylate)	1.4660
32	Poly(vinyl propionate)	1.4665

(continued)

Table 1. (*continued*)

ID	Name	RI
33	Poly(vinyl methyl ether)	1.4670
34	Poly(vinyl acetate)	1.4670
35	Poly(ethyl acrylate)	1.4685
36	Poly(3-methoxypropyl acrylate)	1.4710
37	Poly(2-ethoxyethyl acrylate)	1.4710
38	Poly(isopropyl methacrylate)	1.4728
39	Poly(1-decene)	1.4730
40	Poly(propylene) (atactic)	1.4735
41	Poly(vinyl sec-butyl ether) (isotactic)	1.4740
42	Poly(dodecyl methacrylate)	1.4740
43	Poly(ethylene succinate)	1.4744
44	Poly(tetradecyl acrylate)	1.4746
45	Poly(vinyl formate)	1.4757
46	Poly(ethylene)	1.4760
47	Poly(2-fluoroethyl methacrylate)	1.4768
48	Poly(isobutyl methacrylate)	1.4770
49	Poly(methyl acrylate)	1.4790
50	Poly(oxymethylene)	1.4800
51	Poly(n-hexyl methacrylate)	1.4813
52	Poly(n-butyl methacrylate)	1.4830
53	Poly(2-ethoxyethyl methacrylate)	1.4833
54	Poly(n-propyl methacrylate)	1.4840
55	Poly(ethyl methacrylate)	1.4850
56	Poly(3,3,5-trimethylcyclohexyl methacrylate)	1.4850
57	Poly(vinyl butyral)	1.4850
58	Poly(2-nitro-2methylpropyl methacrylate)	1.4868
59	Poly(1,1-diethylpropyl methacrylate)	1.4889
60	Poly(methyl methacrylate)	1.4893
61	Poly(2-decyl-1,4-butadiene)	1.4899
62	Poly(3-methylcyclohexyl methacrylate)	1.4947
63	Poly(4-methylcyclohexyl methacrylate)	1.4975
64	Poly(vinyl methyl ketone)	1.5000

(*continued*)

Table 1. (*continued*)

ID	Name	RI
65	Poly(sec-butyl α-chloroacrylate)	1.5000
66	Poly(1,2-butadiene)	1.5000
67	Poly(vinyl alcohol)	1.5000
68	Poly(2-bromo-4-trifluoromethyl styrene)	1.5000
69	Poly(ethyl α-chloroacrylate)	1.5020
70	Poly(2-methylcyclohexyl methacrylate)	1.5028
71	Poly(1,1-dimethylethylene)	1.5050
72	Poly(cyclohexyl methacrylate)	1.5066
73	Poly(tetrahydrofurfuryl methacrylate)	1.5096
74	Poly(1-methylcyclohexyl methacrylate)	1.5111
75	Poly(2-hydroxyethyl methacrylate)	1.5119
76	Poly(vinyl chloroacetate)	1.5130
77	Poly(N-butyl methacrylamide)	1.5135
78	Poly(methyl α-chloroacrylate)	1.5170
79	Poly(2-chloroethyl methacrylate)	1.5170
80	Poly(allyl methacrylate)	1.5196
81	Polyacrylonitrile	1.5200
82	Polyisoprene	1.5210
83	Poly(acrylic acid)	1.5270
84	Poly(1,3-dichloropropyl methacrylate)	1.5270
85	Poly(N-vinylpyrrolidone)	1.5300
86	Poly(cyclohexyl α-chloroacrylate)	1.5320
87	Poly[oxy(methylphenylsilylene)]	1.5330
88	Poly(2-chloroethyl α-chloroacrylate)	1.5330
89	Poly(2-aminoethyl methacrylate)	1.5370
90	Poly(vinyl chloride)	1.5390
91	Poly(sec-butyl α-bromoacrylate)	1.5420
92	Poly(cyclohexyl α-bromoacrylate)	1.5420
93	Poly(2-bromoethyl methacrylate)	1.5426
94	Poly(2-bromoethyl ethacrylate)	1.5426
95	Poly(ethylmercaptyl methacrylate)	1.5470
96	Poly(1-phenylethyl methacrylate)	1.5487

(*continued*)

Table 1. (*continued*)

ID	Name	RI
97	Poly(p-isopropyl styrene)	1.5540
98	Polychloroprene	1.5580
99	Poly(methyl α-bromoacrylate)	1.5672
100	Poly(benzyl methacrylate)	1.5679
101	Poly(m-cresyl methacrylate)	1.5683
102	Poly(phenyl methacrylate)	1.5706
103	Poly(2,3-dibromopropyl methacrylate)	1.5739
104	Poly(ethylene terephthalate)	1.5750
105	Poly(vinyl benzoate)	1.5775
106	Poly(o-chlorobenzyl methacrylate)	1.5823
107	Poly(m-nitrobenzyl methacrylate)	1.5845
108	Poly(2-methylstyrene)	1.5874
109	Polystyrene	1.5920
110	Poly(o-methoxy styrene)	1.5932
111	Poly(p-bromophenyl methacrylate)	1.5964
112	Poly(N-benzyl methacrylamide)	1.5965
113	Poly(p-methoxy styrene)	1.5967
114	Poly(pentachlorophenyl methacrylate)	1.6080
115	Poly(o-chloro styrene)	1.6098
116	Poly(phenyl α-bromoacrylate)	1.6120
117	Poly(N-vinyl phthalimide)	1.6200
118	Poly(2,6-dichlorostyrene)	1.6248
119	Poly(chloro-p-xylylene)	1.6290
120	Poly(β-naphthyl methacrylate)	1.6298
121	Poly(2-vinylthiophene)	1.6376
122	Poly[oxy(2,6-diphenyl-1,4-phenylene)]	1.6400
123	Poly(α-naphthyl methacrylate)	1.6410
124	Poly(p-xylylene)	1.6690
125	Poly(α-vinyle naphthalene)	1.6818
126	Poly(N-vinyl carbazole)	1.6830

methods. In all cases Leave-One-Out (LOO) was applied. In Table 3, the QSPR model with the best performance, in terms of coefficient of determination, denoted R^2, and the learning method used to infer it, are shown for each subset.

Table 2. Results of the Feature Selection process, selection method, and cardinality of the six subsets selected.

Subset	Feature selection method	Cardinality
Subset 1	CFS	16
Subset 2	LR	23
Subset 3	NN	9
Subset 4	RF	11
Subset 5	RC	10
Subset 6	KNN	23

In particular, the best models inferred from subsets 2, 3, and 4 achieve performances above 0.90 of R^2 during the training stage. Therefore, we decided to select these models to continue with the next stage.

Table 3. Results of the QSPR Models Inference process to the best models, value of R^2 and learning method employed

Subset	Inference: R^2	QSPR method
Subset 1	0.8438	RF
Subset 2	0.9750	LR
Subset 3	0.9012	LR
Subset 4	0.9118	RC
Subset 5	0.8608	RC
Subset 6	0.8538	RF

QSPR Models Validation

External Validation. For the three QSPR models selected in the previous stage, an external validation was executed using the validation dataset saved at the beginning of the experimentation for this purpose. In Table 4, the results of the external validation for each model, the gap in terms of R^2 between training and external validation performances, the classification of the predictions according to Roy's Method [16], and the model's cardinality are shown.

Considering the performance of the reference model published by Jabeen et al. [6], which reached 0.8820 of R^2 during the external validation and a R^2 Gap of 0.05, we decided to discard the best QSPR inferred from Subset 2. On the other side, both Subset 3 and Subset 4 present similar behaviours in terms of performance, significantly overcoming the reference model [6]. Nevertheless, note that Subset 3 has a negative

Table 4. Results of external validation in terms of R^2, Mean Absolute Error (MAE), R^2 gap, and the classification of performance by Roy's method (Roy's Class) for each model.

Subset (cardinality)	Validation: R^2	MAE	R^2 Gap	Roy's Class
Subset 2 (23)	0.8694	0.0176	0.1056	Good
Subset 3 (9)	**0.9097**	**0.0181**	**–0.0086**	**Good**
Subset 4 (11)	0.9048	0.0161	0.0071	Good

R^2 gap value. This means that the QSPR model associated with this subset obtained a higher R^2 in the external validation than the obtained in the training. In other words, this model has better generalization than the models learned from Subsets 2 and 4. This could indicate that the Subset 3 is the best alternative for modelling the structure-activity relationship among the 6 subsets.

In addition, considering the errors in the model's predictions, an evaluation MAE-based was carried out following Roy's method [16]. This method classifies the model predictions into three categories: Good, Moderate or Bad. All models were classified as Good (see Table 4). In this case, the Roy method was not helpful in discarding models because they all have the same classification. However, it helps to know that they are reliable predictions.

In summary, a rigorous external validation of QSPR models was properly done, and the model inferred from Subset 3 was selected to continue to the next stage. This decision is supported by two main observations: is the one with the lowest cardinality and has a negative R^2 gap value. In other words, the possibility of overfitting is rejected, because the predictive capability does not fall (in fact it increases) for unknown data (validation dataset). From this point, this QSPR model learned from Subset 3 using the LR method will be denoted as our best QSPR model.

Chance Correlation Test. Our best QSPR model is chosen for executing the chance correlation test. In this case Y-Scrambling is used, which consists of randomly mixing the Y (target) values. This test was run 100 times, using the molecular descriptors of Subset 3 and LR method, and finally the mean value performance of the QSPR random models trained in these trials is $R^2 = 0.0258$ and the standard deviation is 0.0422. Thus, we conclude that the QSPR model inferred from Subset 3 is a recommendable option for RI prediction, because it overcomes the reference Jabeen's model performance and, also, the risks of overfitting and chance correlation can be discarded. However, our model has a higher cardinality (9) than the Jabeen's model (4), which is a not desirable characteristic because the models that use several molecular descriptors usually have a more complex physicochemical interpretation.

4   Conclusions

The quality of the data to be used in supervised learning processes always has a central impact on the reliability of the models inferred by these techniques. On the other hand, we know that in the field of material sciences the efforts to provide consistent data sources are high. Nevertheless, the integration of results from different experiments is a complex task, where there are multiple factors that lead to the introduction of mistakes. For this reason, even the databases provided by the scientific literature are subjected to rigorous review processes; they may contain inaccurate or wrong data. Consequently, the checking, curing and cleaning of the databases constitute a mandatory step in any QSPR modelling project in Polymer Informatics.

In this work, we have presented a process of curing and cleaning of a database to be used in the QSPR modelling of the Refractive Index of polymers. In this regard, it should be noted that the generality of the steps proposed for this process allows its application to the curing of other databases in Polymer Informatics. In this sense, the importance of physicochemical knowledge was demonstrated to achieve the correction of chemical structures and their corresponding computational representations guided by an expert in the domain. In this way, it contributes to improving the knowledge extraction process, since the final performance of the QSPR models is increased because of the availability of more reliable data, as can be seen in the models reported in this work.

However, although the models that we have obtained using machine learning methods have achieved a promising performance, it is also true that they constitute "black box" models, i.e., models whose interpretability in physicochemical terms is not always easy to establish. For this reason, it would be desirable that a domain expert could intervene in the process of design of the QSPR model being assisted by supervised learning methodologies. In particular, the use of strategies and tools for active learning may constitute an alternative to take advantage of expert knowledge. In this regard, we plan to use the VIDEAN software [20] in the future. This tool allows us to evaluate the role of alternative subsets of descriptors for the definition of a QSPR model, by using interactive visual analytics. In this way, we can achieve more interpretable models that employ a smaller number of descriptors. Moreover, taking into account the principle of Occam's razor, it is known that the fewer variables used in the definition of a model, the simpler and more generalizable it is the model.

Acknowledgments. This work was partially supported by the *Consejo Nacional de Investigaciones Científicas y Técnicas* (CONICET for its acronym in Spanish) [grant PIP 112–2017-0100829], by the *Agencia Nacional de Promoción Científica y Tecnológica* [grant PICT 2018–04533] and by the *Universidad Nacional del Sur* (UNS), Bahía Blanca, Argentina [grants PGI 24/N042 and PGI 24/ZM17].

References

1. Peerless, J.S., Milliken, N.J., Oweida, T.J., Manning, M.D., Yingling, Y.G.: Soft matter informatics: current progress and challenges. Adv. Theory Simul. **2**(1), 1800129 (2019)
2. Xu, Q., Jiang, J.: Machine learning for polymer swelling in liquids. ACS Appl. Polym. Mater. **2**(8), 3576–3586 (2020)
3. Audus, D.J., de Pablo, J.J.: Polymer informatics: opportunities and challenges. ACS Macro Lett. **6**(10), 1078–1082 (2017)

4. Jha, A., Chandrasekaran, A., Kim, C., Ramprasad, R.: Impact of dataset uncertainties on machine learning model predictions: the example of polymer glass transition temperatures. Model. Simul. Mater. Sci. Eng. **27**(2), 024002 (2019)

5. de Pablo, J.J., et al.: New frontiers for the materials genome initiative. NPJ Comput. Mater. **5**(1), 41 (2019)

6. Jabeen, F., Chen, M., Rasulev, B., Ossowski, M., Boudjouk, P.: Refractive indices of diverse data set of polymers: a computational QSPR based study. Comput. Mater. Sci. **137**, 215–224 (2017)

7. Bicerano, J.: Prediction of Polymer Properties. CRC Press, Boca Raton (2002)

8. Duchowicz, P.R., Fioressi, S.E., Bacelo, D.E., Saavedra, L.M., Toropova, A.P., Toropov, A.A.: QSPR studies on refractive indices of structurally heterogeneous polymers. Chemom. Intell. Lab. Syst. **140**, 86–91 (2015)

9. Xu, J., Chen, B., Zhang, Q., Guo, B.: Prediction of refractive indices of linear polymers by a four-descriptor QSPR model. Polymer **45**(26), 8651–8659 (2004)

10. Khan, P.M., Rasulev, B., Roy, K.: QSPR modeling of the refractive index for diverse polymers using 2D descriptors. ACS Omega 3(10), 13374–13386 (2018)

11. Seferis, J.C. Refractive Indices of Polymers. The Wiley Database of Polymer Properties (2003)

12. Kim, S., et al.: PubChem 2019 update: improved access to chemical data. Nucleic Acids Res. **47**(D1), D1102–D1109 (2019)

13. Molinspiration Cheminformatics: Nova ulica, SK-900 26 Slovensky Grob, Slovak Republic. https://www.molinspiration.com/cgi-bin/galaxy. Accessed 24 Aug 2020

14. Sigma-Aldrich Product Catalog: Polymer Science. https://www.sigmaaldrich.com/materials-science/polymer-science. Accessed 6 Aug 2020

15. Hall, M., Frank, E., Holmes, G., Pfahringer, B., Reutemann, P., Witten, I.H.: The WEKA data mining software: an update. ACM SIGKDD Explor. Newsl. **11**(1), 10–18 (2009)

16. Roy, K., Das, R.N., Ambure, P., Aher, R.B.: Be aware of error measures. Further studies on validation of predictive QSAR models. Chemometr. Intell. Lab. Syst. **152**, 18–33 (2016)

17. Muller, C., et al.: Prediction of drug induced liver injury using molecular and biological descriptors. Comb. Chem. High Throughput Screen. **18**(3), 315–322 (2015)

18. DRAGON for Windows: (Software for Molecular Descriptor Calculations), Talete srl, Version 5.5. Milan, Italy (2007)

19. Topliss, J.G., Costello, R.J.: Chance correlations in structure-activity studies using multiple regression analysis. J. Med. Chem. **15**(10), 1066–1068 (1972)

20. Martínez, M., Ponzoni, I., Díaz, Mónica. F., Vazquez, G., Soto, A.: Visual analytics in cheminformatics: user-supervised descriptor selection for QSAR methods. J. Cheminform. **7**(1), 1–17 (2015). https://doi.org/10.1186/s13321-015-0092-4

The Time-Lagged Effect Problem on (Un)truthful Data, a Case Study on COVID-19 Outbreak

Luis Rojo-González[1,2(✉)]

[1] Facultat de Matemàtiques i Estadística, Universitat Politècnica de Catalunya,
Barcelona, Spain
[2] Department of Industrial Engineering, Universidad de Santiago, Santiago, Chile
`luis.rojo.g@usach.cl`

Abstract. The Coronavirus SARS-CoV-2 (COVID-19) emerged by December 2019, in Wuhan, China; it was reported and, a few months after, in the most of countries we are living a pandemic of an (almost) unknown disease never observed before. For this reason, the importance of a good measurement on the counting of observed cases has a crucial role. This work addresses the time-lagged effect problem via Bayesian analysis supported by a stochastic discrete-event simulation to give an answer to the truthfulness of the data and to validate the obtained results in terms of proportions based on an expected result in the particular case of Spain. Obtained results show that the reported data is untruthful and can make wrong any analysis, but even when the simulating results are as we expected they might be wrong in terms of absolute numbers. However, the most important knowledge we get is related to the fact that the disease might be considered under control because it is more likely that a person gets recover than She/He dies.

Keywords: Bayesian analysis · Stochastic simulation · COVID-19

1 Introduction

The Coronavirus SARS-CoV-2 (COVID-19) emerged by December 2019, in Wuhan, China. It was reported and, a few months after, in the most of countries we are living a pandemic of an (almost) unknown disease never observed before. Despite the drastic, large-scale containment measures promptly implemented by the Chinese government, in a matter of a few weeks the disease had spread well outside China, reaching countries in all parts of the globe [4] resulting in human and economical crisis in most of them.

To decrease the damage associated with COVID-19, public health and infection control measures are urgently required to limit the global spread of the

The author is grateful for partial support from ANID Beca Magíster en el Extranjero, Becas Chile, Folio 73201112.

D. A. Rossit et al. (Eds.): ICPR-Americas 2020, CCIS 1408, pp. 295–307, 2021.
https://doi.org/10.1007/978-3-030-76310-7_23

disease [18], but till now the ongoing pandemic seems not to be overcame unless a vaccine appears. However, the saying: *you cannot manage what you do not measure*, and, therefore, the importance of a well measurement on the counting of observed cases has a crucial role. On this way, the World Health Organization (WHO) [21] has published that the number of real infected individuals are between 10 and 20% of the real ones. Even so, Ioannidis [7] reviews and comments the possibility of a problem that spread even faster than the pandemic, the reliability on data; which results on a clear issue that makes models useless.

So, how COVID-19 pandemic behaves throughout time? this question might be the most important to find an answer for (as soon as possible) in the last time. On this way, it is important to anticipate the number of new cases on a daily-basis to take care about healthcare systems as well as the budget to dedicate to this disease. Several authors has been addresses this question using different models, but the most used has been those ones with dynamic nature such as the Susceptible-Exposed-Infected-Removed (SEIR) model [5,13], Susceptible Unquarantined infected, Quarantined Infected, Confirmed infected (SUQC) model [25], Susceptible Exposed Infected Isolated with treatment Removed (SEIJR) model [14,16], Exponential growth model [9,24], Stochastic simulations of outbreak trajectories [15] and Hierarchical models [3]. Nevertheless, despite of the possibility of working with non-reliable data approximations to transition rates, in the case of dynamic models, as the infection rate, mortality rate and recovery rate has been computed.

This study aims to merge prior knowledge generated at the moment, using the particular case of the observed data in Spain gathered from the Github repository associated with the interactive dashboard hosted by the Center for Systems Science and Engineering (CSSE) at Johns Hopkins University, Baltimore, USA [8], to know the real number of confirmed cases that would be by using Bayesian Analysis assuming the counting cases behave as a Poisson process to fit the necessary parameters to make a simulation throughout Discrete Markov Chains within a specific period.

This work is organized as follows: Sect. 2 establishes the time-lagged effect problem performing a descriptive analysis of the data and then makes the statement of the problem under study. Section 3 shows a Bayesian approach to fit the necessary random variables considering a hypothesis test on the lagged time effect. Section 4 addresses the truthfulness of the data via a stochastic discrete-event simulation to give an answer to the truthfulness of the data and to validate the obtained results in terms of proportions based on an expected result. Finally, Sect. 5 gives the main insights we found and propose to escalate the problem under study.

2   The Time-Lagged Effect Problem

In this section, we perform a descriptive analysis of the counting of each kind of cases in order to motivate the discussion and the way we model the situation that apparently we see. Then, we formally declare the variables in the situation and the equation that support the behaviour through dynamic modelling in its discrete version.

2.1   Descriptive Analysis

From the considered data source, we can recognize that there was an inflexion point before and after a lock-down was established in most of the countries which take that measure. In the particular case of Spain, the lock-down policy was taken in March 14th, and as far as we know the expected effect of this measure was around two weeks after that, by March 28th, an inflexion point must appear on the counting. Motivated on this *belief*, we will work with the data corresponding to March and April given two main reasons: i) these months correspond to the most chaotic in term of how the counting was conducted, and also in this particular country, the beginning of the pandemic; and ii) to consider a larger period would add noise related to the effect of the lock-down on how the disease spread. Figure 1 shows the counting in both months by each kind of case, Confirmed, Recovered and Death cases and two vertical lines that indicate both the day at lock-down was and the expected effect.

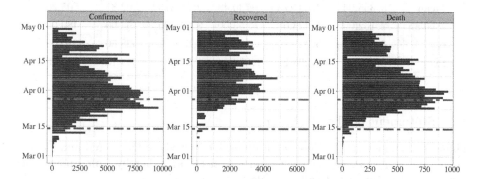

Fig. 1. Counting during March and April by each kind of case, Confirmed, Recovered and Death cases. The vertical lines indicate the day at the lock-down was implemented (March 14th, in red) and at the day its expected effect was (March 28th, in green). (Colour figure online)

It is clear to see that the counting for each kind of case, the so-called curves, do not follow a strict pattern, which can be explained, discarding the lock-down, by the manner, the authorities performed the counting. It affects directly on the Confirmed cases and therefore on Recovered and Death cases, which would show a similar -proportional- pattern after a time to recover or to die.

The later mentioned effect can be supported numerically by the descriptive statistics shown in Table 1. In the case of Confirmed cases, we see that the counting from March to April did not change as much as Recovered and Death cases, where they were (almost) four and two times greater, respectively. However, two simple ways to recognize this effect could be:

– Taking into account that a person must be a Confirmed case before be a Recovered or Death case, if we move these two counting curves some days, they should fit with the counting related to Confirmed cases, which often happens.

- The sum of Recovered and Death cases is lower than the sum of Confirmed cases during the period we are working with.

this informal evidence allows us to motivate the problem formally.

Table 1. Descriptive statistics of the COVID-19 pandemic behaviour during March and April of 2020. Total number of Confirmed, Recovered and Death cases are 225,166, 112,048 and 24,543 people, respectively.

	Confirmed		Recovered		Death	
	March	April	March	April	March	April
Count	104,034.00	121,132.00	22,645.00	89,403.00	9,387.00	15,156.00
Q1	286.50	1,909.25	0.00	2,517.25	7.00	369.75
Mean	3,355.94	4,037.73	730.48	2,980.10	302.81	505.20
Median	2,162.00	4,423.00	53.00	3,256.00	94.00	489.00
Q3	6,621.50	5,425.25	1,396.00	3,655.25	628.50	679.00
SD	3,256.35	2,238.22	1,050.14	1,346.84	344.78	222.43

2.2 Statement of the Problem

These so-called curves are dependent on each other where Recovered and Death cases are lagging product of the Confirmed cases, and thereby from another of non-infected people, the Health people.

The problem under study could be characterized by a dynamic modelling point of view. From now on, we will consider four possible states at which people are at certain time t: i) Health, are the people who have never been positive to the disease before; ii) Confirmed, are the people who currently give positive to the disease; iii) Recovered, are the people who are healthy once they gave positive before; and iv) Death, are the people who have died due to the disease. On this way, it is possible to state the following *Once a person is a Confirmed case there are two possible transitions after a certain elapsed time, that person will get recover or will die*, this situation is such as Fig. 2 shows with their respective transitions.

Formally, we can recognize that curves could be characterized by using Poisson processes with different intensities and elapsed times to change from one state to another. Let's denote the set of states $\mathcal{I}$ and the number of people at time t in each state $N_i(t)$, $\forall i \in \mathcal{I}$. Then, we propose that the number of people $N_i(t)$ is defined by a Poisson process with parameter $\lambda_i t_i$, where λ_i represents the number of new cases (in people per day), intensities from now on, and t_i represents the elapsed time necessary to change from one state to another one, $\forall i \in \mathcal{I}$.

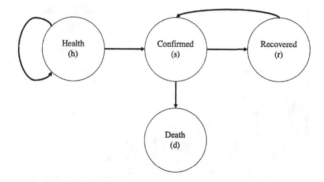

Fig. 2. A general illustration of the dynamic model considering the possibility of the population can get sick again. The states of the model indicate in which category the person is at a certain time t and where She/He from, which is indicated by the arrows.

Thus, considering the elapsed time to get recover or to die as t_r and t_d, respectively, we can define the Poisson processes as Definition 1 shows.

Definition 1 (*Structural Equations*). *Let's consider a person who incubates the disease in t_s days. She/He will get recover or die after t_r or t_d days, respectively. Also, be $Y_i(t_i)$ the cumulative number of cases within a certain days in $[t; t + t_i]$, $i \in \mathcal{I}$. Thus, we can define the total number of Recovered and Death cases depending on the Confirmed cases as follows:*

- *$Y_r(t_r) = N_r(t + t_r) - N_s(t)$, defines the number of total Recovered cases in $[t; t + t_r]$ such that $Y_r(t_r) \sim Poisson(\lambda_r t_r)$.*
- *$Y_d(t_d) = N_d(t + t_d) - N_s(t)$, defines the number of total Death cases in $[t; t + t_d]$ such that $Y_d(t_d) \sim Poisson(\lambda_d t_d)$.*

thus, the equivalence between Confirmed, Recovered and Death cases is given by

$$Y_s(t + t_s) = Y_r(t + t_r) + Y_d(t + t_d) \tag{1}$$

where $Y_s(t + t_s) \sim Poisson(\lambda_s t_s)$ with parameter $\lambda_s t_s$ such that $\lambda_s t_s = \lambda_r t_r + \lambda_d t_d$.

Finally, since a person is different from each other, both the number of new cases λ_i and the elapsed time necessary to change from one state to another one t_i are not fixed parameters, but random variables. Figure 3 shows the idea behind Definition 1 where, without loss of generality, the distribution is a smoothed one of the number of people in each state $N_i(t)$ throughout t. Also, it shows distribution at certain day t without the time-lagged effect, where it is expected that the Confirmed cases are equal to the sum over Recovered and Death cases.

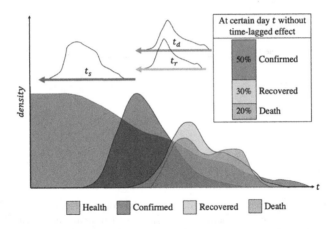

Fig. 3. Illustration of the problem under study. We consider four possible states, where the sum on specific states must be equal to the precedent states, e.g. the density curve that represents the Health people must be equal to the sum over the other density curves without the time-lagged effects t_s, t_r, t_r; which also are random variables.

3   Bayesian Analysis

In this section, we propose a way to overcome the variable fitting via Bayesian Analysis using the package RStan [19]. We state the prior distribution for the later mentioned parameters and perform an interesting hypothesis test related to whether the number of days to change from Confirmed state to another are equal.

3.1   Prior Distributions

As we state in Definition 1 before, the parameters that are involved in Expression (1) are random variables, and to fit them we propose a Bayesian way. On this way, we need to provide priors distributions which draw the knowledge at the moment. These could be obtained from our own knowledge or using other novel researches who also addresses similar problems from an epidemiological point of view.

The case of the intensities (λ_i) is highly dependent on the policies and at which moment they were implemented, so we adopt the knowledge based on how we believe the curves grow throughout time in these two months. Consequently, as intensities represent the number of people in each state at certain day t, it makes sense to think in a Gamma distribution.

On the other hand, the response against the disease has been studied by several authors which report what they found. The literature review shows that the elapsed times t_i distribute as a Gamma for each one of them. The incubation time, t_s, is a Gamma with mean 6.5 days and standard deviation of 2.6 days [2,22]; the elapsed time to get recover, t_r, is a Gamma with mean 10 days but can be up to 14 or 21 days [17]; and the elapsed time to die, t_d, is a Gamma with mean 20 days and standard deviation equal to 10 days [22].

We use such parameterization of the Gamma distribution with expectation defined by $\mathbb{E}(x) = \alpha/\beta$ and variance $Var(x) = \alpha/\beta^2$; where α and β parameters, obtained by solving a system of equation, are shown in Table 2.

Table 2. Parameters that define the Gamma distribution based on the prior knowledge for each random variable to fit.

	λ_s	λ_r	λ_d	t_s	t_r	t_d
α	250	122.50	2.50	26.20	5.00	126.49
β	4.03	0.50	6.32	4.03	0.50	6.32

3.2 Posterior Distributions

Once we already state the prior distributions, we can get the posterior distributions that will show the merge between the prior knowledge and the data. Table 3 shows descriptive statistics from the obtained results, where the $\hat{R}$ statistic shows the fitting process was successfully achieved, and the confidence intervals are not too large respect to the mean. These results are saying that:

- The incubation time, t_s, is about 18 days but can be between 16 and 19 days. The elapsed time to get recover, t_r, is about 11 days but can be up to 9 and 13 days. The elapsed time to die, t_d is 20 days with possibilities to find cases from 17 to 23 days.
- The number of daily Confirmed cases, λ_s is in mean 5,888. The number of daily Recovered cases, λ_r is 3,500. The number of daily Death cases, λ_r is 521.

Table 3. Descriptive statistics for fitted posterior distribution. Notice that the mean and percentile values have been truncated to enhance their interpretation. The $\hat{R}$ statistic shows the fitting process was successfully achieved.

	Mean	Error	5%	95%	$\hat{R}$
t_s	18	0.03	16	19	1.00
t_r	11	0.04	9	13	1.00
t_d	20	0.06	17	23	1.00
λ_s	5,888	306.86	5398	6408	1.00
λ_r	3,500	308.45	2993	4008	1.00
λ_d	521	1.65	452	607	1.00

A graphical analysis of the obtained results can be carried out to see the effects of the selected data which are set out in Fig. 4. It shows the density plot for each posterior distribution. In this, it is possible to distinguish that, although the prior distributions for the elapsed times are evidence from other works, and might be considered informative ones due to their short variances, the data change the landscape in its favour. The most important insight of the results is such a person who gave positive to the disease, it is more likely that She/He gets recover due to a temporal reason. Also, the number of daily Confirmed cases is greater than the Recovered cases which are greater than the Death cases.

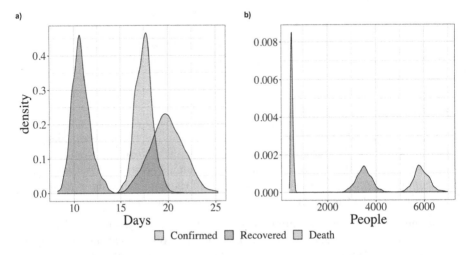

Fig. 4. Density plot for each posterior distribution of elapsed times and intensities. **a)** shows the number of days to change from one state to another and **b)** shows the number of daily cases in mean.

3.3   Differences on the Elapsed Times in the Transitions from Confirmed Cases

An interesting and useful question corresponds to check some conditions of the dynamic, in this particular case whether the number of days to change from Confirmed state to another are equal, in other words, whether the elapsed time to get recover (t_r) is equal to the elapsed time to die (t_d). Taking into account the Definition 1, it is equivalent to the hypothesis test given by

$$H_1 : Y_s(t + t_s) \sim Poisson(\lambda_r t_r + \lambda_d t_d)$$
$$H_2 : Y_s(t + t_s) \sim Poisson((\lambda_r + \lambda_d)t)$$

where the probability of each hypothesis to be true is

$$Pr(H_i|y) = \frac{\pi(H_i)Pr(y|H_i)}{Pr(y)} \tag{2}$$

again, as we are working from a Bayesian point of view, we need a prior probability for each one of the hypothesis.

The literature review did not reveal an answer for this question, but the Bayesian point of view allow us to state a non-informative probability to give to the data an important role. Thus, we use the most non-informative probability, 50%. Doing so, the hypothesis test does not support any of them, indeed the probability to be true for each hypothesis is equal to 50%, it means that there is no difference on the elapsed times in the transitions from Confirmed cases and it is completely at random.

4   Stochastic Discrete-Event Simulation

At the moment, we have proposed the equation and parameters that would enable us to analyze the behaviour of the pandemic. Nevertheless, a crucial question arises and it is still an open problem to overcome. Are we working with (un)truthful data? We show an approach based on a stochastic discrete-event simulation using Markov Chains to give an answer and validate the results obtained previously as an underlying product.

4.1   State Equations

First of all, following the dynamic such as we showed in Sect. 2 and considering the elements given in Definition 1, we can establish a stochastic Markov Chain and the state equations that define it to address a discrete-time simulation.

Let $x_{i,j}(t)$ be the transition rate from state i to state j at time t and $N_i(t)$ be the people in state i at time t, $\forall i, j \in \mathcal{I}$. Thus, for a certain time $t \geq max\{t_r, t_d\}$, the elapsed time to get recover or to die, respectively, the state equations are defined by:

$$N_h(t+1) = x_{h,h}(t)N_h(t) - x_{h,s}(t)N_h(t-t_s) \tag{3}$$

$$N_s(t+1) = x_{h,s}(t)N_h(t-t_s) + x_{r,s}(t)N_r(t-t_s)$$
$$\quad - x_{s,r}(t)N_s(t-t_r) - x_{s,d}(t)N_s(t-t_d) \tag{4}$$

$$N_r(t+1) = x_{s,r}(t)N_s(t-t_r) - x_{r,s}(t)N_r(t-t_s) \tag{5}$$

$$N_d(t+1) = x_{s,d}(t)N_s(t-t_d) \tag{6}$$

this system of equations enables us to simulate the dynamic considering the whole results obtained previously. Equation (3) defines the number of people in Health state at time $(t+1)$ by considering the number of people who were in this state at time t minus the people who incubate the disease at $(t-t_s)$, i.e. since t_s days the person got infected. Equation (4) establishes that the Confirmed Cases at $(t+1)$ is equal to the people who got infected t_s days but never had been in Confirmed state before, and the people who got infected t_s days ago but those ones that at least have been in Confirmed state before, i.e. those people who incubate the disease again; minus the people who get recover since they were a Confirmed case and the people who die. Equation (5) considers the difference between the people who get recover, taking into account that they have been considered as Confirmed cases previously, and those people who incubate the disease again after they got to recover. Finally, Eq. (6) shows the Death state as and absorbing one.

4.2   Implementation, Results and Their Validation

Algorithm 1 shows the implementation of the simulation which aims to compute the number of people at each state $i \in \mathcal{I}$ within a defined period $[t; \hat{t}]$ until the entire population incubate the disease (line 1). It works with the posterior distribution for each parameter as known data, and involve the stochastic component where it makes the intensities sampling for each state $i \in \mathcal{I}$ (line 2–3) and where it makes the difference over the elapsed times to get recover or to die according to a uniform $(0, 1)$ random number (line 4–10). Finally, it performs the simulation by using the state equation defined previously (line 11) at repeat for time $(t + 1)$ (line 12).

Algorithm 1: Stochastic discrete-event simulation

Data: Population, simulation period, posterior distribution of intensities and elapsed times, transition rates and hypothesis test probabilities.

Result: Number of people at each state within a defined period.

---

1   **while** *Population > 0 and $t \leq \hat{t}$* **do**
2   **for** $i \in \mathcal{I}$ **do**
3   get a sample $\hat{\lambda}_i$ from λ_i
4   get a sample $\hat{t}_s$ from t_s
5        $u \leftarrow runif(1)$
6   **if** $|u - Pr(H_1|y)| < |u - Pr(H_2|y)|$ **then**
7   **for** $j \in \mathcal{I} \{s\}$ **do**
8   get a sample $\hat{t}_j$ from t_j
9   **else**
10   get a sample $\hat{t}_{r+d}$ from $t_r \cup t_d$
11   Perform the state equation
12       $t \leftarrow t + 1$

Till now, we already have the almost all parameters we need in order to carry out the later mentioned simulation except for the transition rates values and the number of people in the country. Fortunately, the last value can be obtained from the National Institute of Statistics (INE) [6]; and the transition rates have been reported in epidemiological researches. Velavan and Meyer [20] report the infection rate is 2.2% in average, whereas Anastassopoulou et al. [1] get it is between 1.9% and 1.92%. Zhonghua Liu Xing Bing Xue Za Zhi [26] shows that the mortality rate is equal to 0.3% among the health workers who have become infected and Lai et al. [11] report it decreases from 2.5% to 2% between February 12th and February 14th, when the confirmed cases reached

66,576 globally [23]; finally Yi et al. [23] report a recovery rate of 87%. In the case of the transition rate of incubating the disease again after they got to recover, the literature review has not to evidence about it [10,12].

Thus, considering the whole data mentioned previously, including the posterior distributions for each fitted random variables, we can perform the stochastic discrete-event simulation. The obtained results for the number of people in each state are: i) Confirmed cases are 18,095,738; ii) Recovered cases are 12,048,823; and iii) Death cases are 50,056. These results are different from the reported cases. If we would take the simulation as the true number of people in each state, the reported cases would correspond to a 1.24%, 0.93% and 49.03% for Confirmed, Recovered and Death cases, respectively. However, these last results must be validated.

On this way, as we showed above in particular in Fig. 3, the expected result of the time-lagged effect problem should be such that at some day t the proportion of the cases would correspond to half for Confirmed cases and a half for the sum over Recovered and Death cases, and this must be applied on the reported cases as well as the simulating results to answer the issue related to the truthfulness of the data. Figure 5 shows a stacked column plot of the proportion of people in each state and a dashed horizontal line is overlapped at the middle of the plot to enhance the visualization. It is to say that the reported cases do not behave as we expected, especially at the beginning, but the simulation does despite a couple of atypical counting. Therefore, based on the expected result of the time-lagged effect problem the reported cases, and thus the data, should be considered as untruthful data, whereas the simulation was performed on the right way in terms of proportion, but might be wrong in terms of absolute numbers.

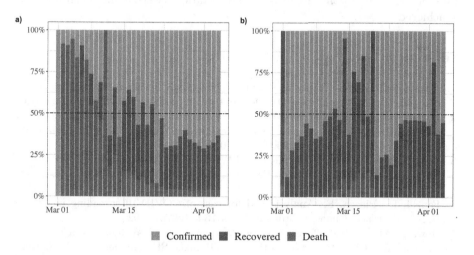

Fig. 5. Validation plot by stacked columns for the proportion of people in each state for **a)** reported cases and **b)** the simulating results. The dashed horizontal line indicates the 50% of the data to enhance the visualization.

5   Conclusion and Future Works

This work addresses the time-lagged effect problem on the COVID-19 pandemic via Bayesian analysis supported by the stochastic discrete-event simulation to give an answer to the truthfulness of the data and to validate the obtained results in terms of proportions based on an expected result. Motivated by this problem, the reported data of Spain is used as a study case. This reported data was characterized by a graphical approach and from a descriptive statistics point of view, where a simple analysis enables us to recognize the proposed problem is set up in the data.

The way we take to fit the established parameters, and therefore to perform the simulation via Markov Chain, is computationally tractable by a desktop computer. Obtained results show that the reported data is untruthful and can make wrong any analysis, but even when the simulating results are as we expected they might be wrong in terms of absolute numbers.

However, despite simulating results are correct in terms of absolute numbers or not, they provide useful knowledge of the pandemic behaviour in this particular study case. The most important ones are related to the fact that the disease might be considered under control, because of it is more likely that a person gets recover than She/He dies, and it behaves differently from other countries according to related studies which are used as prior knowledge on the carried out Bayesian analysis. Nonetheless, the number of daily Confirmed cases is greater than the number of Recovered cases which is greater than the number of Death cases.

For future works, we consider making wider the period modelling the transition rates as random variables and to evaluate the effectiveness of the lock-down through a change of point analysis to find a stationary state and when it would be achieved.

References

1. Anastassopoulou, C., Russo, L., Tsakris, A., Siettos, C.: Data-based analysis, modelling and forecasting of the COVID-19 outbreak. PloS one **15**(3), e0230405 (2020)
2. Backer, J.A., Klinkenberg, D., Wallinga, J.: Incubation period of 2019 novel coronavirus (2019-nCoV) infections among travellers from Wuhan, China, 20–28 January 2020. Euro. Surveill. **25**(5), 2000062 (2020)
3. Fan, J., Liu, X., Pan, W., Douglas, M.W., Bao, S.: Epidemiology of 2019 novel coronavirus disease-19 in Gansu Province, China, 2020. Emerg. Infect. Dis. **26**(6), 1257–1265 (2020)
4. Fanelli, D., Piazza, F.: Analysis and forecast of COVID-19 spreading in China, Italy and France. Chaos, Solitons Fractals **134**, 109761 (2020)
5. How, C., et al.: The effectiveness of the quarantine of Wuhan city against the Corona Virus Disease 2019 (COVID-19): well-mixed SEIR model analysis. J. Med. Virol. **92**(7), 841–848 (2020)
6. Instituto Nacional de Estadística: Cifras de Población (2020). https://www.ine.es/dyngs/INEbase/es/operacion.htm?c=Estadistica_C&cid=1254736176951&menu=ultiDatos&idp=1254735572981. Accessed 14 July 2020
7. Ioannidis, J.P.: A fiasco in the making? As the coronavirus pandemic takes hold, we are making decisions without reliable data. Stat **17** (2020)

8. Johns Hopkins University: COVID-19 Data Repository by the Center for Systems Science and Engineering (CSSE) at Johns Hopkins University (2020). https://github.com/CSSEGISandData/COVID-19/tree/master/csse_covid_19 _data/csse_covid_19_time_series. Accessed 3 May 2020

9. Jung, S.M., et al.: Real-time estimation of the risk of death from novel coronavirus (COVID-19) infection: inference using exported cases. J. Clin. Med. **9**(2), 523 (2020)

10. Kirkcaldy, R.D., King, B.A., Brooks, J.T.: COVID-19 and postinfection immunity: limited evidence, many remaining questions. Jama **323**(22), 2245–2246 (2020)

11. Lai, C.C., Shih, T.P., Ko, W.C., Tang, H.J., Hsueh, P.R.: Severe acute respiratory syndrome coronavirus 2 (SARS-CoV-2) and corona virus disease-2019 (COVID-19): the epidemic and the challenges. J. Antimicrob. Agents **55**(3), 105924 (2020)

12. Ota, M.: Will we see protection or reinfection in COVID-19? Nat. Rev. Immunol. **20**(6), 351 (2020)

13. Prem, K., et al.: The effect of control strategies to reduce social mixing on outcomes of the COVID-19 epidemic in Wuhan, China: a modelling study. The Lancet Public Health **5**(5), 261–270 (2020)

14. Read, J.M., Bridgen, J.R., Cummings, D.A., Ho, A., Jewell, C.P.: Novel coronavirus 2019-nCoV: early estimation of epidemiological parameters and epidemic predictions. MedRxiv (2020)

15. Riou, J., Althaus, C.L.: Pattern of early human-to-human transmission of Wuhan 2019 novel coronavirus (2019-nCoV), December 2019 to January 2020. Euro. Surveill. **25**(4), 2000058 (2020)

16. Shen, M., Peng, Z., Xiao, Y., Zhang, L.: Modelling the epidemic trend of the 2019 novel coronavirus outbreak in China. Innov. **1**(3), 100048 (2020)

17. Singhal, T.: A review of coronavirus disease-2019 (COVID-19). Indian J. Pediatr. **87**(4), 281–286 (2020)

18. Song, F., et al.: Emerging 2019 novel coronavirus (2019-nCoV) pneumonia. Radiol. **295**(1), 210–217 (2020)

19. Stan Development Team: RStan: the R interface to Stan, r package version 2.19.3 (2020). http://mc-stan.org/

20. Velavan, T.P., Meyer, C.G.: The COVID-19 epidemic. Trop. Med. Int. Health **25**(3), 278 (2020)

21. World Health Organization: Coronavirus disease (COVID-2019) situation reports (2020). https://www.who.int/emergencies/diseases/novel-coronavirus-2019/ situation-reports/. Accessed 22 May 2020

22. Wu, J.T., et al.: Estimating clinical severity of COVID-19 from the transmission dynamics in Wuhan, China. Nat. Med. **26**(4), 506–510 (2020)

23. Yi, Y., Lagniton, P.N., Ye, S., Li, E., Xu, R.H.: COVID-19: what has been learned and to be learned about the novel coronavirus disease. Int. J. Biol. Sci. **16**(10), 1753 (2020)

24. Zhao, S., et al.: Estimating the unreported number of novel coronavirus (2019-nCoV) cases in china in the first half of January 2020: a data-driven modelling analysis of the early outbreak. J. Clin. Med. **9**(2), 388 (2020)

25. Zhao, S., Chen, H.: Modeling the epidemic dynamics and control of COVID-19 outbreak in China. Quant. Biol. **8**(1), 11–19 (2020). https://doi.org/10.1007/s40484-020-0199-0

26. Zhi, Z.L.X.B.X.Z.: The epidemiological characteristics of an outbreak of 2019 novel coronavirus diseases (COVID-19) in China. Novel, Coronavirus Pneumonia Emergency Response Epidemiology and others **41**(2), 145 (2020)

Robust Data Reconciliation Applied to Steady State Model with Uncertainty

Claudia E. Llanos[✉] and Mabel Sánchez

Planta Piloto de Ingeniería Química (Universidad Nacional del Sur - CONICET), Camino La Carrindanga km 7, 8000 Bahía Blanca, Argentina
cllanos@plapiqui.edu.ar

Abstract. Different researchers have proposed the treatment of uncertainties in measurements because they interfere in the process state estimation. Data reconciliation procedure improves the information supplied by measurements, minimizing the discrepancy existent between measurements and accurate process model. This problem allows obtaining unbiased estimation when measurements follow exactly a normal distribution. Nevertheless, the presence of outliers do not allow the use of the former procedure, therefore Robust Data Reconciliation is developed. This latter provides accurate solutions when measurements follow approximately the normal distribution. Although many advances have been developed to treat measurement uncertainties in Data Reconciliation framework, there are not research works that consider model and measurement uncertainties simultaneously in presence of outliers. In this work, a Simple robust Method, which takes advantage of temporal redundancy, is applied to benchmarks that contain uncertain parameters. Performances measures are tested for different magnitudes of simulated outliers and compared with the ones provided by a classic Data Reconciliation procedure. Results show that the Robust Data Reconciliation procedure can yield unbiased estimations of measurements and parameters when outliers and parametric uncertainties are present.

Keywords: Robust Data Reconciliation · Model uncertainty · Measurement uncertainty · Parameter estimation · Outliers

1   Introduction

Uncertainties in measurements are given by random events which always are present and produce fluctuations in the observation provided by instruments. These errors can be dealt with procedures as Data Reconciliation (DR). It's well known that DR gives accurate estimations when the assumptions of normality and model accuracy are true. Nevertheless, the presence of systematic errors in measurements or uncertainty in models is a common scenario in an industrial process. Therefore, both should be taken into account for the study of process state.

Systematic errors, as outliers, deteriorate the assumed distribution. Thus, different tests or techniques to detect and eliminate outliers before executing a DR procedure have

© Springer Nature Switzerland AG 2021
D. A. Rossit et al. (Eds.): ICPR-Americas 2020, CCIS 1408, pp. 308–322, 2021.
https://doi.org/10.1007/978-3-030-76310-7_24

been proposed [1]. Using Robust Data Reconciliation (RDR) stands out from the former methodologies since it is able to mitigate the effect of outliers without using auxiliary methodologies [2, 3].

Researchers have proved RDR effectiveness in different operational conditions [2, 3]. This procedure uses M-estimators as objective function instead of the Least Square (LS) function. M-estimators are a generalization of the Maximum Likelihood function and can be classified as monotone or redescending according to their first derivative, which is called Influence Function (IF) [4].

In real-time operation, multiple sources of uncertainties deteriorate the model. This consideration allows avoiding the use of inappropriate models which leads to incorrect state estimation [5]. Some examples are ignored changes of physical flows state, chemical reactions conditions which change kinetics parameters; inaccuracies in model parameters related to energy balances and changes in the process state.

There are few proposals to deal with parametric uncertainty in models using DR. In this sense, [6] and [7] treated linear model which contain error measurements that follow the normal distribution ($\mathcal{N}$). The former research was a semi-empirical technique which used an iterative procedure to weight each constraint regarding to its uncertainties, while the latter research presented two modified DR procedures. Both DR problems provided identical estimations of parameter uncertainties, however the second highlighted because this allowed computing the variables estimation that satisfies the updated model. Furthermore, the latter authors discussed the concepts of redundancy and observability and showed that both diminish when the model contains uncertain parameters. Vasebi et al. [8] demonstrated the importance of selecting uncertainty covariance matrices to treat bilinear system. The three aforementioned research works did not take into account outliers. Recently, the presence of outliers was addressed at the development of a decision support tool when stochastic uncertainty in the demand exists [9]. Although the uncertainty studied was not parametric, this work remarks the importance of obtaining accurate estimation as a first step in the chain of an economic optimization problem.

The papers reviewed show the usefulness of RDR for a posterior step of optimization. This allows proposing the following hypothesis: RDR can treat model with uncertainty when outliers are present. To verify the hypothesis, this research work uses two linear benchmarks which contain parameter uncertainties. A RDR procedure is devised which uses a Simple Method (SiM) developed by [2] to get unbiased estimations of measurements and parameters for two models of error propagation. First, three case studies are presented which compare the results of SiM with the ones obtained by a classic DR problem using a model extracted from [7]. Different magnitudes of outliers are simulated and results of Performance Measure (PM) are presented in performance curves. Finally, a second model is treated with two methodologies which take advantage of temporal redundancy.

This research work is organized as follows. Section two briefly describes the RDR strategy executed for the estimation of measurements and uncertain parameters. Section 3 presents the measurement models considered and Performance Measures (PM) selected to inform the improvement attained. The results in terms of PM for two linear benchmarks are presented in Sect. 4. Finally, Sect. 5 presents the conclusions achieved.

2 Methodology

Data reconciliation is an optimization problem which uses as objective function the LS. This procedure takes advantage of the spatial redundancy provided by the model equation to minimize the discrepancy between measurements and restrictions [10].

In general, the solution of DR is a vector of estimated measured, $\hat{\mathbf{x}}^{LS}$, and unmeasured variables, $\hat{\mathbf{u}}^{LS}$, that strictly satisfy the model.

$$[\hat{\mathbf{x}}^{LS}, \ \hat{\mathbf{u}}^{LS}] = \underset{\mathbf{x}, \, \mathbf{u}}{Min}(\mathbf{y} - \mathbf{x})^T \mathbf{\Sigma}^{-1}(\mathbf{y} - \mathbf{x})$$

st.

$$\mathbf{f}(\mathbf{x}, \mathbf{u}) = 0$$
$$\mathbf{h}(\mathbf{x}, \mathbf{u}) \le 0$$
$$\mathbf{x}^L \le \mathbf{x} \le \mathbf{x}^U$$
$$\mathbf{u}^L \le \mathbf{u} \le \mathbf{u}^U \tag{1}$$

where $\mathbf{\Sigma}$ represents the covariance matrix, while $\mathbf{y}$ stands for the vector of random measurements. Furthermore, $\mathbf{f}$ and $\mathbf{h}$ stand for equality and inequality constraints, respectively. The upper and lower bounds of the optimization variables are represented with the suffix U and L respectively.

Narasimhan et al. (2012) treated parameter uncertainties solving the problem 2:

$$\left[\hat{\mathbf{x}}^{LS}, \hat{\theta}\right] = \underset{\mathbf{x}, \hat{\theta}}{Min}(\mathbf{y} - \mathbf{x})^T \sum{}^{-1}(\mathbf{y} - \mathbf{x}) + \sum_{p}\left(\frac{\theta_p - \hat{\theta}_p}{\sigma_{\theta, \, p}}\right)$$

st.

$$\mathbf{A}_\theta \mathbf{x} = \mathbf{0} \tag{2}$$

where $\hat{\mathbf{x}}^{LS}$ and $\hat{\theta}$ represent the state and parameter reconciled vectors and $\mathbf{A}_\theta$ the model with uncertain parameters [7]. This formulation considers that all the variables are measured. However, if there are unmeasured variables, procedures as QR factorization can provide a reduced problem which includes measured variables [10].

Problem (2) allows address model with uncertainty but does not consider that outliers may exist. Such errors invalidate the assumption of normality, thus the use of classic DR is not suitable. Conversely, Robust Data Reconciliation (RDR) yield unbiased estimation not only when measurements follow exactly the $\mathcal{N}$, but also when they do approximately [4]. Regards uncertain parameters, the presence of outliers can corrupt their estimation, therefore a RDR seems to be a promising strategy to get unbiased estimations of measurements and parameters.

2.1 M-estimators

M-estimator of location, $\hat{x}$, of the i-th variable is the solution provided by the following minimization problem:

$$\sum_{i=1}^{n} \rho\left(y_i - \hat{x}\right) = Min! \tag{3}$$

where ρ symbolizes the robust estimator. Based on this function the mathematical expressions of the IF (ψ) and Weight functions (W) are developed.

$$\sum_{i=1}^{n} \psi(y_i - \hat{x}) = 0 \tag{4}$$

$$\sum_{i=1}^{n} W_i(y_i - \hat{x})(y_i - \hat{x}) = 0 \tag{5}$$

where $\psi = d\rho(x)/dx$ and W is defined in terms of ψ [4].

According to their IF, M-estimators are classified as monotone or redescending. The former are convex functions, therefore they always converge to the optimal solution. The latter W tend or are equal to zero, because of that redescending estimators are more efficient for heavy-tailed data. However, they require a good starting point to ensure attaining the global solution. Examples of monotone functions are the Huber (HU) and Fair functions, while classical redescendig estimators are the Biweight (BW) and Hampel functions [3]. Taking into account the former analysis, researchers proposed a Simple Method (SiM) which takes advantage of the M-estimators characteristics [2].

2.2 Simple Method

Different strategies have been developed to solve the RDR problem, a comparison among their performances established that SiM can be applied for the measurement treatment in real-time optimization loops because it provides a good balance between the estimation accuracy and the computational load. The SiM comprises two sequential steps, which are the following:

Step 1: A robust median $\tilde{y}_i$ of the i-th variable, $(i = 1: I)$, is calculated using a moving window of measurements, which length is N, and the BW function.

$$\tilde{y}_i = Min \sum_{p=j-N+1}^{j} \rho_{BW}\left(\frac{y_{ip}-\tilde{y}_i}{\sigma_i}\right) \tag{6}$$

where σ_i stands for the measurement standard deviation and j represents the actual time.

Equation (6) takes advantage of the temporal redundancy supplied by a window of observations and eliminates the outlier effect because BW is a redescending function. This problem is initialized with the solution of the j-1 time for $j > 1$ or the mean of a data moving window if $j = 1$.

The BW function is defined as follows:

$$\rho_{BW} = \begin{cases} 1 - [1 - (a/c_{BW})^2]^3 & if \ |a| \leq c_{BW} \\ 1 & if \ |a| > c_{BW} \end{cases}$$

$$a = \left(\frac{y_{ip}-\tilde{y}_i}{\sigma_i}\right) \tag{7}$$

where c_{BW} stands for the tuning constant calculated to fix the asymptotic efficiency of the BW estimator.

Step 2: The state of the system at the time interval j, $\hat{\mathbf{x}}_j^R$, is obtained by solving the following optimization problem

$$[\hat{\mathbf{x}}_j^R] = \underset{\mathbf{x_j},\, \mathbf{u_j}}{Min} \sum_{i=1}^{I} \rho_{HU}\left(\frac{\tilde{y}_i - x_i}{\sigma_i}\right)$$
$$st.$$
$$\mathbf{Ax = 0}$$

(8)

where ρ_{HU} and a corresponds to:

$$\rho_{HU} = \begin{cases} a^2 & if \quad |a| \leq c_{HU} \\ 2c_{HU}|a| - c_{HU}^2 & if \quad |a| > c_{HU} \end{cases}$$
$$a = \left(\frac{\tilde{y}_i - x_i}{\sigma_i}\right)$$

(9)

where c_{HU} is the tuning parameter of the HU function and a stands for the standardized measurement error. The solution achieved is optimal, because this step applies a monotone estimator. The computation time of the RDR is reduced initializing the problem with a good starting point that is the robust median obtained in Step 1.

2.3 RDR with Uncertainty Parameters

Fluctuations around a state or variations of operative conditions generate uncertain models since they affect the model equations. In this first approach, the parametric uncertain is tackled.

The following development is based on a maximum likelihood procedure which assumes independence between measurements and model with uncertain parameters, which are distributed following $\mathcal{N}(0, \sigma_{\theta,p}^2)$ [6].

A RDR problem is formulated which allows obtaining the measurement and parameter estimations for process operating in steady-state. Equation (10) presents the RDR problem formulated at the j-th time:

$$\left[\hat{x}_j^R,\ \hat{\boldsymbol{\theta}}_j\right] = \underset{\mathbf{x_j},\, \boldsymbol{\theta_j}}{Min} \sum_{i=1}^{I} \rho_{Hu}\left(\frac{\tilde{y}_i^R - x_i}{\sigma_i}\right) + \sum_{p=1}^{P} \left(\frac{\theta_p - \hat{\theta}_p}{\sigma_{\theta,p}}\right)^2$$
$$s.t.$$
$$\mathbf{g(x, \boldsymbol{\theta}) = 0}$$

(10)

where $\hat{x}_j^R$ and $\hat{\boldsymbol{\theta}}_j$ stand for the reconciled measurements and parameter vectors, respectively, $\mathbf{g(x, \theta)}$ represents the model with p, $p = 1:P$, uncertain parameters.

The problem (10) uses as initial point the robust median presented in Eq. 6. The major differences between the problem devised in this work and [6] are the use of a robust estimator to mitigate the presence of outliers and the temporal redundancy that is taken into account at the robust median computation. It is expected that this formulation allows achieving accurate estimations independently of the lack of redundancy of some variables [11].

3 Implementation

The RDR problem of (Eq. 10) is implemented for two measurement models which include measurements that follow the $\mathcal{N}$ and measurements which contain outliers.

3.1 Measurements Models

Random Errors. They are caused by unknown and unpredictable fluctuations which may occur in the measuring instruments or in the operational conditions. These errors generate inconsistencies between the measurements and the process model.

The measurement model at the j-th time of the i-th variable, y_{ij}, is the following:

$$y_{ij} = x_i + e_{ij} \tag{11}$$

where x_i stands for the true value of the variable and e_{ij} represents the random error.

Outliers. These are extreme values deviated from the bulk of observation. A low probability (p_G) of outliers occurrence is defined since they are presented with a lower frequency than random errors. The outlier model equation is:

$$y_{ij} = x_i + e_{ij} + K\sigma_i \tag{12}$$

where K stands for the magnitude of the outlier.

3.2 Performance Measures

Mean Square Error (MSE) and Root (RMSE). These PMs are related to the variance and bias achieved with the estimated values. Such parameters can be computed globally and individually for each variable as follows:

$$MSE = \frac{1}{I\,J} \sum_{j=1}^{J} \sum_{i=1}^{I} \left(\frac{\hat{x}_{ij} - x_i}{\sigma_i} \right)^2 \tag{13}$$

$$RMSE_{\rho,\,i} = \sqrt{ \frac{1}{J} \sum_{j=1}^{J} \left(\frac{\hat{x}_{ij} - x_i}{\sigma_i} \right)^2 } \qquad \rho = LS,\ RDR \tag{14}$$

where $\hat{x}$ stands for the estimated value and J stands for the number of trials simulation.

Ratios. Different ratios are evaluated and compared to analyze the improvement or deterioration achieved with the DR procedure. The root mean square error ($RMSE_y$) of measurements is calculated as follows:

$$RMSE_{y,\,i} = \frac{1}{J} \sum_{j=1}^{J} \left(\frac{y_{ij} - x_i}{\sigma_i} \right)^2 \tag{15}$$

Ratios are calculated as follows:

$$R_{\rho,\,i} = \frac{RMSE_{\rho,\,i}}{RMSE_{y,\,i}} \tag{16}$$

Spatial Redundancy (SR). This PM is computed following the procedure presented by Maronna and Arcas (2010).

$$SR_i = 1 - h_i \tag{17}$$

where h_i is the i-th diagonal element of the hat matrix **H**. This matrix allows calculating the reconciled variable values as a linear transformation of the measurements. It is defined in terms of the matrix that represents the process model and the measurement covariance matrix [12].

Parameter Mean ($\tilde{\theta}_p$) and Standard Deviation (std ($\hat{\theta}_p$)). Both are calculated for each parameter using the estimations of J trials.

$$\tilde{\theta}_p = \frac{\sum_{j=1}^{J} (\hat{\theta}_{p,j})}{J} \quad p = 1 : P \tag{18}$$

$$std(\hat{\theta}_p) = \sum_{j=1}^{J} \left(\frac{\hat{\theta}_{pj} - \tilde{\theta}_p}{J} \right) \quad p = 1 : P \tag{19}$$

3.3 Useful Consideration for the RDR Implementation

– A matrix of measurements of size $[I \times J]$ is generated to treat the same set of data for all the simulations of a benchmark.
– The global probability of outliers is fixed on $p_G = 0.1$
– Outliers magnitudes are simulated on the range $[0:10]$
– Linear models are transformed into nonlinear because of uncertainty
– Equation (16) is used for computing the SR of variables of nonlinear models. Such models have been previously linearized

4   Results

Two linear benchmarks are treated with the proposed technique. The first one is extracted from [7]. Its PMs obtained with SiM and the procedure used in [7] are compared at different scenarios. The second model compares the DR solution when temporal redundancy is considered.

4.1   Model 1

This linear model comprises 12 measured variables, which are interconnected by six balance equations. Three balances include uncertain parameters, these are highlighted in Fig. 1. The values of variables and parameters considered in [7] are reported in Tables 1 and 2.

Three cases of study are proposed. Case 1 tests an accurate model. Case 2 estimates the uncertain parameters using as initial point the values informed by [7]. Finally, Case 3 formulates a model that fixes the uncertainty in the values reported at [7]. This case

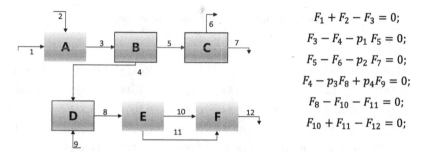

$$F_1 + F_2 - F_3 = 0;$$
$$F_3 - F_4 - p_1 F_5 = 0;$$
$$F_5 - F_6 - p_2 F_7 = 0;$$
$$F_4 - p_3 F_8 + p_4 F_9 = 0;$$
$$F_8 - F_{10} - F_{11} = 0;$$
$$F_{10} + F_{11} - F_{12} = 0;$$

Fig. 1. Model 1 with 4 uncertain parameters

Table 1. Flows true values

I	1	2	3	4	5	6	7	8	9	10	11	12
F	70	30	100	60	40	20	20	91	31	44	47	91
σ_i	3.5	1.5	4.9	3.0	0.4	0.15	0.1	1.7	1.5	1.05	0.5	1.65

is analyzed to show the deterioration estimation when uncertainties exist but are not considered. Results of PMs are reported in Tables 3, 4, 5 and Figs. 2, 3 and 4.

Table 3 present the SR and Ratios (R) for all the variables of the first model. The comparison of SR between Cases 1 and 2 shows that the lowest values of SR are obtained for Case 2, thus estimations accuracy diminishes. This fact represents an increment of RMSE for the model with uncertainty.

Table 2. Parameters and standard deviation values

θ_p	p1	p2	p3	p4
Value	1.8651	2.3325	0.7493	1.4603
$\sigma_{\theta,p}$	2	0.8	2	1.6

Ratios are computed as the division between $RMSE_{LS}$ or $RMSE_{SiM}$ in relation to $RMSE_y$. Ratios below 1 mean that the estimation procedure achieves more accurate results than measurements. Let analyze the PMs of variable F_4 at Case 2. Its redundancy is $SR_{4,2} = 0$, while its ratios are $R_{LS,2} = 1$ and $R_{SiM,2} = 0.33$, $R_{SiM,2} < R_{LS,2}$. These results show that LS is not able to enhance estimation, however SiM allows correcting the measurements. This is due to the fact that SiM uses temporal redundancy. A general ratios comparison shows:

– LS: the R of Cases 1, 2 and 3 confirm that uncertainty deteriorates the accuracy of the estimation procedure. Estimation of Case 2 provides better results than raw measurements, but non-redundant variables ($SR = 0$) are not corrected. This is in

Table 3. *SR* and *Ratios* for random measurements

F	Case 1			Case 2			Case 3	
	SRi,1	$R_{LS,1}$	$R_{SiM,1}$	*SRi,2*	$R_{LS,2}$	$R_{SiM,2}$	$R_{LS,3}$	$R_{SiM,3}$
1	*0.7402*	0.5184	0.1606	*0.3181*	0.8267	0.2644	7.9353	13.7292
2	*0.1359*	0.9197	0.2857	*0.0584*	0.9653	0.3007	3.5026	0.3077
3	*0.9252*	0.2737	0.0885	*0.6235*	0.6077	0.1972	6.6046	9.6614
4	*0.8022*	0.4368	0.1412	*0.0000*	1.0000	0.3332	11.5925	12.1756
5	*0.8316*	0.4137	0.1327	*0.0000*	1.0000	0.3293	44.1636	66.6664
6	*0.1183*	0.9345	0.2981	*0.0000*	1.0000	0.3241	17.5425	0.3214
7	*0.0526*	0.9788	0.3104	*0.0000*	1.0000	0.3174	27.3919	0.3077
8	*0.7838*	0.4739	0.1533	*0.7618*	0.4964	0.1636	0.4994	0.1624
9	*0.3018*	0.8294	0.2722	*0.0000*	1.0000	0.3254	1.0368	0.3017
10	*0.4386*	0.7562	0.2467	*0.4003*	0.7790	0.2565	0.7825	0.2554
11	*0.0995*	0.9391	0.3032	*0.0908*	0.9448	0.3065	0.9452	0.3061
12	*0.7705*	0.4888	0.1581	*0.7472*	0.5119	0.1687	0.5150	0.1675

concordance with theory. Case 3 presents the worst results, therefore not treating the uncertainty is not a smart decision.

- SiM: this procedure shows similar tendency than the obtained for LS. However, for Cases 1 and 2, the most accurate results are attained with SiM. The use of *SR* and temporal redundancy allows achieving more accurate results for redundant and non-redundant variables.

In Case 3 there is not a clear tendency between the ratios of SiM and LS, however this is a limit case since both estimation procedures present high deterioration.

The *MSE* of the measurements (y) and the three cases studied are reported in Table 4. A rough comparison lets arrive at similar conclusions than the ones obtained with R of Table 3. However, Case 3 shows that the highest deviation is reached with SiM.

The ability to mitigate the presence of outliers and get accurate parameter estimations are analyzed for Case 2. Results are presented in Table 5, and Figs. 2, 3 and 4. The *MSE* results demonstrate that the most accurate results are attained with SiM for all the K considered. Least Square estimations are deteriorated whit the increment of K, while the robust procedure does not present significant changes.

The selection of four variables with different let analyze this PM influence, these are F_2, F_5, F_8 and F_{10}. Their RMSE curves are presented in Fig. 2. Figures 2.a and Figures 2.b stand for variables non-redundant (F_5) and with low redundancy (F_2), while Fig. 2.c and Fig. 2.d represent variables with *SR* equal to 0.40 (F10) and 0.76 (F_8). As it was expected, it is observed that LS achieves better results when *SR* increase, however this estimator is not able to enhance the estimation of non-redundant variables. On the

Table 4. MSE of cases studied for random measurements

	Case 1		Case2		Case 3	
y	LS	SiM	LS	SiM	LS	SiM
	0.9913	0.4955	0.0508	0.7463	0.0785	271.6773	405.7392

Table 5. MSE for different outlier magnitudes

K	0	2	4	6	8	10
y	0.9913	2.4091	5.0099	9.6123	16.9779	27.9915
SiM	0.0785	0.1058	0.1192	0.0909	0.0885	0.0912
LS	0.7463	1.0628	1.9339	3.4379	5.5103	8.2281

other hand, SiM results show that all the variables can give accurate results independently of their SR.

Figures 3 and 4 present the means and standard deviations (stds) estimated with the LS and SiM for different outlier magnitudes.

- LS: The first parameter (p_1) presents the highest deviation, while the lowest is obtained for p_4. The stds of the first parameter p1 takes values between [0.1073–0.3415] and p_4:[0.0166–0.0545]
- SiM: Similar to LS, the std of p_1 and p_4 present the highest and lowest std($\tilde{\theta}_p$). However, these parameters take values between [0.0351–0.0431] and [0.0054–0.0065] respectively.

A comparison of both methods shows that the means calculated for SiM and LS are similar, nevertheless the stds computed with LS are bigger than the ones corresponding to SiM. This tendency is observed for all the K tested.

4.2 Model 2

This case of study allows compares the behavior of M-estimators and LS function when temporal redundancy is taken into account. The classic DR problem is initialized with the mean of a measurement data window that is called LSw.

Model 2 is a linear model composed of 23 flows related by 15 process units Fig. 5. The 6 units highlighted in Fig. 6 are the ones which have uncertain parameters, their initial values are presented in Table 6. Differently to Model 1, Table 7 presents the suggested values for parameters and standard deviation. These are improved by solving the DR or RDR problem for $K = 0$. Then, the upgraded parameters are fixed as initial values for the following simulations. Results are displayed in Table 8, Figs. 6 and 7.

Table 8 shows that both procedures achieve more accurate results than the ones provided by the measurements. The LSw presents the most accurate estimation for K

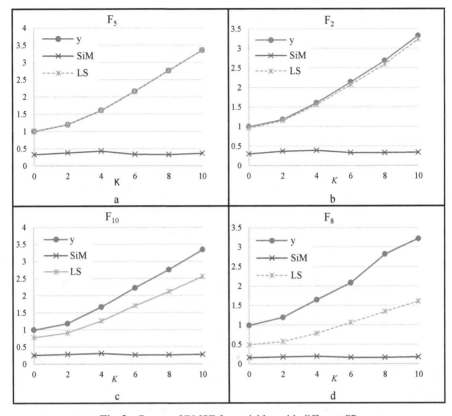

Fig. 2. Curves of RMSE for variables with different *SR*.

≤ 2 for $K > 2$ the MSE increases until reaching the maximum value for $K = 10$. The deterioration is smaller than the observed in the first example, this is as a result of the use of temporal redundancy. The SiM gets similar MSE for all the range of K analyzed.

Figures 6 and 7 display the means and stds of uncertain parameters. The lowest stds are achieved by LSw when measurements follow the $\mathcal{N}$ ($K = 0$). However, this PM deteriorates with the increment of K, achieving stds that are 3 or 6 times bigger than the ones corresponding to SiM for $K = 10$.

In conclusion, Fig. 6 shows that the stds of all the parameters get worse with the increment of K, therefore the means of the parameters also deteriorate. Opposite, Fig. 7 shows that uncertain parameters do not present changes with the increment of K. This is because of the use of a robust estimator in the DR procedure.

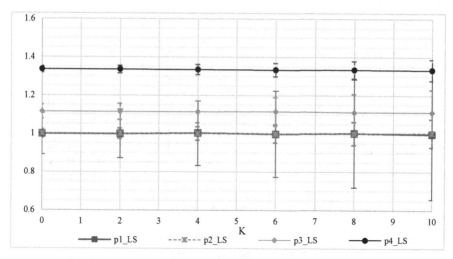

Fig. 3. Parameters' mean and standard deviation estimated using LS

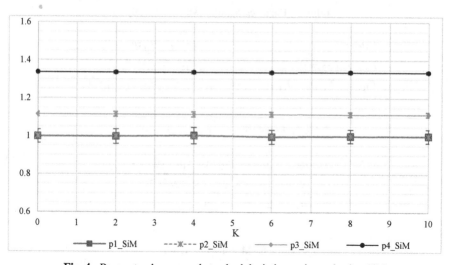

Fig. 4. Parameters' mean and standard deviation estimated using SiM

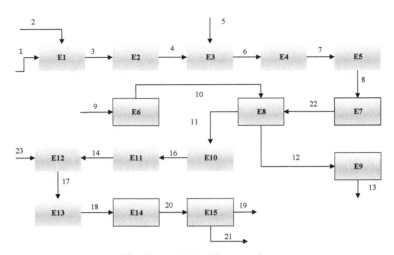

Fig. 5. Model 2 with uncertainty

Table 6. Balances equation with uncertainty

$F_9p_1 - F_{10}p_2 = 0$;	$F_{12}p_3 - F_{13}p_4 = 0$
$F_8 - p_6F_{22} = 0$;	$F_{18} - F_{20}p_5 = 0$;
$F_{10}p_2 + F_{22}p_6 - F_{12}p_3 - F_{11} = 0$;	$F_{20}p_5 - F_{19} - F_{21} = 0$;

Table 7. Initial values of parameters and std

$\hat{\theta}_p$	p1	p2	p3	p4	p5	p6
Value	1,8	1,5	2	1,8	1,5	1,2
$\sigma_{\theta,p}$	2	0,8	2	1,6	2	0,8

Table 8. MSE for different K

K	0	2	4	6	8	10
y	0.9991	2.3946	4.9611	9.5848	16.9928	27.8381
SiM	0.0639	0.0865	0.0932	0.0738	0.0716	0.0711
LSw	0.0620	0.0865	0.1505	0.2720	0.4342	0.6162

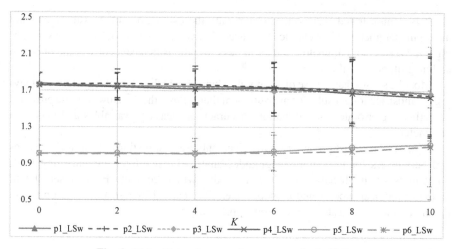

Fig. 6. Water Network parameter means and std of LSw

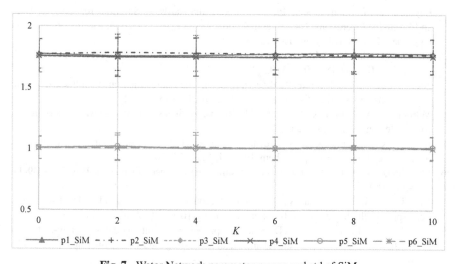

Fig. 7. Water Network parameter means and std of SiM

5   Conclusion

In this first approach, two linear models with uncertainty were treated with RDR and a classic DR formulation, the analysis of results allows to conclude that:

Robust data reconciliation can treat models with uncertainty when outliers are present. In this sense, the Simple Method (SiM) can achieve accurate estimations of parameters and variables when measurements follow exactly or approximately the $\mathcal{N}$.

Uncertainty diminishes the SR. The analysis of RMSE curves for variables with different redundancy degree showed that SiM provided accurate estimation for all the types of variables. Because SiM takes advantage of temporal redundancy.

It is demonstrated that outliers deteriorate the estimations of measurements and uncertain parameters when classic DR is used. Conversely, the use of SiM allows obtaining good results for parameters and variables estimations independently of the presence and magnitude of outliers.

Model 2 shows that the use of temporal redundancy in the LS problem allows achieving the most accurate results for measurement that follows the $\mathcal{N}$. However, the presence of outliers deteriorates the estimations accuracy of measured variables and uncertain parameters.

Regards the robust procedure, SiM gives unbiased estimations for all the variables and parameters for the outlier magnitudes analyzed. Therefore, implementing RDR procedures is a suitable alternative to treat models with uncertainty. Future research will address nonlinear models and explore uncertainty caused by dynamics and uncertain parameters when measurements follow approximately the $\mathcal{N}$.

References

1. Bagajewicz, M.: Smart Process Plants: Software and Hardware Solutions for Accurate Data and Profitable Operations: Data Reconciliation, Gross Error Detection, and Instrumentation Upgrade. McGraw-Hill, New York (2010)
2. Llanos, C.E., Sanchez, M.C., Maronna, R.A.: Robust estimators for data reconciliation. Ind. Eng. Chem. Res. **54**(18), 5096–5105 (2015)
3. Nicholson, B., López-Negrete, R., Biegler, L.T.: On-line state estimation of nonlinear dynamic systems with gross errors. Comput. Chem. Eng. **70**, 149–159 (2014)
4. Maronna, R.A., Martin, R.D., Yohai, V.J.: Robust Statistics: Theory and Methods (with R). John Wiley and Sons, Oxford (2019)
5. Kravaris, C., Hahn, J., Chu, Y.: Advances and selected recent developments in state and parameter estimation. Comput. Chem. Eng. **51**, 111–123 (2013)
6. Maquin, D., Adrot, O., Ragot, J.: Data reconciliation with uncertain models. ISA Trans. **39**(1), 35–45 (2000)
7. Narasimhan, S., Narasimhan, S.: Data reconciliation using uncertain models. Int. J Adv. Eng. Sci. Appl. Math. **4**(1–2), 3–9 (2012). https://doi.org/10.1007/s12572-012-0061-3
8. Vasebi, A., Poulin, É., Hodouin, D.: Selecting proper uncertainty model for steady-state data reconciliation–application to mineral and metal processing industries. Min. Eng. **65**, 130–144 (2014)
9. Galan, A., de Prada, C., Gutierrez, G., Sarabia, D., Grossmann, I., Gonzalez, R.: Implementation of RTO in a large hydrogen network considering uncertainty. Optim. Eng. **1**(30), 1161–1190 (2019). https://doi.org/10.1007/s11081-019-09444-3
10. Narasimhan, S., Jordache, C.: Data Reconciliation and Gross Error Detection. Gulf Publishing Company, Houston (2000)
11. Llanos, C.E., Sanchéz, M.C., Maronna, R.A.: Robust estimation of nonredundant measurements and equivalent sets of observations. Ind. Eng. Chem. Res. **58**(42), 19551–19561 (2019)
12. Maronna, R., Arcas, J.: Data reconciliation and gross error diagnosis based on regression. Com. Chem. Eng. **33**(1), 65–71 (2009)

Author Index

Printed in the United States
by Baker & Taylor Publisher Services